D0394633

Partners *in* Science

PARTNERS *in* SCIENCE

Foundations

and

Natural Scientists

1900–1945

ROBERT E. KOHLER

The University of Chicago Press
Chicago and London

Robert E. Kohler is professor of the history and sociology of science at the University of Pennsylvania.

The publication of this book has been supported in part by a generous grant from the Rockefeller Foundation. The contents of the book, however, do not necessarily reflect the views of the foundation.

The University of Chicago Press, Chicago 60637
The University of Chicago Press, Ltd., London

Printed in the United States of America

00 99 98 97 96 95 94 93 92 91 5 4 3 2 1

Library of Congress Cataloging-in-Publication Data

Kohler, Robert E.
Partners in science : foundation managers and natural scientists. 1900–1945 / Robert E. Kohler.
p. cm.
Includes bibliographical references and index.
ISBN 0-226-45060-0 (cloth).—ISBN 0-226-45061-9 (pbk.)
1. Science—History—20th century. 2. Science—Endowments—History—20th century. 3. Science—United States—Endowments—History—20th century. 4. Scientists. 5. Naturalists. I. Title.
Q125.K62 1991
507.9—dc20 90-43520
CIP

♾ The paper used in this publication meets the minimum requirements of the American National Standard for Information Sciences—Permanence of Paper for Printed Library Materials, ANSI Z39.48–1984.

This book is printed on acid-free paper.

For Frances
mea cum confuge maius perfectum

Contents

Illustrations

Tables

Figures

Preface

My interest in foundations and patronage dates from about 1974, when I wandered into the newly opened archives of the Rockefeller Foundation, which at the time was shoehorned into temporary quarters in a second-story warehouse on Manhattan's West Side. A recent convert to history from chemistry and biochemistry, I came looking for material on the history of heavy isotope research in biochemistry, which I knew the foundation had sponsored. I was not disappointed: half-a-dozen files of correspondence detailed a behind-the-scenes story that, for an ex-scientist used to working only with published sources, was a revelation of what doing history could be. Even more eye-opening were the vast files in the foundation's "program and policy" series, to which I was pointed by the foundation's archivist, Dr. William Hess—the first of many times he was to point me in a fruitful direction. Here was revealed an almost daily record of the activities of a small group of people who, together with groups of academic scientists, were creating new relationships between science and society: reshaping their institutions to accommodate extramurally sponsored research. And, astonishingly, these records had been seen by almost no one since they had been written. That day I was hooked on excavating archives and was committed, willy-nilly, to an institutional approach to the history of science.

Gradually, through working with the records of foundation managers, I became aware (as many others were at the time in other ways) that science was a complex social system with many actors, in which securing resources, negotiating with patrons, creating departments and disciplines, competing for talents, designing products and services, and projecting public images were no less essential than bench research. More precisely, I became aware that these aspects of science were proper subjects for history. As an apprentice chemist I had seen that this was how science really worked, but that did not seem to be what the most respected historians of science wrote about. Rummaging through the letters, diaries, and reports of foundation officers and their academic allies, I was drawn from a history of science, that is, of finished intellectual products, to a history of scientists, that is, of science as social process.

The history of *scientists:* the idea was at the time a reproach and rebuff to those who wanted to enlarge the horizons of their discipline.[1] But it became for me the emblem of an actor-centered, pluralistic history, which dealt evenhandedly with everything that scientists did; that valued equally the social and intellectual products of science; that gave equal weight to the aims and perceptions of all the diverse social actors, who gave their own meanings to science and in doing so made the system work. Science and scientists mean something quite different to university presidents, congressmen, businessmen, and foundation officers, and their views are no less valid and worthy of study than the view from the laboratory bench. All are essential actors. None has a privileged historical value. Science is the sum of their views and activities.

When my first articles on the history of foundations were published in the mid-1970s, only a handful of historians were working on the subject, most notably Stanley Coben, Donald Fisher, Nathan Reingold, and (in partnership) Barry Karl and Stanley Katz. Since I sat down to write this book about five years ago, there has been a veritable flood of work on foundation activities (thanks in part to the Rockefeller Archive Center's generous program of research grants-in-aid). And the flood continues. As I revised, I learned of John Heilbron's work on the diffusion of cyclotron technology, Finn Aaserud's on the role of foundations in Niels Bohr's embracing of nuclear physics, Pnina Abir-Am's on the Cambridge theoretical biologists, Larry Owens's on Warren Weaver's activities in World War II, and Thomas Glick's on Rockefeller funding of physics in Spain—topics central to the story I hoped to tell. Studies appear almost monthly that throw new light on some aspect of foundation patronage that I missed or set aside. My aim in writing this book was to give a comprehensive view of foundation patronage of natural scientists: how the partnership was constructed, and how it worked and evolved in different regions of the scientific world. It will not be long, I think, before the interpretations I offer will face informed scrutiny from many directions.

Anyone who has tried to write history that deals with many scientific disciplines and national institutions knows how easy it is to make mistakes. I am grateful to colleagues who read chapters and set me right on

[1]See, e.g., Leonard Wilson's review of my *From Medical Chemistry to Biochemistry*, *J. Hist. Sci.* 38 (1983): 462–464.

matters both technical and contextual, especially John Heilbron, Tom Glick, Finn Aaserud, Nathan Reingold, Pnina Abir-Am, and Robert Olby. I am grateful, too, to sponsors who supported the researching and writing of this book: the Commonwealth Fund, the Rockefeller Foundation, the Rockefeller Archive Center, and the National Science Foundation (SE81-19490). Helen Weaver and Martha Thorkelson Riddell Kohler kindly provided family recollections and photographs. Last but not least, I would like to acknowledge the many archivists and records officers on whose knowledge I and all historians depend. I want especially to thank the remarkable staff of the Rockefeller Archive Center, above all Dr. J. William Hess, Dr. Joseph W. Ernst, Thomas Rosenbaum, Emily Oakhill, Harold Oakhill, Melissa Smith, and Dr. Darwin Stapleton. Without their expert and enthusiastic participation, doing history would be a lot less fun.

Chapter One

Systems of Patronage

The importance of patronage in modern science is almost too obvious to dwell upon. Since the late nineteenth century, competing in fast-paced research fronts has become ever more dependent on the ability to command resources. It takes money to improve facilities, compete for research talent, keep up with changing laboratory technologies, and produce papers as fast as the competition. Also, it takes organizational know-how to use resources effectively and to keep the knowledge produced in circulation. For scientists in universities, material resources and organizational skills have come increasingly from extramural patrons. The bread-and-butter service of university scientists—collegiate teaching—produces neither the money nor the organizational experience that modern science requires. Thus, for academic scientists, the practice of science has meant a more complex social relationship with extramural sponsors, whose institutional values and agendas are not quite the same as the values and agendas of their clients. Often, sponsors were skeptical of conventional boundaries between academic departments and disciplines. They had their own ideas of how science should be organized and practiced. Getting and managing grants, and managing sponsors, has gradually become a constitutive part of doing research and making careers in academic science, along with teaching, department and discipline building, or professional service.

Who would argue? And yet for some historians of science (and many scientists), grant-getting and the organizational work of science seem far less interesting and important than the production of knowledge as such. These activities are often regarded as necessary evils, as noncreative distractions from the real business of bench research. This seems too narrow a view of the complex reality of scientific careers and practices. It does an injustice to scientists who find satisfaction in solving problems of organization. It rends apart the production process, in which mobilizing resources, building disciplines, and maintaining boundaries are integral parts. Assembling the material and human resources for doing research is no less a part of the creative process than doing experiments. Good ideas are essential (obviously), but so too are the skills that infuse good ideas into research teams and schools. Ideas are often made good by the process of securing resources, enlarging disciplinary boundaries or tearing them down, lining up allies. Fundability has become, for better or for worse, an important element in the system by which credibility is allocated to researchers and their work.[1]

The production process of science is a complex social system, and we will not understand it better if we impose upon its elements an artificial hierarchy of values. It will not pay to make arbitrary distinctions between insiders and outsiders. It is not just the people who work in laboratories who do science, but everyone who takes part in sponsoring, producing, justifying, or making use of scientific knowledge. Foundation managers were doing science no less than the scientists whose work they helped to make possible. We need, as Bruno Latour exhorts us, to take a holistic and inclusive view.[2]

The hardest part of working on the history of patronage is knowing where to stop, how to keep it from becoming a history of science in toto. How much need one say, for example, about the institutions and disciplines that benefited from foundation largess? How far should one go in describing the work done with foundation grants and in assessing the influence of foundation patronage on research fashions? How can the subject of patronage be contained without leaving out some essential part of the seamless web?

I have tried to do so by focusing on the social system of patronage itself: the evolving partnership between patrons and recipients. I am interested in the process of giving and getting grants, in how the social machinery of sponsorship was constructed, in the people who kept the machinery in running order. There is something about bench research in this book, but there is far more about how research methods and agendas figure in interactions of patrons and clients. I have avoided the question of the influence of foundation patronage on scientific disciplines (I think it is an unanswerable question), but disciplines figure as key elements in my account of how systems of patronage worked. I want to show how workable systems of sponsorship were constructed and

[1]Bruno Latour and Steve Woolgar, *Laboratory Life* (Beverly Hills, Calif.: Sage, 1976), ch. 5. Latour, *Science in Action* (Cambridge, Mass.: Harvard University Press, 1987).

[2]Latour, *Science in Action,* pp. 157–162.

TABLE 1-1

Rockefeller and Carnegie Philanthropies in Science

Organization	*Founded*	*Active in Science*
Rockefeller Institute	1901	1902–
Carnegie Institution of Washington (CIW)	1902	1902–
General Education Board (GEB)	1902	1923–30
Carnegie Corporation of New York (CC)	1911	1919–31
Rockefeller Foundation (RF)	1913	1919–
Laura Spelman Rockefeller Memorial	1918	1922–28
International Education Board (IEB)	1923	1923–28

how they were embodied in institutions and in the conventions, expectations, and rituals of the patron-client relation. The social system itself is the object of study here, not its effects on something else.

Choices of what to put in and what to leave out in this volume derive from this basic principle. I focus on university scientists, because it was with them that the tensions between patron and recipient were sharpest and most interesting. The activities of the research institutes endowed by John D. Rockefeller and Andrew Carnegie are outside my scope, except for their extramural grants programs. This book deals with only the largest foundations, the seven founded by Carnegie and Rockefeller (see table 1.1), because they were the most self-conscious and programmatic in their relations with scientists. They created the system and accounted for 85–90 percent of foundation expenditures on science in the interwar period.

I have also confined myself to the natural sciences, omitting the social sciences, medicine, and public health. This is perhaps less defensible, not just because the Rockefeller and Carnegie boards spent vast sums in these fields but because these programs preceded those in the natural sciences and constituted a body of practical experience upon which the later programs drew. (Also, the social and health sciences offer opportunities for social history that the natural sciences cannot.)[3] The trouble is that the social bases of the natural, social, and

[3]Barry Karl, "Philanthropy, policy planning and the bureaucratization of the democratic ideal," *Daedalus* 105 (1976): 129–149. Karl and Stanley N. Katz, "The American private philanthropic foundation and the public sphere, 1890–1930," *Minerva* 19 (1981): 236–270. Karl and Katz, "Foundations and ruling class elites," *Daedalus* 116 (1987): 1–40.

health sciences are so different as to make it impracticable to do all of them at once. With one group it becomes possible to look at the fifty years in which foundations were their chief sponsors, to see how the creators of the system meant it to work, how it did in fact work, and how it changed through being worked.

Finally, I have tried to maintain an even balance in this history between ideas and practices. Chapters tend to alternate in a dialectical fashion between internal foundation politics and policymaking, and what foundations actually did in the field. The relation between boardroom policy and field practice is about as straightforward as that between party platforms and practical politics. Patronage, like politics (or science itself), is the art of the possible. New modes of patronage evolved out of existing modes, in the field, just as new scientific practices derived from existing ones, at the bench. Projects were constructed with an eye to what officers knew from experience would survive the process of review by university officials and foundation trustees. Grant-givers and grant-getters were no less constrained by academic subcultures and conventions of academic careers. To understand who got what, we need always to keep in mind how the process of grant-giving really worked as well as how it was supposed to work.

The following chapters are divided into three groups, which correspond to three periods or stages in the evolution of the private system of sponsorship. The first group (2–5) begins with an examination of the extramural grants program of the Carnegie Institution of Washington. It is a story of misunderstanding and conflict between trustees, managers, and scientists, and it reveals most clearly the basic assumptions and limitations of the nineteenth-century system of individual grants-in-aid. It also sets the stage for the invention of a new system of patronage, in which the National Research Council mediated between foundations and individual scientists and universities. Just after World War I, the Carnegie Corporation and the Rockefeller Foundation provided a handsome endowment fund and grants for national research fellowships to the National Research Council. I will focus on these key decisions and show how, in the process of negotiation, basic assumptions of the old system of patronage were discarded for a new rationale for large-scale patronage. I will show also how in practice this was a limited relationship and how efforts to expand it failed.

The second group of chapters (6–9) focuses on the activities of the International and General Education Boards. These educational foundations, unlike the Carnegie Corporation and Rockefeller Foundation, were direct operating agencies and created a system of large institutional grants to university science departments, a system that was quite unlike earlier individual grants and much more expandable. The emphasis of this segment of the story will be less on policy than on field experience, which was rich and diverse. I give a systematic account of this experience, both in Europe and in the United States, and try to explain large-scale patterns of distribution—who did well and why. It is an ecology of patronage, which shows how the distribution of resources was shaped by the social systems of science in which foundation officers operated.[4] In the case of institutional grants, who got what depended on location in national educational systems—old elite universities and the up-and-coming; those in cosmopolitan or provincial centers and on undeveloped peripheries. This group of chapters concludes with the reorganization of the Rockefeller Foundation in 1928, which brought the era of institutional grants to a close and set the stage for a new system of grants for individual or group projects.

The third group of chapters (10–13) concerns the natural science programs of the Rockefeller Foundation in the 1930s, managed by Warren Weaver, the quintessential manager of science. (The Carnegie Corporation abandoned natural science in the 1920s.) Again the approach is ecological—who got what and why is the question. But the story turns not on institutional systems but on disciplines, research schools, and laboratory practices. A group of projects in experimental biology was shaped by the structure of discipline communities (or their lack of structure). Projects in "molecular biology" were inspired by Weaver's desire to encourage transdisciplinary careers in science, careers in transit from physical to biological science or vice versa. A third group of projects, in biophysics, was patterned by the technical and social characteristics of different kinds of laboratory instruments, like cyclotrons and radioisotopes. In the 1930s system of projects, grant-giving and grant-getting was a more intimate and personal experience, and to understand it we need to look at strategies of making careers, the ways in which scientific territory is divided up, and the conventions of laboratory practice.

As a byproduct of the ecological approach, many things will be said about science between the wars: about university reform movements, the emergence of leadership elites, efforts to carve out space for new disciplines like general physiology and biophysics, the synchronous appearance of novel instruments for biomolecular research, and holistic movements in biology. Systems of patronage took their shape from

[4]Charles E. Rosenberg, "Toward an ecology of knowledge: On discipline, context, and history," in Alexandra Oleson and John Voss, eds., *The Organization of Knowledge in Modern America, 1860–1920* (Baltimore: Johns Hopkins University Press, 1979), pp. 440–455.

large-scale structures in the world of science; they amplified long-term trends. One of the pleasures of studying the history of patronage is the opportunity it affords to cross boundaries of expertise that divide historians of the different disciplines and different national contexts.

The common thread of all these diverse experiences, however, is the process of making projects, the dynamic partnership between grant-giver and grant-seeker. Each project is an occasion to see a system of patronage being constructed or renegotiated. On the foundation side, the key players are the small cadre of officials—the managers of science—who gave substance to official policies by making projects in the field. On the university side are the entrepreneurial scientists who created common ground between disciplinary and philanthropic agendas. The principals of my story are middling people, people on the shop floor of science and philanthropy. Bench researchers and department heads figure more prominently than university presidents and deans. On the foundation side, John D. Rockefeller remains entirely offstage, and Rockefeller, Jr., and Andrew Carnegie appear only as minor players whom the managers of their fortunes had to steer tactfully onto new tracks. It is the first two or three generations of foundation managers whom I want to bring to life, people who left academic research or administration for the unfamilar world of large foundations.

Most of the material for this book comes from the records left by foundation managers, especially from the extraordinary wealth of correspondance, diaries, applications, and field reports left by the officers of the Rockefeller boards. I have tried to see the system of patronage through the eyes of grant-getter as well as grant-giver. (It is hard not to see both sides where so much of the documentation has to do with negotiation.) But as the managers of science are the central characters in this story, they will have the most lines. In fact, a good part of the excitement and fascination of philanthropic records is that they enable us to see the enterprise of science through the eyes of people who were participants but not bench scientists. While I would not claim any privileged status for their point of view (or for any actor's view), I know of no participants in the system of science who had a more comprehensive view of science between the wars. Foundation officers approached science and scientists with an equal blend of sympathy and skepticism, idealism and common sense, and that made them unusually canny observers. They made it their business to see the whole show, in every region of Europe and the United States, and in a dozen disciplines. I know of few people with a more varied and cosmopolitan experience, and none whose experience offers historians a better chance to see how science really works in the modern world.

The Nineteenth-Century System

We need first, however, to take a brief look at how the system of sponsorship worked before the large foundations appeared on the scene. In the United States, academic scientists operated in a system (such as it was) of genteel poverty and laissez-faire. They admired and envied the accomplishments of European academics with their state-supported institutes, but they feared the strings attached to government support. Scientific leaders like Harvard astronomer Edward C. Pickering had been propagandizing the well-to-do for decades, hoping to divert some of the vast wealth created by the new mass-production industries into academic science, but with little success.[5] Hence the sensation when, in 1902, steel magnate and philanthropist Andrew Carnegie announced that he was giving $10 million to found a new institution for promoting science—the Carnegie Institution of Washington (CIW). Carnegie's benefaction, following hard upon John D. Rockefeller's endowment of the Rockefeller Institute for Medical Research in 1901, seemed to open a new era in the patronage of academic science.

No one, however, seemed to know exactly what to do with a windfall of $400,000 a year. Everything seemed possible, but there was little in the experience of late nineteenth-century academic scientists to suggest what a large-scale system of sponsorship should look like. Certainly Carnegie did not know: his deed of gift gave his trustees a blank check to recruit youthful talents, sponsor individual research projects, help federal scientists expand their professional horizons, and aid publication and diffusion of knowledge. Invited to share their dreams openly in the pages of *Science,* academics poured forth plans for special professorships and European-style institutes for research elite, small grants-in-aid to the underprivileged academic rank and file, fellowships for graduate students, great national research institutes to rival the Royal Institution and Prussian Academy, and so on.[6]

One provision of Carnegie's deed of trust especially captured the

[5]Howard Plotkin, "Edward C. Pickering and the endowment of scientific research in America, 1877–1918," *Isis* 69 (1978): 44–57.

[6]Nathan Reingold, "National science policy in a private foundation: The Carnegie Institution of Washington," in Oleson and Voss, *Organization of Knowledge,* pp. 313–341. David Madsen, "Daniel Coit Gilman at the Carnegie Institution of Washington," *Hist. Educ. Quar.* 9 (1969): 154–186. Nathan Reingold and Ida H. Reingold, eds., *Science in America: A Documentary History* (Chicago: University of Chicago Press, 1981), ch. 1, reprints many relevant documents. Excerpts from scientists' letters to the editor of *Science* were printed in *Science* 16 (1902) and 17 (1903).

imagination of both elite and middling academics: "To discover the exceptional man in every department of study whenever and wherever found, inside or outside of schools, and enable him to make the work for which he seems specially designed his life work."[7] "Discover the exceptional man"—the phrase resonated with deeply held social values: American particularism and manifest destiny, a belief in self-improvement and upward mobility, and a sense that everyone should be able to participate in elite cultural activities, including scientific research.

The condition of American colleges and universities circa 1900 made Carnegie's plans for the "exceptional man" especially resonant. Since the 1870s, colleges and universities had proliferated in response to local demands for higher education, unrestrained by state regulation or effective gatekeeping by the old Eastern elite institutions. This made the American system of universities very broad based, but also very uneven in quality. Many small colleges announced "graduate" programs in the 1890s in the hope of capitalizing on the growing market for better-quality teachers, as Johns Hopkins, Bryn Mawr, and a few others were doing. Many institutions aspired to university status but lacked the means to support research. Many college teachers were inspired by the research ideal, but few could hope to realize that ideal. Even in universities like Harvard or Princeton, research was valued mainly for its use in teaching college teachers.[8] The history and political economy of American universities strictly limited internal resources available for research, while at the same time they encouraged a large rank and file to hope they too might one day make the research ideal a reality for themselves. Thus, in 1900 there was an exceptionally large number of undiscovered geniuses in American colleges and universities. Some had arrived at their prime years knowing nothing but a frustrating struggle for minimal resources; others were entering their careers hoping that things would be easier for them than for their elders. To all, Carnegie's plan to rescue "the exceptional man" from obscurity seemed to speak directly and personally.

Scientists' expectations of Carnegie's and Rockefeller's millions reflected their very limited prior experience with extramural patronage. What small endowment there was for research was held in trust by national scientific societies like the National Academy of Sciences and the

[7]CIW *Yearbook* 1 (1902): xiii. T. C. Chamberlin to R. S. Woodward, 11 Feb. 1910, CIW.

[8]Owen Hannaway, "The German model of chemical education in America: Ira Remsen at Johns Hopkins, 1876–1913," *Ambix* 23 (1976): 145–164. Larry Owens, "Pure and sound government: Laboratories, gymnasia, and playing fields in nineteenth-century America," *Isis* 76 (1985): 182–194. Roger L. Geiger, *To Advance Knowledge* (New York: Oxford University Press, 1986), ch. 1. R. E. Kohler, "The Ph.D. machine: A view from the shop floor," *Isis* 81 (1990): forthcoming.

American Academy of Arts and Sciences. It was derisory by later standards—the several research funds of the National Academy added up to $94,000 in 1890 and produced an annual income of less than $4,000. These modest funds were doled out in small grants—a few hundred dollars on average—to deserving individuals selected by committees of academicians.[9] For scientists, this early form of peer review had the advantage of limiting intervention by overly zealous gatekeepers into individual decision making. But the system had the great disadvantage that it was not an attractive object for wealthy philanthropists. Aiding obscure academics to pursue their miscellaneous private researches was not competitive with traditional forms of civic philanthropy—colleges, museums, libraries, churches—that delivered real public services. The only benefactors of the National Academy and American Academy were their own members, and even the wealthiest of them were far from wealthy on the scale of the great industrial families.

There was one minor exception to this pattern: in 1884 a New York philanthropist, Elizabeth Thompson, created an independent endowment of $25,000 to support scientific researches. (In 1878 she had given $10,000 for research on yellow fever.) In its operations, however, the Elizabeth Thompson Science Fund followed the pattern of the scientific societies. Mrs. Thompson had hoped that the fund would be managed by the American Association for the Advancement of Science, but when that fell through the endowment was put in the charge of a group of Harvard scientists led by embryologist Charles S. Minot and the ubiquitous Edward Pickering.[10] In effect, this group acted as a peer review panel, though it did not represent a national society. It had no definite program. The fund's grants were extremely eclectic: obviously, the committee did not want to be criticized for favoring particular disciplines or places. Grants were small, mostly $50–$250 with a cap for individuals of $300.

The managers of the Elizabeth Thompson fund did make some effort to use grants as carrots to make American scientists more competitive in disciplines where Europeans dominated: making grants to Europeans, for example, was meant to put Americans on their mettle. Minot and Pickering also used grants as a stick. Recipients who failed to publish in due course were publicly named in the fund's brochures; some were requested to give the money back, and the names of those sinners too were made public. Pickering lost no chance to boast of the

[9]Howard S. Miller, *Dollars for Research* (Seattle: University of Washington Press, 1970), ch. 6.

[10]Ibid., pp. 127–129. C. S. Minot, "A new endowment for research," *Science* 6 (1885): 144–145.

93 percent "success rate" achieved by these methods.[11] Pickering would have gone even further in using extramural funds to get scientists to be more business-like in their research, less individualistic and more focused on important problems. He hoped that wealthy industrialists might be more inclined to make large endowments for research if they could see that academic scientists shared their own business habits and values.

Pickering's hopes were doomed on both counts, however. Scientists resented his efforts to organize them. They regarded with deep suspicion Pickering's guileless but insensitive suggestion that Harvard's board of overseers would be an ideally disinterested body to hold research endowments for the nation.[12] After one of Pickering's organizing committees collapsed, a colleague observed: "The whole subject was filled with dynamite and he [Pickering] didn't know it."[13] Wealthy benefactors responded to Pickering's pleas for Harvard's observatory (a familiar kind of gift to a local scientific cause), but none stepped forward to endow research in general.

Pickering was tactless and heavy-handed, but his frustrations highlight the basic limitation in the nineteenth-century system of patronage: scientists resented any interference in their work by outside agencies; philanthropists saw little point in aiding academic scientists in their private and individualistic endeavors. It was not that philanthropists undervalued scientific research; quite the contrary, they thought research too important to leave to part-time academics.

Both Carnegie and Rockefeller opted for freestanding research institutes, against the advice of their academic advisors, who advocated institutes attached to universities. Carnegie decided against a national university because he feared it would only compete with nearby Johns Hopkins.[14] Rockefeller's scientific advisors, like William Welch, favored an institute connected with Columbia University. But Rockefeller, Sr. (or Jr., it is not clear), and Starr J. Murphy, a lawyer engaged by the Rockefellers to assess the options, preferred an independent institute.[15] The philanthropists believed in a functional division of labor: the raison d'être of medical schools was to teach and of hospitals to care; both could do research that pertained specifically to

[11]"Elizabeth Thompson Science Fund," Jan. 1909. Pickering to F. T. Gates, 14 July 1909, GEB 1.4 612.6479.

[12]Plotkin, "Pickering," pp. 46–49.

[13]S. C. Chandler to Wolcott Gibbs, 7, 13 Mar. 1901, Gibbs Papers, Bache Fund, Franklin Institute, Philadelphia.

[14]Reingold, "National science policy."

[15]George W. Corner, *A History of the Rockefeller Institute 1901–1953* (New York: Rockefeller Institute Press, 1964), pp. 26, 39–42, 583–584.

these missions. But only a separate institute with research as its sole mission could "take up the problems where the medical schools leave them, and treat them in their broadest aspects." In Murphy's view, "[H]ospitals and the medical schools, so far as they carry on research work, will lead up to and be feeders for the Institute, which will be the crown of the whole system."[16] The same argument applied to universities and research in the natural sciences: research could best be done separated from teaching, and any effort to combine them would only compromise both. In this view, grants to academic scientists were not an appropriate use of endowments earmarked for research. That was how the trustees of the CIW saw the matter and why they too, like the founders of the Rockefeller Institute, chose to devote most of their income to in-house research departments rather than to programs of extramural grants.

It is not surprising, then, that scientists' eager anticipation of their new patrons soon turned to disappointment and conflict. These conflicts, symptomatic of fundamental changes in the system of patronage, are the subject of the next chapter.

[16]Starr J. Murphy to Rockefeller, Jr., 13 Feb. 1902, C. W. Eliot Papers, 104.47, Pusey Library, Harvard University.

PART I

Creating a System, 1900–1920

Chapter Two

Troubled Beginnings: The Carnegie Institution

The CIW's grants program has gone down in history as a footnote to the history of the institution's research departments and as a failed experiment in the patronage of science.[1] In fact, it was the first significant experiment in large-scale sponsorship of academic science. On average, over $100,000 per year was spent on individual grants between 1903 and 1920.[2] This was twenty times the expenditure of any other grants fund and more than all such funds put together.[3] The CIW's grants program was not a failed but a troubled experiment, marked by conflict and misunderstanding between academic recipients of grants, the CIW's board of trustees, and the CIW president, Robert S. Woodward.

The unprecedented wealth of Carnegie's endowment, plus the vagueness of his directives, brought to the surface fundamental conflicts about the purpose of research endowments and how they should be managed. Within the CIW there was competition for authority between the president and the board of trustees. The first president, Daniel Gilman (emeritus president of Johns Hopkins), resigned over this issue, and Woodward spent a decade battling the executive committee for control of programs and budgets. No less divisive was the competition for authority between the president and the CIW's extramural scientific constituents. (With the heads of the in-house departments, conflict was minimal.) The extramural grants were the source of

[1]Geiger, *To Advance Knowledge*, pp. 61–67.

[2]Financial data are in CIW *Yearbooks*.

[3]"Endowments for research in the United States," in Report of the Executive Committee, 1902, pp. 247–269, CIW. Callie Hull, "Funds available in 1920 in the United States of America for the encouragement of scientific research," National Research Council, *Bulletin* no. 9 (1921).

chronic, grinding friction between grant-givers and grant-getters. Woodward was constantly complaining about the irresponsibility of his academic clients in not abiding by the standards of practice of the CIW's research departments. Academics never ceased to believe that Woodward was trying to impose his personal agendas on the scientific community and subverting funds earmarked for individual grants to the CIW's departments.

These conflicts have usually been taken as signs of personal failure or mismanagement; in fact, they are symptoms of a new system of patronage in the making—more precisely, of a nineteenth-century system displaying its inability to work in twentieth-century conditions. The sudden scaling-up of the patronage system created conflicts because it was not accompanied by any change in practices or expectations. Academics used to the laissez-faire of grants committees run by friends and colleagues were unprepared for a granting agency that had its own agenda and its own ideas of how academic science should be practiced. So, too, boards of trustees were unaccustomed to full-time scientist-managers exercising an independent authority. In hindsight, Woodward was the first modern manager of science, and his experience is a window for historians on a new patron-client partnership as it was beginning to take shape.

In this chapter we will examine in turn the policies and politics behind the grants program, then Woodward's conflicts with the board, and finally the troubled relations between Woodward and the CIW's academic constituents. We will see how expectations from the nineteenth-century system of patronage resulted in mutual misperceptions on both sides, and how conflict cleared the way for a fundamentally new system of patronage by the end of the 1910s.

The CIW Grants Program: Origins

It was clear from the beginning that the CIW would have both research departments and an extramural grants program. The only question was what the balance would be. Grants overshadowed departments in the first few years because organizing departments took more time and effort.[4] Eager to do something while advisory committees pondered and politicked, the trustees appropriated several hundred thousand dollars for a system of "minor grants" (up to $3,000) for academic scientists. (A few larger grants were also made for expensive but once-only

[4]Reingold, "National science policy."

undertakings like field expeditions.)[5] The early grants program of the Rockefeller Institute was likewise partly a stopgap while in-house programs were organized, though it remained much smaller than the CIW's program (it never topped $15,000 per year).[6] The CIW's grants program survived the organization of the departments because it was the vehicle for Carnegie's dream of discovering the "exceptional man."[7] At least, that is how the all-powerful executive committee saw it.

The key people on the board's executive committee were the medical politician, John Shaw Billings, and Charles D. Walcott, the chief of the U.S. Geological Survey. They virtually ran the CIW during its first few years. Aging, gentlemanly, and uncertain of his new role, Daniel Gilman was no match for ambitious and energetic people like Billings. Relations between Gilman and the executive committee became increasingly strained, and Gilman resigned after only two years, tired of being pushed around by Billings and his friends.[8]

As chairman of the executive committee, Billings was the single most influential person in determining what kind of institution the CIW would become. His preeminence in the library world (he more or less created the National Library of Medicine and the New York Public Library) gave him great authority with Carnegie, the great benefactor of public libraries. Carnegie's deed of trust to the CIW echoed memoranda prepared for him by Billings. It was Billings who persuaded Carnegie to overrule Gilman's counsel to emphasize education over research. He also took credit (perhaps undeservedly) for putting the idea of the "exceptional man" in Carnegie's mind. In any case, Billings was certainly the strongest advocate of using minor grants to bring obscure talents to light.[9] Billings saw the CIW not simply as a research institute but as a general trust fund for encouraging research throughout the United States, complementing the funds of the national scientific societies.[10] He was an unrelenting protector of the grants

[5]CIW *Yearbook* 4 (1905): 17–18. Billings, memorandum to the executive committee, 1 Feb. 1902, JSB box 58 Memoranda.

[6]Corner, *Rockefeller Institute,* pp. 43–46. The grants had been vestigial for some years when they were finally curtailed in 1914.

[7]Billings, remarks to board, 8 Dec. 1903, board minutes, vol. 1, pp. 178–181, CIW.

[8]Madsen, "Daniel Coit Gilman," pp. 176–181, 185.

[9]Billings, "Memorandum," 22 Nov. 1901; "Memorandum handed to Mr. Carnegie 22 Nov. 1891" [*sic* for 1901]; "Memorandum," 25 Nov. 1901; memorandum to executive committee, 1 Feb. 1902; memorandum, 7 Feb. 1903; all in JSB box 58 Memoranda. Madsen, "Daniel Coit Gilman," pp. 156–159. Reingold, "National science policy," pp. 317–320.

[10]Billings, "Memorandum," 25 Jan. 1902, JSB box 58 Memoranda. Billings memorandum in executive committee minutes, 26 Oct. 1903, p. 2, JSB box 59.

program, and that made him Robert Woodward's arch antagonist.[11]

Billings's vision of patronage was rooted in a lifetime's experience of science politics. Sixty-three years old in 1901, Billings had spent his entire career in public institutions. From his base in the Army Surgeon General's Office, he had had a hand in most of the campaigns of the previous thirty years to organize national institutions of science and medicine. His ambitions were vaulting and eclectic, and he liked to shape events from behind the scenes by controlling executive committees.[12] Billings believed in organized, programmatic research, but he took a middle ground between the laissez-faire ideals of most academics and the mission science of the federal bureaus:

> No doubt many of the younger scientific men . . . have beautiful dreams as to what might be done by collectivism if they could have the direction of it, set the questions and assign them to special investigators. No doubt also the tendency of most of the older men is conservative, and to think that what has been done in science . . . by individual workers following their own bent is a good argument for aiding such workers now. But there are also a number of young and old men who think that much can be done by cooperation. . . . They think it wiser to give to the exceptional men the means of working where they are. . . . They believe that the present policy of the Carnegie Institution should be cooperative and not collectivism. Out of this cooperation, a form of collectivism may ultimately develop, but it does not seem wise to try to *create* it at first.[13]

Billings was arguing not for the invisible hand of laissez-faire but for the visible hand of a managed marketplace—to borrow a phrase from historian Alfred Chandler. A program of individual grants seemed to him the most natural way of improving a highly dispersed and disparate community of scientists.

Not surprisingly, academic scientists did not behave like ideal citizens of Billings's cooperative commonwealth. Most of them—and they were legion—simply hoped to get a bit of Carnegie's largesse, which they took to be unlimited and theirs by right. The publicity in *Science* and the press produced an avalanche of applications. No fewer than 1402 were received by November 1903, requesting a total of

[11]Billings, memorandum to executive committee, 1 Feb. 1902, JSB box 58 Memoranda. Billings remarks to the board, 8 Dec. 1903, board minutes, vol. 1, pp. 178–181, CIW.

[12]Silas Wier Mitchell, "Memoir of John Shaw Billings," *Biog. Mem. Nat. Acad. Sci.* 8 (1917): 375–383. A. Hunter Dupree, *Science in the Federal Government* (Cambridge, Mass.: Harvard University Press, 1957), pp. 230–231.

[13]Billings, memorandum, 29 Jan. 1903, JSB box 58 Miscellaneous.

$2,200,398—one-fifth the CIW's entire endowment. Ten expert advisory committees appointed to design intramural departments came up with projects that would cost $911,500 more.[14] Applicants resented any effort to impose criteria of selection and took for granted that if the money ran out Carnegie would provide more from his deep pockets. In short, academics expected the CIW's patronage to be just like the individual grants-in-aid programs of the national scientific societies, only lots bigger.

Growing Conflicts

Charles Walcott and Billings, who managed the extramural grants, quickly became disillusioned with academic grant-seekers. When Woodward succeeded Gilman as president in December 1904, he was warned by Billings and Walcott to expect trouble. A few months later, Walcott wrote to Billings: "I am very much interested and somewhat amused to hear of the experiences Dr. Woodward is having with University and other men. I told him that it was a duplication of what you and I had been through and that he would probably come to the same conclusions before long."[15] Woodward soon did, recalling later: "No vagaries of fiction could surpass the realities of the unrealizable ideals and of the dreams of avarice developed in this wave, which culminated in 1905–06 and is only now [1911] slowly subsiding."[16]

Woodward was in fact an old hand in the grants business when he came to the CIW. As treasurer of the American Association for the Advancement of Science (AAAS), he had managed its grants program smoothly for twelve years and expected that the CIW's grants program could be run in the same way.[17] He soon realized how wrong he was. After only six weeks on the job, he wrote to his friend George Ellery Hale that he was "almost buried under an avalanche of importunities and good advice to say nothing of bad advice." Every recipient of a grant expected it to be continued, whether or not they had produced results. Many of those who received grants lacked the facilities to carry out the research, and those who did not get grants denounced the CIW's policies and damned Woodward's management. Woodward's letters, to Harvard chemist Theodore W. Richards, for example, reveal an in-

[14]CIW *Yearbook* 2 (1903): li–lii.

[15]Board minutes, 10 Dec. 1910, vol. 1, pp. 650–651, CIW. Walcott to Billings, 26 May 1905; Woodward to Billings, 1 May 1905; both in JSB reel 20.

[16]CIW *Yearbook* 10 (1911): 12.

[17]Board minutes, 11 Dec. 1906, vol. 1, pp. 564–565; Woodward to Walcott, 18 Oct. 1904; both in CIW.

creasingly harassed and embattled man.[18] The grants program consumed three-fourths of his time, he reckoned, though it accounted for only one-fourth of the CIW's expenditures.[19] It reminded him altogether too much of the political patronage that plagued the lives of Washington scientists. "Many of the evils of the 'spoils system' already confront us," he wrote. "Some applicants file claims; many are impatient for speedy action; and many . . . speak in the possessive case with respect to grants long before they are awarded." Some, indeed, came armed with demanding letters from their congressmen.[20]

Woodward knew all too well about the "spoils system," having spent the first two decades of his professional career as a federal scientist with the U.S. Lake Survey (1882–1884), the U.S. Geological Survey (1884–1890), and the U.S. Coast and Geodetic Survey (1890–1893). Woodward had left the Geological Survey when Congress demolished its Irrigation Survey (as chief geographer in charge of the division of mathematics he was especially vulnerable). His departure from government service for a chair of mathematical physics and a deanship at Columbia University in 1893 coincided with the political dismemberment of the rest of John Wesley Powell's scientific empire.[21] With this diverse experience behind him, Woodward came to the CIW with a belief in managed research and an expectation that his reception as a patron would be a cordial one. There was no reason to expect that managing the CIW's extramural grants would be any different from the experience of the much smaller AAAS and National Academy programs. Hence, no doubt, his intense and personal reaction against academics when the CIW grants began to sour.

Woodward was surprised and dismayed when some colleges and universities cut internal funds for research when their faculty received grants from the CIW.[22] It is easy to see why this happened: heads of small colleges and universities naturally viewed outside grants as contributions toward their institutional mission, which was to teach and provide some minimal opportunity for faculty research. From Woodward's perspective, they were selfishly subverting the CIW's mission of *increasing* funds for research. Until academics understood the CIW's special place in the system of scientific institutions, he warned, grants

[18]Woodward to Hale, 5 Feb. 1905, GEH reel 38. Woodward to Richards, 26 Mar. 1906, TWR 1.5 box 3. Reingold and Reingold, *Science in America,* pp. 27–33, 40–45.

[19]Board minutes, 11 Dec. 1906, vol. 1, p. 562, CIW.

[20]CIW *Yearbook* 4 (1905): 29. Board minutes, 20 Jan. 1920, vol. 2, p. 881, CIW.

[21]Fred E. Wright, "Robert Simpson Woodward 1849–1924," *Biog. Mem. Nat. Acad. Sci.* 19 (1938): 1–24.

[22]Woodward to E. L. Mark, 24 Nov. 1905, CIW Castle. Woodward to Richards, 1 Dec. 1905, TWR 1.5 box 3.

might actually diminish local resources for research.[23] Whether or not the abuse was as serious as Woodward alleged (probably it was not), it was a good argument against the grants.

What happened, apparently, was that a sudden increase in wealth from outside collegiate walls disrupted the moral economy of a community accustomed to genteel poverty. Where extramural grants were occasional and small, college and university administrators were not tempted to make individual grants part of their financial calculations. When such grants became larger and more systemmatic, however, it was hard not to take the windfall and forget that the purpose of granting agencies was not identical with those of academic institutions. A gap opened up between expectations, based on experience in the nineteenth-century system, and the new reality of a more extensive and active patronage. Just what were the mutual obligations of patron and client? Where grants were few and small, it was not an urgent question; but the sudden scaling-up of the CIW's grants program brought it into sharp focus.

Within a year Woodward had concluded that the conflict of interest in the grants program could never be resolved. Research belonged in institutions whose mission was research; colleges and universitites should be left to educate. "Except in a small way, it will prove impracticable to cooperate with other institutions organized for other purposes," he wrote to Richards. The CIW had to "work out its own destiny."[24] That destiny was to take two forms: in-house departments, and a small number of full-time extramural research associates.

Woodward's plan was to give grants only to individuals who had already demonstrated their capacity for productive research and who had institutional backing. Fewer grants would be made, but they would be larger and for longer periods of time. Recipients would be in effect employees of the CIW, not attached to any particular department but salaried by the CIW and accountable to it. Woodward proposed to give individuals so honored the title "research associate," to remind them that they would be held up to the same standards of productivity and efficiency as researchers in the CIW's departments. Woodward expected that associates would devote their whole time to research: they would be a special elite of research professors or, from another point of view, external research staff of the CIW.[25]

[23]CIW *Yearbook* 4 (1905): 29–30; 5 (1906): 30–37. Board minutes, 11 Dec. 1906, vol. 1, pp. 562–569, CIW.

[24]Woodward to Richards, 26 Mar. 1906, TWR 1.5 box 3.

[25]CIW *Yearbook* 5 (1906): 30–37. Board minutes, 11 Dec. 1906, vol. 1, pp. 562–569, CIW.

Well aware that this plan might not be looked upon favorably by the trustees and recipients of minor grants, Woodward took no hasty action.[26] He prepared a statistical report on the minor grants program, which showed that only about 60 percent of recipients had published any results, a poor showing in comparison with the AAAS grants.[27] An opinion poll of academic scientists and administrators showed that a majority (60 percent) liked the present system of minor grants, although some of the most eminent university presidents did not.[28] Woodward gathered numerous letters from academic scientists applauding his idea of research associates, especially from those who already had grants and had been encouraged by Woodward's hints that they would be strong candidates for long-term support.[29]

The response of the executive committee to Woodward's plan was not what he hoped. Billings and Silas Wier Mitchell (a Philadelphia physician, physiologist, and author) prepared a rebuttal to his report, opposing the research associates scheme and arguing for the minor grants to "exceptional men."[30] The stage was set for a confrontation with the full board at their meeting in December 1906.

In his presentation to the board, Woodward dwelt on the inefficiency of the minor grants: they were not as productive of research as the departments and were an unproductive use of his own time. Fruitless interviews and correspondence with hopeful or disappointed applicants kept him from attending to the productive business of the CIW's departments. The only way to ensure that the CIW got a good return from its investment in individual grants, Wooodward contended, was to clearly separate research from education. Research associates had to be selected by the same standards as regular department staff: not for youthful enthusiasm and promise but for a track record of productive research on important problems. Associates would be forbidden to delegate research to student assistants, and they had to be employed in universities where research had already become an accepted requirement of academic careers.[31]

The fatal flaw in Woodward's case was that it entailed giving up the

[26]Woodward to Richards, 6 Oct. 1906, TWR 1.5 box 3.

[27]I have not found a copy of this report, but the results are summarized in Woodward to Billings, 13 Nov. 1906, JSB reel 21. Of 98 minor grants, 61 resulted in publication, 28 did not, 9 failed completely.

[28]Board minutes, 11 Dec. 1906, p. 564, CIW

[29]Woodward to Richards, 1 Dec. 1905, 6 Oct. 1906, TWR 1.5 box 3. Woodward to T. C. Chamberlin, 16 Nov. 1906; Woodward to E. L. Nichols, 9 Oct. 1906; A. A. Noyes to Woodward, 5 Dec. 1906; all in CIW.

[30]Board minutes,, 11 Dec. 1906, p. 581, CIW.

[31]CIW *Yearbook* 5 (1906): 35–36. Board minutes, 11 Dec. 1906, vol. 1, pp. 565–566, CIW. Reingold and Reingold, *Science in America,* pp. 47–49.

search for "the exceptional man." For believers in the "exceptional man," Woodward's language of productivity and efficiency was simply a reminder that the minor grants were designed not to produce research but to widen participation in the process of research by assisting disadvantaged young scientists. Woodward knew that he might have to fight hard for his research associates plan. He was not prepared, however, to have it summarily dismissed.

Billings was not present at the critical meeting, owing to a recent operation, and it was Mitchell who presented the case against the research associates scheme. Mitchell minced no words: he sympathized with the inconveniences of administering a large number of gants, but his own experience suggested that Woodward exaggerated the hardships. Besides, that was what he was paid to do. Carnegie had established the CIW to discover exceptional men, and Woodward's job was to carry out the wishes of the founder and his trustees. He and Billings disliked the idea of supporting only "men of middle age and assured competence by success in many investigations," observing that he himself would not have qualified for support at a time when he could have best used it. Mitchell hoped there would be "no serious warfare between the President and Executive Committee," and on that note the board moved on to other matters.[32] The president was put in his place as the servant of the board, not its master. Woodward neither forgot nor forgave it.

President versus Board

The confrontation in December 1906 was just one in a series of skirmishes between Woodward and the executive committee, in which both sides tried to assert their authority. In selecting Woodward to succeed Gilman, Billings and Walcott may have been trying to make sure they kept control. Woodward was their hand-picked candidate, and they lobbied vigorously on his behalf against trustees who argued that Woodward's record as administrator failed to show "any executive capacity whatsoever." The Boston faction, businessman Henry L. Higgenson and Harvard zoologist Alexander Agassiz, favored either Ira Remsen or Columbia's Henry Pritchett, both of whom were less accomplished scientists than Woodward but more experienced administrators.[33] In fact, Pritchett was offered the presidency of the CIW, but he declined it, apparently because it became clear to him that

[32]Board minutes, 11 Dec. 1906, vol. 1, pp. 581–582, CIW. Reingold and Reingold, *Science in America,* pp. 49–50.

[33]Higginson to Billings, 25 Nov. 1904, 19 Oct. 1904; W. N. Frew to Billings, 6 Dec. 1904; Mitchell to Billings, 9 Dec. 1904; all on JSB reel 20.

he would be expected to do the bidding of the executive committee.[34] Probably Billings and Walcott expected that Woodward would accommodate more easily to their way of doing things; he was, after all an old Washington hand.

For a time, Woodward did dance to the tune of the executive committee, communicating regularly with Billings and Walcott, informing them of what he was doing, and seeking their advice. That changed, however, as Woodward gained experience. The flow of information dwindled. In October 1905 Woodward requested that he replace Walcott as secretary of the executive committee, to Walcott's dismay, and in November he asked for a seat on the board of trustees, obviously with the intention of cutting Walcott out as middleman between himself and the board. A few months later Walcott resigned as secretary.[35] Woodward proved more independent than expected, using his command of routine administration effectively to divert authority from the executive committee to his office. This silent internal struggle continued for some years and was potentially a factor in every policy decision, especially in the matter of minor grants.

The grants program gave the executive committee regular opportunities to challenge Woodward's control, because he needed their approval to spend money. He would screen grant applications and prepare a summary form and a recommendation for the monthly meetings of the executive committee. Accustomed to deciding on grants themselves, Billings, Mitchell, and Walcott apparently continued to take a personal interest in the applications. Too personal, Woodward thought: he described the process as a "species of Havana lottery, with monthly drawings, in which the inexperienced and inexpert man is almost as likely to receive a prize as the expert and the experienced man."[36] It appears that Woodward's recommendations were generally accepted, but occasional episodes reminded him that he was not in control. Once, for example, Mitchell and Billings were incensed when Woodward failed to consult them in advance about an application from a medical scientist of whom they disapproved. Occasionally, disgruntled applicants would go over Woodward's head to Billings.[37]

34A. Flexner, *Henry S. Pritchett* (New York: Columbia University Press, 1943), pp. 157–158. Mitchell to Billings, 14, 16 Dec. 1904, JSB reel 20. Woodward to T. C. Chamberlin, 11 July 1904, in Reingold and Reingold, *Science in America,* pp. 24–25.

35Walcott to Billings, 24 Oct, 5, 20, 28 Nov. 1905, JSB reel 21.

36Board minutes, 11 Dec. 1906, vol. 1, p. 563, CIW. Executive committee minutes, 16 Oct. 1905, JSB box 59.

37Mitchell to Billings, 16 May 1907; Billings to E. Root, 3 June 1907; Woodward to Billings, 20 Dec. 1905, 17 May, 1907; all on JSB reel 21.

Woodward also had to come to the executive committee for approval of expenditures of the CIW departments. There, too, he could be challenged and occasionally was; Billings balked, stalled, and queried before signing checks. He could harass Woodward but not diminish his power to take the initiative, especially in the CIW departments, where Woodward had a close personal relation with directors like George Ellery Hale.[38] It was the monthly "Havana lottery" of the minor grant where the struggle for authority ground Woodward down.

For Woodward, the Geophysical Laboratory and the Mt. Wilson Solar Observatory were the ideals of organized research: institutions with clearly defined missions and controlled by leaders like the geophysicist Arthur L. Day and George Ellery Hale, who combined scientific vision and administrative efficency, and who were clearly accountable to the CIW for the funds they spent. The extramural grants were doomed to failure, in Woodward's view, in institutions where research was frosting on the cake of college teaching and where recipients of CIW grants could not be made accountable. Woodward's vision of a system of "associates" employed by the CIW and accountable to it was an extension of his experience with the CIW's department heads—not surprising, since that relation had proved easy and productive, while the extramural grants were nothing but trouble.

In public Woodward was philosophical about the trustee's rejection of his plan for research associates, but privately he felt personally affronted.[39] He determined to put his plan quietly into practice without the formal approval of the board. In fact, as he wrote to Hale, the system was more or less in operation even as he sought the board's approval.[40] For several years he had been allowing minor grants to lapse while renewing those to eminent and productive scientists for longer periods of time. Few new applications were approved. The board continued to appropriate large sums for minor grants, but Woodward simply did not spend them. In 1906 expenditures were half what they were in 1904, and in 1908 they were cut by a third. The number of grants in force declined dramatically, from a peak of seventy-five in 1905 to a steady state of about thirty-two by 1909. The net loss was sharpest in 1906 and 1907, when thirty-seven grants were terminated and only five begun.[41] It is no wonder that Woodward's relation with his academic clients reached a nadir in these years! In 1910 Wood-

[38]Woodward to Billings, 24 Jan., 26 Feb. 1906, JSB reel 21.

[39]Board minutes, 10 Dec. 1907, vol. 1, pp. 648–651, CIW. Woodward to Hale, 12 Dec 1906, GEH reel 38. Woodward to Richards, 7 Dec. 1907, TWR 1.5 box 4. Woodward to Chamberlin, 21 Jan., 8 Mar. 1910, CIW.

[40]Woodward to Hale, 12 Dec. 1906, GEH reel 38.

[41]Data are from CIW *Yearbooks*.

ward wrote to his friend, the geophysicist Thomas C. Chamberlain, that his research associates scheme was a practical reality, although the trustees still refused to make grantees official employees of the CIW.[42]

Woodward also moved forcefully ahead with the organization of new departments. Six had been approved before he arrived, and between 1905 and 1907 he shepherded four more through the board. But for financial constraints, there might have been more: for example, Woodward and T. W. Richards often talked longingly of a department of chemistry for work on precise measurement of atomic weights. But as existing departments expanded their operations and budgets, the limit of income from Carnegie's endowment was soon reached.[43] Woodward repeatedly warned the board that the minor grants were threatened by declining real income. No doubt that was so, but it is also apparent that Woodward was squeezing Carnegie's favorite program in the hope of persuading him to make a large addition to the CIW's endowment.[44]

The ploy worked: Carnegie gave a second gift of $10 million in January 1911. Woodward gave the credit to department chiefs like Hale, who had built the magnificent temples of science that so impressed the founder. But Woodward, too, had been working relentlessly on Carnegie. He visited him in Skibo, took up golf, even adopted (briefly) Carnegie's simplified spelling, and suffered patiently the founder's headstrong and whimsical ways.[45] The minor grants were expanded somewhat, but most of the new funds went to the departments.

Carnegie's addition to the endowment was one of several things that, in the early 1910s, gave Woodward a greater sense of security and confidence in his role of manager of science. The most important thing, he thought, was the acceptance by academic scientists of the CIW departments. The prestige of operations like the Mt. Wilson Observatory vindicated his belief that research institutes were a superior way of organizing research and that criticisms of the CIW by academics were misguided and self-serving. This became a recurrent theme in Woodward's reports after 1911.[46] Probably Woodward exaggerated academic opposition to the CIW departments, but even so it is clear that he was finally feeling somewhat more comfortable in the role of manager of science.

[42]Woodward to Chamberlin, 21 Oct. 1909, 21 Jan., 8 Mar. 1910, CIW.

[43]CIW *Yearbook* 7 (1908): 26; 8 (1909): 39; 10 (1911): 12. A department of physical chemistry was discussed in 1917. Woodward to Pritchett, 23 May 1917, CIW.

[44]CIW *Yearbook* 9 (1910): 23; 10 (1911): 7–8.

[45]Hale to Woodward, 30 Mar. 1910; Woodward to Hale, 11 June, 21 Nov., 18 Dec. 1910, 6 Feb., 19, 22 Dec. 1911; all on GEH reels 38 and 39.

[46]CIW *Yearbook* 11 (1912): 10–12; 12 (1913): 11; 13 (1914): 12–13; 14 (1915): 11–17. Woodward to Hale, 19 Dec. 1911, GEH reel 39.

Woodward's relations with the board also improved after 1910. Oddly, Woodward's closest personal friend among the trustees was Silas Wier Mitchell, once Billings's chief ally in the fight for the "exceptional man." A large and warmly personal correspondence with Mitchell reveals the octogenarian author and medical statesman in the role of confidante and conciliator in Woodward's relations with the board. They exchanged reflections and opinions about the philosophy of science and literary matters. Woodward felt Mitchell's death in 1914 as a personal loss: of all the trustees, he reflected, Mitchell had best understood him as a person.[47] Billings, however, continued to be a thorn in Woodward's flesh right up until his death in March 1913. Billings harassed him in the monthly ritual signing of checks, and in 1911 there was a violent confrontation, in which Woodward seemed to have reached the limits of his patience.[48] Only when Billings was safely buried could Woodward afford to be philosophical. He wrote to Hale: "Senator Root characterized him fitly as we were on our way to Arlington [Cemetary], as a man without malice but with at times much cussedness."[49]

The more tolerant and clear-cut division of authority between president and board did not, however, resolve the issue of the minor grants. That continued to be the one source of conflict with the executive committee.[50] To quiet Woodward's complaints about the inordinate amount of time he had to spend administering the minor grants, the board offered to hire an assistant to take charge of the program. Woodward refused, however, presumably because that would have strengthened a system that he still hoped radically to change.[51] Woodward continued to press the board to create a division of research associate with funds for fifty to 100 new full-time extramural researchers. It never happened: no doubt the trustees sensed that it would mean the end of the CIW's quest for the "exceptional man."

It is no accident that the struggle between Woodward and the board was keenest in the question of the extramural grants. Woodward's sys-

[47]Mitchell to Woodward, 3 Oct. 1910, 25 Mar. 1911; Woodward to Mitchell, 28 Nov. 1910, 20 Nov. 1911; Woodward to Cleveland Dodge, 14 Jan. 1914; all in CIW.

[48]Woodward to W. H. Welch, 28 Nov., 2 Dec. 1911; Welch to Woodward, 30 Nov. 1911; all in CIW.

[49]Woodward to Hale, 15 Mar. 1913, GEH reel 39.

[50]Woodward to Hale, 13 Jan. 1912, GEH reel 39. Woodward to Mitchell, 24 Nov. 1911, CIW.

[51]Cleveland Dodge to Woodward, 12 Jan. 1912; Woodward to Dodge, 13 Jan. 1912; both in CIW. Hale to Woodward, 24 Jan. 1911, GEH reel 39. Also discussed and rejected was a plant to assign one or more members of the executive committee to assist Woodward on the minor grants. Mitchell to Woodward, 23 Nov. 1911; Woodward to Mitchell, 25, 27 Nov. 1911; all in CIW.

tem of research associates would have enlarged his authority as manager, allowing him to set the terms of appointment and determine how the work of associates was organized. Grants-in-aid to "exceptional men," in contrast, limited Woodward's discretionary authority. No wonder, then, that Billings saw eye to eye with Woodward's clients. Power was not the only thing at stake, but it had to be an element in every point of policy at a time when the powers of board and manager were still up for grabs.

Manager versus Constituents

Woodward's relations with recipients of minor grants were no less troubled than his relations with the board, even more so. Woodward and the board at least operated in a common institutional context with common rules. With his academic clients, Woodward often found himself talking at cross-purposes. Academics believed that a grant from the CIW had no strings attached and did not carry with it any obligation to change academic ways of doing things. Woodward expected recipients of grants to be accountable just as if they were employees of the CIW's departments, though he never said so openly. Grant-giver and grant-getter operated in different frameworks and thus tended to treat the other's behavior as incomprehensible and deviant.

It is significant that the most heated exchanges occurred not with disappointed grant-seekers but with people like Arthur Noyes and Thomas C. Chamberlin, Woodward's close friends and exemplars of what he thought research associates should be. Their differences remained submerged but surfaced occasionally, usually in connection with some minor administrative matter. Such episodes were painful reminders to both parties of how differently they saw the patron-client relation. They are also fascinating glimpses of that new partnership in the making.

Scientists were ambivalent about the research associate scheme from the start: that is clear from the letters Woodward received in his opinion survey in 1906. Cornell physicist Edward L. Nichols applauded Woodward's desire to eliminate grants to "the immature, the dilettante, and the tyro" (who could disagree?) but thought small grants a useful way of promoting science. Hale declared that he was "a great believer in the efficiency of small grants," pointing to the good results he had obtained with the National Academy's William Hale Fund. (He acknowledged, however, that small grants-in-aid might be practicable only on a small scale.) Noyes approved the research associates idea as a superior method of "finding the exceptional man"—not exactly what Woodward had in mind. Chamberlin agreed that the CIW should stay out of education but lectured Woodward that small individual grants were better suited than permanent research institutes to exploit unexpected "bonanza" research problems. Engineer and metallurgist Henry Howe made a classic case for small grants-in-aid, unaware of the political mine field into which he stepped.[52] "Look at the facts," Howe wrote.

> There are scattered through the world a very great number of men of very great ability who are willing to give their time for nothing for making scientific investigations, and are able to use the laboratories, buildings and collections of their institutions, and the services of the members of their staffs, and of the advanced students working under them, all for this purpose. They have everything to carry on the investigation . . . everything but money. . . . Money sent to them is like seed sowed in a farm already plowed, manured and irrigated, and with the purchase price and taxes already paid. For work which you do at the Carnegie Institution you have to pay the whole expense of the salaries of your men, the installation and maintenance of the laboratories—in short you not only have to provide your seed, but pay for your farm, plow, manure and irrigate it.[53]

Howe was surprised there was any question about which was the better way.

Woodward claimed his opinion poll supported his plan for research associates. In fact, there was real agreement only on motherhood issues like quality and professionalism. More significant is the general uneasiness with the underlying assumption of Woodward's plan: namely, that research could be practiced in the same way in universities as it was in research institutes. The political economy and culture of academic science were different from those of the CIW departments. Research was part of a more complex professional role, which most professors liked and wanted to preserve. Limited resources and egalitarian mores discouraged a two-class system of research and teaching faculty. Universities had always distributed resources in a parliamentary way among departments, never strictly by merit and productivity. Woodward seemed not to appreciate these differences.

Support for student assistants was the most frequent cause of contention. Woodward expected "associates" to work at the bench and not

[52]E. L. Nichols to Woodward, 11 Oct. 1906; Noyes to Woodward, 5 Dec. 1906; Chamberlin to Woodward, 24 Nov. 1906; H. M. Howe to Woodward, 23 Aug. 1906; all in CIW. Hale to Woodward, 20 Dec. 1906, GEH reel 38.

[53]Howe to Woodward, 23 Aug. 1906.

delegate research to inexperienced juniors, whose main purpose was to acquire a degree. For Woodward, paying graduate research assistants with CIW funds was an abuse, even though being an "associate" of the CIW hardly relieved professors of their obligations to teach and provide problems for Ph.D. dissertations. Where Woodward saw a conflict of interest, academics saw a familiar and effective symbiosis between training and research.

Ironically, dependence on student assistants was most prevalent among academics who had begun to learn how to organize large-scale programmatic research. Professors who did occasional researches, just to keep their hand in, could manage to work at the bench. Not so those who brought a more entrepreneurial business style to academic science—people like Richards, Noyes, Nichols, or Thomas Hunt Morgan, all of whom managed large groups that systematically exploited specific lines of research. The more academics behaved like scientists in bureaus or research institutes, the more efficient they had to be in exploiting every kind of local resource, and that meant the cheap labor of graduate students. Thus, they were more likely than isolated individual researchers to run afoul of Woodward's prejudice against combining research and training.

Noyes and Richards especially stumbled on that hidden root of discord. In 1906 Noyes was supervising the research of fifteen apprentice researchers, two or three of whom he thought might qualify as "exceptional men" for long-term CIW support. Richards, too, used students extensively for research in his private laboratory. They made no secret of their practice to Woodward; indeed, they went out of their way to emphasize how both sides benefited from it. Universities spared the CIW the cost of building new laboratories; the CIW gave universities what their alumni would not, money for graduate training and research. The CIW got good research cheap; Harvard and MIT enjoyed an intellectual leaven they would otherwise not have been able to afford.[54] To Woodward, however, it was false economy. He dismissed Richards's argument as "fine sentiment," asserting that in his experience teachers and students were not productive researchers. He wanted to invest in those who directed research, not those who were directed.[55] He did nothing to stop Richards or Noyes from using student assistants, however, probably because they were doing outstand-

[54]Richards to Woodward, 11 Dec. 1905; 14 Apr. 1908; both in TWR 1.5 box 3. Noyes to Woodward, 5 Dec. 1906, CIW.

[55]Woodward to Richards, 30 Mar. 1906, 16 Apr. 1908, TWR 1.5 box 4. Woodward to C. L. Jackson, 29 June 1906; Richards to W. A. Noyes, 23 Nov. 1905; both in TWR 1.5 box 3.

ing work and because he knew and trusted them as individuals. But the underlying difference in principle was like a land mine waiting for someone to make a false step.

Noyes unwittingly made such a step in early 1909, when he sent Woodward a copy of a printed Ph.D. dissertation in which the CIW's aid was acknowledged. Outraged, Woodward demanded an explanation of why funds had been diverted to help a student get a degree, hinting ominously that "it is certain to make trouble for the Institution, and I fear may make trouble for you."[56] Taken aback, Noyes explained that the student had been given the choice of working as a graduate student for $50 a month or as Noyes's professional assistant for twice that amount; choosing the former, the young man got a degree, and Noyes more than doubled the value of the CIW's grant. Moreover, graduate students could be counted on to stay on the job for three years while experienced assistants, even with annual increases in pay, often departed after a year or two, leaving Noyes with the problem of training inexperienced recruits in the middle of a project.[57]

Noyes had assumed that Woodward understood and approved such standard practices. In any case, he hoped Woodward would not be more concerned with procedure than results:

> The question for the administration of the Institution to decide seems to me to be whether, when grants are made to professors at educational institutions, it will hold them responsible for *results*, making the continuance of their grants depend on this, or whether it desires to restrict by specific regulations the freedom of its grantees in the use of the funds assigned to them because of the fear of abuses by individuals or of criticism by outside persons. In view of the fact that such minor grants are now made only to a few carefully selected investigators, it would seem to me best for the Institution to pursue the methods that will lead to the greatest effectiveness, avoiding such hampering restrictions as are sometimes necessary in large administrative organizations, like the government, where direct personal responsibility and supervision are impracticable; and that it is wisest to deal with individual abuses individually and to ignore the ill-founded criticisms that may be made by outside persons who may be uninformed or disgruntled. But these are, of course, questions for you to decide.[58]

[56]Woodward to A. A. Noyes, 2 Mar. 1909, CIW.
[57]Noyes to Woodward, 6 Mar. 1909, CIW.
[58]Ibid., pp. 4–5.

Not at all pleased by Noyes's lecture on administrative philosophy, Woodward retorted that it was indeed for him to decide, and that he must decide on general principle and not particular cases. While Noyes may have saved $2,000, ten times that amount was wasted by others who delegated research to inexperienced students. As Woodward saw it, Noyes was arguing for what would most benefit universities, ignoring entirely the CIW's quite different goals.[59] Stung by Woodward's accusation of academic self-interest, Noyes again explained that the common goal of producing good research seemed to him to transcend any differences in the way research was done. "Is it not very probable," he asked, "that the best methods of securing such results may be quite different when the work is to be carried on by professors in educational institutions . . . [and] when it is carried on in a laboratory devoted exclusively to research?"[60]

The proximate cause of this spirited set-to was easily remedied. The offending acknowledgment was omitted, and Woodward grudgingly allowed that Richards and Noyes could pay graduate assistants with CIW funds, if they did it discretely. The fundamental difference, however, was not resolved. Noyes failed to understand that Woodward, as head of a research institute, was bound to see grants as an extension of his institution's mission. Obsessed with the mechanics of the selection process, Woodward failed to understand that a workable large-scale system of grants would require accommodation with the realities of academic work. Noyes offered him a practical rationale for the relation between sponsor and client. Woodward heard it as special pleading.

The other side of the same issue surfaced a year later in an exchange with Thomas C. Chamberlin over what the CIW owed to its "associates." Chamberlin had received grants every year since 1903 and had been an "associate" for several years. He had mistakenly interpreted his new status as a guarantee of automatic renewal. In January 1910, however, he was shocked to learn that his grant had not been renewed. After more or less recovering his composure, he wrote Woodward to complain. As an experienced and productive investigator in the advanced stage of a large project, Chamberlin felt humiliated at being compelled to beg for annual renewals. He had assumed that his elevation to "associate" status was equivalent to academic tenure. Such security was essential for the relation between patron and researcher: "Otherwise," he wrote, "research is really placed at a disadvantage as compared with university service, and research is rather humiliated than honored, for its basis is made precarious and even hazardous."[61]

[59]Woodward to Noyes, 10 Mar. 1909, CIW.
[60]Noyes to Woodward, 11 Mar. 1909, CIW.
[61]Chamberlin to Woodward, 14 Jan. 1910, CIW.

Woodward replied with some heat that Chamberlin had no right to assume that he had a permanent grant or that he was excused from giving detailed specifications of research needs and purposes. Woodward did tacitly acknowledge that the title "research associate" was misleading but blamed the trustees for the subterfuge. (In fact, Woodward was as much to blame for adopting a misleading name in the hope that it would in time become a reality.) Chamberlin was one of Woodward's oldest and closest friends. They worked on similar problems in mathematical geophysics, and professional kinship made Chamberlin's criticism seem like a personal betrayal. Woodward complained bitterly that even old friends were not understanding and considerate of the difficulties he labored under, carping when they should be giving constructive advice.[62]

Whether or not Woodward really wanted advice, Chamberlin gave it to him, sixteen pages of it. Realizing from Woodward's letter just how profound their misunderstanding was, Chamberlin put his personal grievance aside and sought to make Woodward understand why the grants program looked so different to his academic clients. The basic problem, he thought, was that Woodward seemed to be operating under his own rules, not those set down in the CIW *Yearbook.* The stated purpose of the program seemed to be to enable people with an exceptional capacity for research to have secure, full-time careers in research. In practice, however, Woodward seemed to be more concerned with particular projects and research lines: "No exceptional man has been found in America or elsewhere," Chamberlin complained, "unless it be done indirectly in connection with some project." Academics had expected the CIW to offer a "distinctive inspiration" to researchers, that is, a degree of freedom in research that academic or professional jobs could not afford—not even in the CIW departments.[63]

Chamberlin tried to make Woodward see how academics had misunderstood the purpose of his shift to "research associates." The "minor grants," they had assumed, were a way of sifting out the researchers who were most fit for long-term support. Woodward's announcement in 1906 that grantees would in future be "associates" of the CIW was taken by the survivors of the sifting to mean that they had won a tenured connection with the CIW. Having carefully reread what Woodward actually put down in black and white, Chamberlin had to admit that no such promise had been made. However, Woodward's letters and conversations had given him the definite impression that individuals like himself, or Richards, or mathematical astronomer Forest R.

[62]Woodward to Chamberlin, 21 Jan. 1910, CIW.
[63]Chamberlin to Woodward, 11 Feb. 1910, p. 3, CIW.

Moulton—"masters of research"—had left their probationary status behind for a more independent and permanent position.[64]

As Chamberlin saw it, the minor grants had worked well in discovering "the fittest." Where Woodward had failed was in persuading academic scientists of the legitimacy of a selection process in which they were themselves not a part: "The real difficulty is merely in *a competent jury to pass on the fact* [of fitness]."[65] The obvious solution, it seemed to Chamberlin, was for Woodward to appoint a body of eminent scientists to advise and approve of his selection of research associates. Such a "judicial auxiliary" would present "a wall of defense" against the inevitable resentment and criticism, which Woodward had to bear entirely by himself, "for the simple reason that no one else was known to be responsible."[66] Chamberlin reminded his friend that they themselves, as members of the CIW's advisory committee in geophysics in 1902–1903, had strongly advocated using expert advisors to select researchers and projects. The committee system had worked well, and Chamberlin had hoped it would become a permanent feature of the CIW. It had not, however, and the result was the personalized criticism that Woodward daily had to endure.[67]

Chamberlin's long letter was a thoughtful effort to understand the evolving relationship between patron and client. Chamberlin tried to see both points of view in a way that Woodward never did. He did not completely succeed, of course: he was no more free from his own professional self-interest than was Woodward. His experience as a tenured professor in an elite university was a misleading model for his relation to the CIW, just as Woodward's relation with the CIW's in-house staff was a misleading model of his relation to the "associates." Chamberlin took his privileges too much for granted. He was too much the nineteenth-century individualist to fully grasp the new rules of large-scale patronage.

A system of peer review would have been fine for a national scientific society. But the moral economy of the National Academy's grants-in-aid could not obtain in the scaled-up grants program of the CIW. Large-scale patronage had to have policies and projects. The mechanics of applications, appraisals, and accountability were inescapably part of the large-scale patronage of science. So too was the "project" mode of organization. Faced daily with the problems of managing grants, Woodward learned some of these lessons faster than his academic colleagues. Lacking experience as a recipient of grants, he was slower to appreciate the

[64]Ibid., pp. 4–5, 11–13.
[65]Ibid., p. 5.
[66]Ibid., pp. 7–8.
[67]Ibid., p. 5–7.

importance of flexibility, tolerance, and a light touch. Also, he never was able to acknowledge how much confusion he had sown by calling grantees "associates" when they were not and could never be employees of the CIW. In the absence of a new vision of the patron-client partnership, the two sides talked at cross-purposes.

To Woodward, Chamberlin's reflections were just another display of academic "fallacies"—misguided idealism, ignorance of practical administration, failure to give constructive advice, and ingratitude.[68] He summarily dismissed Chamberlin's suggestion of expert advisory committees, citing the "bitter complaints against what were called cliques and favoritism generated by such boards and committees." The fundamental limitation of such committees was that they could not be held accountable, and thus were free to recommend grandiose schemes for spending far more than the CIW had to spend. He recalled with a deep sense of shame, he said, how he and Chamberlin had done precisely that on the geophysics committee, thus sowing the seeds of his own tribulations. His subsequent experience with academic advice had been the same: it was parochial, impractical, or self-interested, none of it of any real use.[69]

Woodward was especially irked by Chamberlin's invocation of the infamous "exceptional man":

> Like thousands of other correspondents, you quote Mr. Carnegie's happy frase about the exceptional man; but [you do not] . . . see that that happy frase was destind to work no end of difficulty to the Institution. Why can you not see the fact . . . that there are almost as many different views of the application of that frase as there are individuals desiring its application to them. The quotation to me of this article from the Founder's Deed of Trust could be tolerated in its voluminous triteness if I could even occasionally encounter a correspondent who would do so much as to suggest a mode of general application.[70]

Obviously, Chamberlin had caught his friend at a bad moment. Billings was being especially irritating, and Carnegie was ignoring his pleas for more endowment. Harassed and tired, Woodward was unable to see that Chamberlin's advice was constructive, even wise.

Obviously, Woodward's exasperation with Noyes and Chamberlin had a personal dimension; however, similar exchanges with other long-term "associates," like chemists Gregory Baxter and Henry Sherman,

[68]Woodward to Chamberlin, 8 Mar. 1910, CIW.
[69]Ibid., p. 1–5. CIW *Yearbook* 11 (1912): 10–12.
[70]Woodward to Chamberlin, 8 Mar. 1910, pp. 3–4.

reveal that the conflicts were not personal but constitutive of the relation between patron and client.[71] Little changed in the CIW grants program between 1914 and Woodward's retirement in 1920. Woodward still seized every opportunity to complain about academic "fallacies," even though the grant files reveal less and less real controversy and more and more the routine of annual reports and renewals. Woodward's ritualized complaints became increasingly fatalistic and misanthropic. It became too much even for Richards, who urged his old friend to be more forgiving of human frailties.[72]

A First-Generation Manager of Science

In his swan song as president, Woodward reflected ruefully on the limitations of human action:

> I have sometimes said to my wife and friends that gods or even angels could not have made it much different from what it has been. One who has studied the doctrine of evolution and who has read history carefully, particularly the history of science, must come to believe I think that all of the agencies, the organizations, the developments of mankind are in a sense very largely determined not simply be environment but by the long course of history out of which we and our organizations have come.[73]

The history of the CIW grants program lends support to Woodward's view. This experiment in large-scale patronage of academic research, the first ever, was shaped by interests and expectations rooted in different historical experiences.

In the 1880s and 1890s, scientists in colleges and universities had improvised a role for research by adapting it to departments and collegiate teaching. Naturally, they looked to make the most of their historical advantages: especially their monopoly in the training of scientists, who in their apprenticeship years constituted a pool of cheap and eager research labor. Academic scientists were untroubled by their improvised, dual-purpose role. In contrast, research institutes like the CIW or Rockefeller Institute were created and overseen by businessmen, who were accustomed to the functional organization of work

[71]Woodward to H. C. Sherman, 23 Sept. 1911, 28 Aug., 13 Sept. 1915; Sherman to Woodward, 29 Sept. 1911, 31 Aug. 1915; G. P. Baxter to Woodward, 15 Nov., 19 Dec. 1912, 24 Mar. 1915; Woodward to Baxter, 19 Dec. 1912, 26 Mar. 1915; all in CIW.

[72]CIW *Yearbook* 14 (1915): 11–18. Board minutes, 10 Dec. 1920, vol. 3, p. 10, CIW. Woodward to Richards, 22 Mar. 1918; Richards to Woodward, 6 Apr. 1918; both in TWR 1.8 box 4.

[73]Board minutes, 10 Dec. 1920, vol. 3, pp. 48–49, CIW.

in large corporations. Leaders of research institutes also sought to make the most of their historical advantages: endowments specifically for research, business organization, and a new role for professional, full-time researchers. No wonder Woodward was so misunderstood and resented.

Many scientists had managed funds before—Pickering, George Minot, Billings, Woodward himself—but they had done so as volunteers and as members of universities or professional societies. Woodward was the first to make patronage his full-time occupation, and as the representative of the CIW he came to embody the managerial ideals of his institution. Psychologist and editor James McKeen Cattell recalled how, as colleagues at Columbia, he and Woodward had complained about the bureaucratic methods of Billings and Walcott, and how they had hoped that Woodward would do more as president to enrich scientific careers and promote cooperation and solidarity. To Cattell, it seemed that Woodward had only made the bureaucratic machinery more efficient and oppressive: "I am . . . compelled, to my great distress, to say the things about you that we used to say about Billings and Walcott."[74] The scaling-up and professionalization of science patronage had the effect, initially, of widening the breach between patron and academic clients. Academics began to see Woodward as an outsider and interloper in their domestic business rather than as an insider with delegated authority to distribute funds. Woodward came to feel that he had lost cast among his former colleagues. "I would rather write a chapter in supplement to [Laplace's] *Mechanique Céleste* than be president of the C. I. of W.," he wrote Hale. "The latter job necessarily brings disappointment to the great majority of my contemporaries, if not friends, while the former disconcerts no one and may even live to the permanent advantage of our race."[75]

Woodward's managerial practices also reflected the larger social context in which he operated. His years as president of the CIW overlapped almost exactly with the high tide of progressivism and the American public's romance with the scientific method. Although he thought Frederick Taylor's "scientific management" was dangerous (a public that expected miracles from science must eventually be disappointed and turn against it), as a manager of science he took heart from the growing belief that planning and business methods should be applied to science. "Research," he wrote, "like architecture and engineering, is increasingly effective in proportion as it is carefully

[74]Cattell to Woodward, 9 Mar. 1907, JMC box 46.

[75]Woodward to Hale, 23 July 1913, GEH reel 39. Woodward to Cattell, 11 Aug. 1913, JMC box 158. Woodward to Richards, 9 Oct. 1913, TWR 1.8 box 1.

planned and executed in accordance with definite programs."[76] For Woodward, programs and business methods were what made the CIW superior to universities: they were its raison d'être; they became his raison d'être as a professional manager of science.

Woodward's annual call for budgets from his department heads became a ritual occasion to exhort them to be efficient and business-like: the CIW could win the trust of the public only by showing that it was more responsive to public issues than the National Academy, and more efficient and business-like than colleges and universities in using public resources. If research institutes could do that, there would be plenty of funds available; if not, they only expect a "chronic poverty."[77] Woodward never lost the feeling that he had to prove himself as a businessman to Carnegie and the practical businessmen on the board. Thus were managerial ideals of planning and cost-effectiveness transferred from the corporate sector to science.

Woodward attempted through the extramural grants to teach business ideals to academic scientists. He berated academics on their unbusinesslike ways and on the need for definite programs of research and careful budgeting.[78] No doubt business methods were a practical necessity for administering a large grants program, but the strong moral flavor of Woodward's exhortations suggests that it was more than that. Extending the ideals and practices of the CIW research departments to universities was for Woodward an important part of his personal and institutional mission.

In retrospect, individual research grants were not an appropriate vehicle for changing academic science. They were the most obvious vehicle in the early 1900s, familiar from the programs of the national scientific societies and the Elizabeth Thompson fund. It was natural for the new patrons of science to assume that they could simply be scaled-up. But as so often happens in the growth of complex organizations, what works on a small scale does not work on a larger one. Individual research grants led academics to assume that the relation between patron and client would be the same with the CIW as with the National Academy or AAAS. It was not. Scaling-up from $2,000 to $100,000 entailed a more intensive style of management and a more intense, even agonistic relationship between grant-giver and grant-getter, because their aims in giving and getting were not identical. The terms of a new

[76]CIW *Yearbook* 13 (1914): 16; 14 (1915): 7.

[77]Woodward to Hale, 23 July 1913, GEH reel 39. See also Woodward to department heads, 18 Dec. 1908, 8 July 1910, 16 Dec. 1910, 15 Oct. 1912, GEH reels 38 and 39.

[78]CIW *Yearbook* 13 (1914): 12–14, 16; 14 (1915): 7.

partnership had to be renegotiated, and Woodward's stormy relations with his academic clients were symptomatic of that process.

Woodward's troubles with the grants program, most scientists believed, was the result of his overly personal style of management. Certainly, Woodward could be insensitive and heavy-handed. He was patronizing and dismissive in his relations with academics and took criticism as a personal attack. He caricatured opinions that differed from his own and ridiculed them as "fallacies."[79] But it was more than personalities. The perception of Woodward as intrusively "personal" was symptomatic of a stage in the history of science patronage when perceptions lagged behind practice. Academics expected the CIW grants to resemble the passive, even-handed distributions of the national societies; when they proved to be more directed and programmatic, academics perceived Woodward as a usurper of the familiar committee of peers. Unaccustomed to an activist manager with his own institutional agenda, Woodward's critics saw an excess of personal ambition.

Pickering too had stirred suspicion and resentment when he tried to use grants as a vehicle for making academic research more businesslike. But whereas Pickering was acting as an individual, Woodward was acting as the representative of a sponsoring institution. His agenda was not his own but that of the CIW. As secretary of the AAAS grants committee, he would have indeed been irresponsible to act as aggressively as he did for the CIW. But as manager of the CIW grants, he would have been irresponsible not to take an activist role. Old perceptions thus conflicted with the new reality of a more institutional, managed system of patronage. Individual grants were an appropriate vehicle for small-scale patronage in a moral economy of scarcity, but less for the large-scale patronage in the richer and more competitive world of twentieth-century science.

The CIW grants program continued through the 1920s and 1930s with little change. It was overshadowed, however, by the much larger and quite different programs of the Carnegie Corporation and the Rockefeller boards. Individual research grants went out of fashion in the 1920s, as large foundations turned to institutional grants for facilities, research endowments, and fellowships. The purpose of the new system of patronage was to build infrastructure and develop national scientific communities.

This retreat from individual research grants was partly due to the experience of the CIW grants program. Its troubled history was widely

[79]Woodward to Richards, 30 Mar. 1906; Richards to Woodward, 2 April 1906; both in TWR 1.5 box 3.

known, and in foundation circles was taken to be an object lesson in the dangers of direct aid to academic scientists. Woodward remained an outspoken advocate of separating research from teaching and, as an eminent personage and trustee of the Carnegie boards, he was well-placed to influence policy. Memories of Woodward's experience made it easier for new modes of patronage to emerge as philanthropists became interested in science after World War I.

Chapter Three

Reluctant Patrons: The Rockefeller and Carnegie Boards

It makes sense that the large-scale patronage of science was pioneered by research institutes, since research grants were obviously relevant to their institutional mission. The potential for growth of extramural programs, however, was limited. In the competition of resources, all the advantages lay with the in-house departments: predictability, productivity, accountability. Only the CIW's peculiar internal politics allowed the grants program to survive. The Rockefeller Institute's extramural grants became, after only a few years, a mere vestige of the embryonic phase of its research departments. The dozen or so smaller medical endowments that appeared in the early 1900s were exclusively for in-house research.

The potential for science patronage was even less, however, in most of the foundations that appeared between 1900 and 1914. The Russell Sage Foundation, Rockefeller's General Education Board (GEB), and others were operating agencies, created to deliver services in some specific field of public health, education, or social welfare. In principle, basic science could have been done as an overhead on the foundations' social services. The people in charge of the Sage Foundation and the GEB believed in experiment and investigation. But their investigations were surveys and statistics, and their experiments were field demonstrations. What mattered to this new generation of "scientific" philanthropists was the efficient diffusion of existing knowledge, not the creation of new knowledge. The science in "scientific philanthropy" meant system and business methods, not the science of universities and research institutes.[1]

[1]Ernest V. Hollis, *Philanthropic Foundations and Higher Education* (New York: Columbia University Press, 1938), chs. 2–5.

More promising patrons for natural scientists were the large general purpose foundations that appeared after 1910: especially, the Carnegie Corporation (1911) and Rockefeller Foundation (1913). Unlike the operating foundations, these organizations were not restricted to particular lines of work. The last and largest of Rockefeller's and Carnegie's philanthropies, they were also the most open-ended, leaving the choice of programs to succeeding generations of trustees. If operating foundations like the GEB were the philanthropic equivalent of manufacturing corporations, the general purpose foundations were the equivalent of trusts or holding companies. As Carnegie's biographer wrote: "[They] were as natural a development in the field of philanthropy as Standard Oil and United States Steel were in the field of manufacturing."[2]

This resemblance was not lost on contemporary observers, who liked the idea of a philanthropic trust about as well as they like standard Oil, U.S. Steel, and other new corporate giants. The Rockefeller Foundation and Carnegie Corporation were born in controversy, especially the foundation, since Rockefeller's decision to seek a federal charter opened the door to congressional investigation and partisan politicking. Antitrust sentiment was in full flood (Standard Oil had been broken up just a few years before), and foundations were caught up in the political storms that split the Progressives from the Republican party and brought the "new freedom" Democrats to power.[3] The idealism and political conflicts of high progressivism also shaped the policies of the philanthropic trusts.

The Rockefeller Foundation (RF) and Carnegie Corporation (CC) were envisioned as nurseries of new philanthropic programs. With their great wealth (ten times the CIW endowment) and open-ended charters, they could transcend any one kind of social service. The idea was to use these great endowments not to monopolize a field but to demonstrate new kinds of organized philanthropy that smaller foundations could take as their own. A combination of philanthropic think tank and investment corporation—that was the ideal: the reality was considerably more complex.

In practice, the options open to the new philanthropic trusts were restricted by many things: by politics and especially by Rockefeller's and Carnegie's previous philanthropies. The CC was regarded by its founder and its board as a holding company for Carnegie's other institutions.

[2]Joseph F. Wall, *Andrew Carnegie* (New York: Oxford University Press, 1970), p. 833, also pp. 880–884. Raymond B. Fosdick, *The Story of the Rockefeller Foundation* (New York: Harper, 1952), chs. 2–3.

[3]Fosdick, *Rockefeller Foundation,* ch. 2. John W. Chambers II, *The Tyranny of Change: America in the Progressive Era, 1900–1917* (New York: St. Martin's Press, 1980).

The heads of the Carnegie Institute, Institution, Endowment, Foundation, and Hero Fund sat on the corporation's board and saw to it that the bulk of its income went to their institutions. Like a holding company, the CC in this view had no product or service of its own.

A similar pattern can be seen in the Rockefeller Foundation, though less dramatically. The nucleus of the RF was the International Health Commission (IHC), and its staff could marshal powerful arguments against any new and competing programs. The hookworm campaign in the Southern United States gave them prestige and authority, and they had the backing of powerful trustees like Frederick Gates. Large grants were also made to the Rockefeller Institute, though the institute never dominated the RF as Carnegie's institutions did the CC. Unlike Carnegie, the elder Rockefeller never meddled in the affairs of his boards (in 1917 he even relinquished control of $2 million set aside each year for his personal, predominantly religious charities). On the whole, the RF succeeded a good deal better than the CC in living up to the ideal of a nursery of philanthropic ideas. But decisions about new programs were heavily weighted toward public health, in which the IHC had a proven and unassailable track record. In both boards, the politics of the decision-making process discouraged radically new ventures.

These contending forces produced a distinctive pattern of initiatives and retreats that began in earnest about 1913 and petered out in the early 1920s. One after another, new programs were debated and set aside, and the CC and RF emerged from this process not very different from what they had been when the process began. Natural science was one of the many causes contending for support and, surprisingly, one of the very few that got it. In 1919 two large projects were consummated: The CC's gift of $5 million to the National Research Council (NRC), and the RF's grant to the NRC for a system of national research fellowships. A new system of patronage was created in 1919 that was quite different from the prewar system of the CIW grants. To understand how this system of science patronage took shape, we need first to understand the philanthropic trusts. We need to look at the people who made decisions and how the decision-making process worked.

There were three key players: the trustees, of course; the scientist-entrepreneurs who came knocking on their doors; and the foundation officers who dealt with the one and reported to the other. These middle managers, because they did the work, were strategically placed to shape decisions. They were a distinctive generation, these professional philanthropists, whose careers happened to intersect with the appearance of the philanthropic trusts. Not natural scientists like Woodward, most came either from the social sciences or from an emerging class of university middle managers—"university handymen" or "maintenance men," as some jokingly called themselves.[4] The successors to the great innovating presidents, or their lieutenants, they were the systematizers who made universities, and foundations, into twentieth-century bureaucratic organizations. What these people believed and did is the subject of this chapter.

Jerome Greene: "A University of Human Need"

In the RF the person most responsible for developing programs was its secretary, Jerome D. Greene. A scion of an old Boston family, Jerome was brought up in a tradition of Christian social service. (His parents were missionaries in Japan, and his elder brother, Roger S. Greene, was the first head of Rockefeller's China Medical Board.) Following his graduation from Harvard in 1896, Jerome became part of the expanding administrative machinery of the college. He helped found the *Harvard Alumni Bulletin* in 1898 and served as Charles W. Eliot's personal secretary (1901–1905) and as secretary of the Harvard Corporation (1905–1910.) Greene's job was to strengthen ties with civic and alumni constituencies and manage endowment funds—a typical apprenticeship for his generation of university handymen.

Greene entered the Rockefeller circle in 1910, as business manager of the Rockefeller Institute. Two years later he became personal secretary to John D. Rockefeller, Jr., and for the next year, with Starr J. Murphy, was in charge of the arduous campaign for a federal charter for the RF. In May 1913 he became the new foundation's first secretary.[5] Rockefeller, Jr., was president, but he had little taste for administration and acted more like a chairman of the board. In practice, Greene was the chief executive officer, analogous to Woodward in the CIW, only without the title and prestige of president. He was well placed to shape the RF's program. Idealistic, ambitious, and efficient, Greene brought to his new role a decade's experience with the business side of nonprofit institutions.

Greene's vision of the RF's mission came directly from this experience. He envisioned the foundation as a "university of human need": comparable to a university with its comprehensive array of disciplines and its dual purpose of advancing and diffusing knowledge. So, too, Greene hoped to build within the RF departments devoted to every important field of philanthropy and reform, each of which would spon-

[4]Charles S. Johnson, "Phylon profile X: Edwin Rogers Embree," *Phylon* 7 (1946): 317–334, see p. 326. I am grateful to Vanessa Gamble for calling my attention to this source.

[5]"Jerome D. Greene," RF 903 1.21.

sor research on social, economic, or health problems as well as demonstrations and service operations in the field.[6] Just as a premier university could not neglect any important field of learning, so the premier foundation must explore every important area of philanthropy.

In 1914 Green drew up a comprehensive statement of the possibilities, many of which were already being supported by the Rockefeller family. The most important ongoing program was, of course, the public health work of the IHC. Surveys of health and education in China had been underway since 1909 and would soon result in the organization of the China Medical Board (CMB). (Like the IHC, it had its own board but received funds from the RF.) Psychiatrist and mental health reformer Thomas W. Salmon had been borrowed from the National Committee on Mental Hygiene to survey and report on opportunities in that field. An Industrial Relations Commission was being organized by Mackenzie King, and an institute for economic research had been proposed. Greene also envisioned a department of charities and correction, which would bring together a number of projects supported by the Rockefellers. A Division of Social Hygiene could have been formed in the same way as the IHC and CMB, by assimilating the Bureau of Social Hygiene, recently organized by Rockefeller, Jr., to investigate prostitution and vice in New York City. Other possible fields were public administration, municipal and rural charities, pensions, industrial hygiene, asylum administration, housing, immigration, fine arts, and wildlife conservation. Greene recommended to the trustees that the first exploratory conferences be held in alcoholism, mental hygiene, economics and sociology, venereal disease, and education and medicine in the Near and Far East.[7]

Taken item by item there was little new in Greene's list. (Indeed, Greene professed that his grand plan was no more than a schematization of what was already being done in an unsystematic way by the Rockefeller family.) Greene's distinctive contribution was to bring all together in a single "university of human need." He was, in effect, applying the RF's ideals of large-scale organization to its own organization. Greene's faith that any branch of philanthropy could be done better as part of a comprehensive whole was the same faith that had created huge corporations like Standard Oil or modern universities, with their diverse faculties and professional schools.

Greene did not intend the foundation to set up operating programs in all these fields but only to reveal in a systematic way where philanthropic investment would be most effective. The unique role of the

[6]Greene, "Principles and policy of giving," 22 Oct. 1913, pp. 20–21, RF 900 21.163.

[7]Ibid. Greene, "Future organization of the Rockefeller Foundation," 1914, RF 900 21.163. Greene to Eliot, 2 June 1914. RF 900 21.159.

RF, Greene felt, was to investigate, coordinate, and set standards. Thus, surveys had a special place in his "university of human need." Surveys did not commit the board to any operating program and did not create false expectations. They could be stopped at any point without losing the knowledge already gained and could be taken up again at any time. Cheap and flexible, they left all doors open and produced knowledge that might serve as a guide to other foundations, if the RF chose not to enter the field.[8] While surveys of particular fields were customary preliminaries for all foundations and voluntary health agencies, only the RF could take investigation as its special mission. That, in Greene's view, was what made the RF like a national university of philanthropy.

Greene's method of inaugurating new departments in the "university of human need" was to appoint the best experts on a temporary basis and set them loose to study, plan, try out a program on a small scale, "play with it," and prove to the trustees that their special field was worthy of large-scale investment. If they proved their mettle, these temporary recruits would become directors of operating divisions like the IHC or CMB. Greene intended that Salmon and King would become directors of divisions of mental health and industrial relations.[9] Steps were also taken toward assimilating the Bureau of Social Hygiene, presumably with Raymond B. Fosdick continuing as head.[10]

Opposition to Greene's Program

Opinion in the Rockefeller circle was divided on Greene's philanthropic university. One group favored expanding beyond the safe and proven fields of medicine and public health. Raymond B. Fosdick was the most constant advocate of expansion, first as trusted advisor to Rockefeller, Jr., and after 1921 as an increasingly influential trustee. A lawyer and the younger brother of social gospeler Harry Emerson Fosdick, Raymond represented the secular and pragmatic style of Progressive reform. He was an alumnus of Lillian Wald's Henry Street Settlement and was active in the reform of municipal government and the penal system. As commissioner of accounts for New York's reform mayor John P. Mitchell, he rooted out graft and corruption in the city bureaucracy. As Rockefeller, Jr.'s chief man in the Bureau of Social Hygiene, he surveyed the police systems in the United States and Europe. In the

[8]"Notes on the conference," 7 Dec. 1915, pp. 9–10, RF 200 15.158.

[9]Ibid., pp. 13–14.

[10]Ibid. Fosdick, "A plan for the development of the Bureau of Social Hygiene," 23 Oct., 27 and 31 Dec. 1915; "The Bureau of Social Hygiene," n.d., is an unfinished report probably written by Greene for eventual presentation to the board; all in RF 200 15.158.

1920s he was an ardent supporter of the League of Nations.[11] His views on unions and the balance of power between labor and capital were liberal, and he worked hard to wean Rockefeller, Jr., away from the family's hidebound conservatism. Raymond's brother Harry wrote that he was glad to see the elder Rockefeller's "old Tory" entourage being replaced by more liberal men and thought Raymond's influence on Rockefeller, Jr., was crucial.[12] Raymond's desire to make the RF a force in social reform was shared by Starr Murphy, Abraham Flexner (who investigated European prostitution for Rockefeller, Jr.), and trustee Arthur Woods, a former commissioner of police. In 1915 Fosdick and Flexner envisioned a comprehensive program in criminology, alcoholism, drug addiction, feeblemindedness, venereal disease, family structure, incomes policy, and delinquency. These were the activists.

On the other side were those who thought the RF should stick to medicine and public health, in which it had a solid reputation. The leader of this group was the redoubtable Frederick T. Gates, the elder Rockefeller's most trusted advisor on business and philanthropy, and still a force to be reckoned with on the board. Gates opposed Greene's program with his own vision of a comprehensive program in medicine and public health, uniting the Rockefeller Institute, IHC, CMB, and the General Education Board (GEB). In Gates's scheme, the Rockefeller Foundation would be a "great treasury" to be drawn upon at strategic moments to meet special opportunities encountered by the four operating boards—a kind of holding company but a passive not an active, one, a purse for Rockefeller's other institutions. Gates's scheme had every advantage over Greene's: it was comprehensive, coherent, and complete; its programs were tried and true; it limited claims on the foundation's income—a feature especially appealing to the business-minded trustees.[13]

With his usual eloquence and absolute conviction, Gates caricatured Greene's proposal (without referring to it directly) as a policy of "scatteration," unsystematic giving to every worthy cause great or small, frittering away capital with no programmatic touchstone. Gates tarred Greene's "university of human need" with the same brush as he did old-fashioned individual charity. Quite unjustly, of course: the essential point of Greene's plan was that it brought business methods to philanthropy. Turning the tables on Gates, Greene wondered if it was really a good use of the RF's funds to apply known methods to one disease after another in one region of the world after another. Would that

[11]Raymond B. Fosdick, *Chronicle of a Generation* (New York: Harper, 1958).

[12]R. B. Fosdick to Rockefeller, Jr., 5 Sept. 1919; R. B. Fosdick to H. E. Fosdick, 29 Sept. 1919; H. E. Fosdick to R. B. Fosdick, 8 June 1921; all in RBF box 8.

[13]Greene, "The policy of the Rockefeller Foundation," 23 May 1916, RF 900 21.163.

not simply relieve communities form routine work that they should support themselves? The RF's distinctive role, Greene argued, was to discover and promote new philanthropic ideas: "It is not through the multiplicity of scattered efforts of minor importance . . . that the Foundation would render its greatest service, but rather through the discovery of a very few lines of work having a very great strategic importance." Thus did Greene hope to hoist Gates with his own rhetorical petard.[14]

Gates and Greene were not just disputing philanthropic philosophy: at stake was the balance of powers of officers and board. Greene's idea of a "university of human need" gave the power of initiative to the professional officers and their expert advisors. It was the ideal of an emerging class of philanthropic middle managers. Greene took for granted that the RF would someday be run by a corps of professional administrators. Gates was no less committed to the view that power should remain with the board and the inner circle of Rockefeller family advisors, among whom he presided. Experienced generalists, Gates and his allies viewed with disdain the pretentions of university handymen like Greene.

None of the projects initiated by Greene ever came to fruition. Not because of Frederick Gates: there is little evidence that he blocked Greene's plans in any systematic way.[15] Gates's vivid personality and his undoubted influence on the elder Rockefeller tempt one to see his hand behind every move of the Rockefeller philanthropists. But the evidence suggests that his influence was waning as the RF became more independent of the Rockefeller family. Practical realities were more important than ideology in frustrating Greene's dream: the growing appetite for funds of the two operating boards, the IHC and CMB; the failure of Mackenzie King and others to live up to expectations; the cutting short of all new initiatives by war relief projects; and by Greene's departure.

Public health and medical missionary work in China were everybody's top priorities. The hookworm commission was the one venture of John D. Rockefeller's of which the public approved. Wickliffe Rose, the IHC's vigorous director, was eager to recapitulate that success in other regions of the world. The Rockefellers' deep interest in China was probably the single most important reason for creating the RF. The survey of 1909 provided a definite plan of action, years ahead of any other potential field. Greene stated that medical education and public health in China were "the most tempting field for philanthropic investment in

[14]Ibid., pp. 4–5. Green to Vincent, 13 Jan. 1917, RF 900 21.163.

[15]Gates did oppose the proposed institute for economic research. Gates to Rockefeller, Jr., 19 Mar. 1914, FTG 3.58.

the world today."[16] The IHC and CMB were what everyone most wanted; what no one forsaw was how expensive they would be and how, like cuckoo babies, they would starve other hatchlings. With the Peking Union Medical College, the Rockefellers broke their golden rule and created an institution that was dependent entirely upon their support and in perpetuity. (Serious miscalculations of building costs compounded the problem.) So, too, in public health, no one quite realized how fast Rose would expand his foreign operations and how expensive they would become.

Other unanticipated problems arose from the attack launched on the RF by Senator Frank Walsh's Commission on Industrial Relations, following the notorious "Ludlow Massacre" in 1914.[17] The Walsh hearings did not result in legislative restrictions on private endowments, as some had feared, but the scandal did contribute to the narrowing of the RF's horizons. Most immediately, the public rage against anything bearing the Rockefeller name slammed the door shut on the nascent program in industrial relations. Mackenzie King lost interest, and by 1920 it was generally agreed that the project was a washout.[18] Hopes for projects in economic and government research also evaporated in the heat of the Walsh commission's attacks. It became clear that any information, much less propaganda, coming from a Rockefeller organization would be taken by the public as self-interested and thus would do more harm than good. Henceforth, the RF steered clear of such controversial subjects. With some reluctance. Industrial relations, economics, and public administration were the great public issues of the early 1910s, and it was a bitter pill for Greene and others in the RF that these areas were out of bounds to them.

America's entry into World War I also contributed to narrowing the RF's scope. Anticipating large expenditures for emergency relief in Europe, the trustees shelved all new programs. Large grants to the American Red Cross and other relief organizations mounted to over $20 million by the time the fighting stopped.[19] The most important ca-

16[Greene], "Memorandum," 12 Aug. 1913; Greene to Rockefeller, Jr., 12 Nov. 1913; Greene, "Educational and other needs in Far East," 22 Oct. 1913; all in RF 900 21.159. Mary B. Bullock, *An American Transplant: The Rockefeller Foundation and Peking Union Medical College* (Berkeley: University of California Press, 1980).

17Graham Adams, Jr., *The Age of Industrial Violence, 1910–1915* (New York: Columbia University Press), 1966. James Weinstein, *The Corporate Ideal in the Liberal State 1900–1918* (Boston: Beacon Press, 1968), ch. 7. Howard M. Gitelman, *Legacy of the Ludlow Massacre* (Philadelphia: University of Pennsylvania Press, 1988).

18Gitelman, *Ludlow Massacre.*

19Merle Curti, *American Philanthropy Abroad* (New Brunswick, N.J.: Rutgers University Press, 1963). RF *Annual Report* 1917–1919.

sualty of mobilization was the proposed department of social hygiene. During the war Raymond Fosdick was busy with the War Department's campaign against prostitution and venereal disease in army camps, and after the war the government's Interdepartmental Social Hygiene Board (1918–1921) co-opted much of what Fosdick had hoped the RF would do.[20]

Greene Out, Vincent In

No less disruptive was Greene's unexpected resignation in July 1916 to make way for the RF's first president, George Vincent. A shake-up was inevitable. The roles of both Greene and Rockefeller, Jr., were makeshift: the only reason Greene enjoyed such authority as secretary was that Rockefeller, Jr., was a part-time figurehead trying to do a full-time job and not liking it at all. He wanted out, and Greene, though he never said so, undoubtedly wanted up. Greene was one of the most vocal in pressing the trustees to appoint a full-time president, with the authority to organize and adjudicate among the departments of the university of human need.[21] Rockefeller, Jr., and Gates wanted a full-time president too, but for different reasons. They were concerned that the foundation be headed by a person of national stature. However energetic and competent, Greene was not a public personage, and Rockefeller was determined to have someone "more mature, more experienced and of larger calibre."[22] Visibility and social connections were qualities that Greene, the university handyman, could never acquire. Rockefeller, Jr., and his circle knew the value of talented middle managers like Greene or Murphy, but they never thought of them as material for top executive positions.[23]

In the wake of the Walsh committee scandal—a public relations debacle—Rockefeller, Jr., and Gates hoped that an eminent personage would reassure the public that there was no conspiracy afoot in the Rockefeller Foundation. Their perception was not unfounded: some part of the public's suspicion of the RF was the result of its tendency to do its business in secret. To dispel such suspicions, Greene had proposed to establish a public advisory council made up of representatives

[20]Allan Brandt, *No Magic Bullet* (New York: Oxford University Press, 1985), chs. 2, 4. The Social Hygiene Board was the only nonagricultural agency empowered to make research grants in private institutions.

[21]Greene, "Future organization of the Rockefeller Foundation," 1914, p. 10, RF 900 21.163.

[22]Rockefeller, Jr., to H. P. Judson, 19 June 1915, RF RG3 Boards 33.333. I am grateful to Dr. William Hess for calling my attention to this group of documents.

[23]Rockefeller, Jr., to John D. Rockefeller, 14 June 1915, RF RG3 Boards 33.333.

of the states and major universities. Rockefeller, Jr., and Gates quashed the idea, fearing politics and logrolling. They preferred to make decisions in the privacy of their offices and looked to a highly respectable figurehead to represent the RF in public.[24]

The first choice of Rockefeller, Jr., and Gates was Harry Pratt Judson, president of the University of Chicago. A nationally known educator, possessed of great dignity and presence, socially well connected, and a skilled speaker, Judson seemed ideal. It soon became apparent, however, that Judson was less interested in the challenge of the new job than in the prestige and salary that came with it. (He fancied the title of "Chancellor" and "artfully" suggested a salary twice what Rockefeller, Jr., felt proper.)[25] Gates finally asked Judson bluntly if his wide experience and reading had revealed to him "a program for the Foundation; a scheme . . . of world philanthropy so important, so inviting, so full of promise and hope . . . that he could feel justified, nay compelled . . . to resign his presidency . . . [of the university.]" Judson did not rise to the challenge, replying that "he had no such program in mind, . . . no plan, and no compelling sense of duty arising from any plan."[26] After that, opinion in the RF was that Judson was too important to his university to be spared.

George E. Vincent, Rockefeller, Jr.'s second choice, was remarkably like Judson. He, too, had been associated for most of his career with the University of Chicago, as graduate student in sociology (Ph.D. 1896), professor (1904), dean of Junior Colleges (1900) and of Arts and Sciences (1907). Disappointed in his hopes for the University of Chicago presidency to which Judson was appointed, Vincent moved on to become president of the University of Minnesota in 1911.[27] He was a master of public relations. Son of the founder of Chautauqua, Vincent had long been active in the movement, starting as literary editor of the Chautauqua Press (1886) and working up to president (1907–1915) of the Chautauqua system. Possessed of a commanding public presence, quick intelligence, and great personal charm, Vincent was known as the best orator in the Middle West. It was his silver tongue and the wide social connections he acquired on the Chautauqua lecture circuit that caught the attention of Rockefeller, Jr.: "Dr. Vincent would be able to

[24]Greene to Rose and Rockefeller, Jr., 10 July 1913; Rockefeller, Jr., to Greene, 14 July 1913; both in RF 900 21.159.

[25]Rockefeller, Jr., to Rockefeller, 29 June 1915; Rockefeller, Jr., to Gates, 27 July, 14 Aug. 1915; both in RF RG3 Boards 33.333.

[26]Gates to Rockefeller, Jr., 19 Aug. 1915, RF RG3 Boards 33.333. A memorandum by Judson suggesting possible programs is notable for its windy high-minded generalities. Judson, "Suggestions of program for the Rockefeller Foundation," 1913, RF 900 21.163.

[27]E. W. Burgess, "George Edgar Vincent: 1864–1941," *Amer. J. Soc.* 46 (1941): 887.

render a great service to the Foundation by speaking on its behalf as extensively as opportunity offered, thus bringing the public into both an understanding of, and closer sympathy with, the Foundation's work."[28] Vincent was the "large calibre" figure who could repair damages done to the RF's public image in its controversial early years.

Some people did wonder what was behind the brilliant presence. One referee thought Vincent's one fault was a "fondness or at least a willingness, to permit himself to be called upon to make speeches." Another noted that "many people feel that when they have heard him once they have heard all that he has to say," and a third wrote, "He has been and is an unusual publicity man, both for the University and the State, and incidentally for himself."[29] However, Vincent had a record as an efficient academic administrator, self-confidant and unafraid of controversy. He had moved so decisively (some thought precipitately) in clearing out deadwood at Minnesota that he had acquired a controversial aura and some powerful enemies—a good sign, it was thought in the RF.[30]

Vincent and Judson both belonged to that generation of university heads who tidied up after dynamic, expansive innovators like Charles W. Eliot or William Raney Harper. Whereas Harper was always starting something new, running in the red and blackmailing Rockefeller to bail him out, Judson—and Vincent too, it turned out—were good at cutting back, eliminating programs, and balancing budgets. Judson is remembered as an unpopular president who did little, a hiatus between Harper and the dynamic Robert Hutchins. It was from this generation of trimmers that the foundations had to recruit their leaders—one reason, perhaps, for the "crisis" in leadership in the 1920s.

As president of the RF, Vincent would of course take over Greene's duties in formulating policy and program, leaving the secretary with administrative routine. Vincent felt embarrassed at offering Greene a job for which he was clearly overqualified, but Rockefeller, Jr., not wishing to give Vincent the smallest reason to decline, wrote that he would "take care of Greene." Disappointed and embarrassed, Greene resigned at once. With no immediate prospect of a job, he seemed concerned only that it not appear that he had left out of pique or in disgrace.[31]

[28]Rockefeller, Jr., to R. B. Fosdick, 10 Aug. 1916, RF RG3 Boards 31.311.

[29]A. R. Rogers to W. S. Richardson, 27 May 1916; Richardson to Rockefeller, Jr., 31 May 1916; D. D. Dayton to Richardson, 5 June 1916; all in RF RG3 Boards 31.311

[30]Judson to Rockefeller, Jr., 30 May 1916; Dayton to Richardson, 5 June 1916; both in RF RG3 Boards 31.311.

[31]Greene to Rockefeller, Jr., 1 and 6 Sept. 1916; Vincent to Rockefeller, Jr., 5 July 1915; Rockefeller, Jr., to Vincent, 11 July 1916; all in RF RG3 Boards 31.311. Greene went into investment banking and in 1931 returned to Harvard as secretary to the corporation.

Greene's sudden departure meant the end of his dream of a philanthropic university. The hiatus in administrative leadership was complete. Vincent did not come for six months, and in the interim the business of the foundation was run by Greene's young and inexperienced successor, Edwin Embree. When Vincent did arrive, he was immediately swamped with emergency war work. In these crucial months and years, the impetus of Greene's vision was lost. The medical barons of the IHC and CMB, Wickliffe Rose and Roger Greene, expanded their operations unrestrained by competition from the president's office, until it began to seem doubtful that the RF could afford any new programs at all.

Transition: Edwin Embree

Like his predecessor, Edwin Embree was one of the "university handymen," coming to the RF from a post as assistant secretary of Yale University.[32] He came from an old Quaker family, with a long history of reform activism. He spent his formative years in the household of his maternal grandfather John G. Fee in Berea, Kentucky, a community founded as a demonstration in racial integration. An old-time abolitionist and head of Berea College, Fee managed to imbue his grandson with his deep social idealism, though not with his rigid piety and puritanical self-discipline. Handsome and fun living, Edwin acquired a wordly facade and a taste for sophisticated company and good living that only partly covered a youthful and provincial naiveté. That naiveté was a good part of his charm, but it deprived him of practical judgment, and he suffered for many years from a lack of self-confidence.[33] He combined the temperament of a bon vivant with the ambition and moral commitment of a crusader.

Embree had little sense of how to manage a career. After a brief stint as cub reporter for the New York *Sun,* he took a job with the *Yale Alumni Weekly.* Reorganized by Clarence Day (of *Life with Father* fame), the *Weekly* served as a vehicle for the "literary renaissance" at Yale, a movement that united cultural improvement with social activism.[34] (At Harvard, a similar movement turned young Walter Lippmann to Socialism and the *Nation.*) As assistant to Yale's secretary, Anson Phelps

[32]Embree, "Rockefeller Foundation," pp. 5–6, ERE box 1. The post of secretary was first offered to Edwin's brother, William, then chief council for the Voluntary Defenders League, where he met Rockefeller, Jr. Edwin was appointed in January 1917; Vincent arrived in July 1917.

[33]Johnson, "Embree."

[34]George W. Pierson, *Yale College: An Educational History 1871–1921* (New Haven: Yale University Press, 1952).

Stokes, Embree was responsible for strengthening the social connections of Yale College, building alumni networks, administering scholarships, and finding employment for graduates.[35] Following Clarence Day to New York in 1917, Embree was drawn into the cultural life of the big city. Day was cultivated, an avid theatergoer, a brilliant conversationalist, and a literary lion; Embree became his adoring disciple.[36] He would develop a similar relationship with George Vincent.

Unsure of himself, Embree tended to imitate personages who possessed assurance and cultivation, as he himself noted in a remarkably candid (and uncharitable) autobiographical sketch: "As an officer at Yale I felt terribly young," he wrote. "Stokes . . . and the Deans seemed to me to be in a world completely beyond me and I literally trembled in the presence of President [Arthur] Hadley. . . . So, when I went to New York I was again a great deal impressed by the importance and maturity of the men with whom I was associated."[37] So similar in background to Jerome Greene, Embree had little of his confidence and practical knack for administration.

It would have been difficult for anyone to keep Greene's program alive in early 1917. Alone in the president's office and totally without experience in foundation work, Embree was little more than a caretaker. He later reproached himself with his indecisiveness:

> If I had been another kind of man I might have seized a great deal of authority during the first six months, or during the early years, but I could not persuade myself that I knew enough to make final judgments or that I should exercise authority until I had lived with the job for a long time. I think this is good philosophy from the standpoint of the job, but it is not the American practice.[38]

Unaware of the varied options Greene had been weighing, Embree quite naturally shaped his views of what the RF should do on what he saw it doing—that is, what the medical barons were doing.

George Vincent: Opportunities Forgone

Briefing his new boss in September 1917, Embree painted a picture of a foundation devoted to medicine and public health (including mental health) and projected little expansion beyond these activities for the

[35] Johnson, "Embree," pp. 325–326.
[36] Ibid., pp. 322–324.
[37] Embree, "Rockefeller Foundation," n.d. [1930], pp. 7–8, ERE box 1.
[38] Ibid., p. 8.

foreseeable future. He anticipated some new projects in social reconstruction after the war, but these too were mainly in health and medical education.[39] The immediate task was to maintain the IHB (the IHC had become the International Health *Board*) and the CMB. As chief executive, Vincent felt his first responsibility was to balance the budget. As it proved impossible to restrain Rose and Roger Greene, his job soon degenerated into a perpetual battle to prevent new programs from being created.

At the first peacetime meeting of the board, Vincent presented a sobering financial picture. Of the RF's annual income of $6.5 million, $5 million was committed for the indefinite future: the IHB and CMB took $1.5 million each, medical education and schools of public health $.75 and $.3 million, the institute $.15 million, $.8 million for administration and sinking fund. That left only $1.5 million for new ventures, ruling out any large projects. Vincent proposed that the RF (as distinct from its quasi-independent subsidiary boards) limit itself to surveys and temporary demonstrations of philanthropic methods.[40] It was Jerome Greene's vision of a philanthropic nursery, but without the prospect of surveys leading to new operating divisions. By restricting the president's office to investigation, Vincent unintentionally ensured that the power of initiative would remain with the IHB and CMB.

Except for a vague interest in promoting international understanding, Vincent confined himself almost entirely to medicine and public health. His list of possible fields included mental hygiene, drug addiction, physiological causes of insanity, public health, dispensaries, voluntary health agencies, and international health information services. To prepare specific proposals, he selected a group of advisors who were bound to press the cause of medicine and public health: among them William Welch, Simon Flexner, and New York Public Health Commissioner Herman Biggs. The board had a long list of fields to consider, but only the items related to public health were actually discussed. Care for drug addicts was regarded as too social; medical research was the province of the Rockefeller Institute; social hygiene campaigns were ruled out as moral and political propaganda; industrial health (suggested by Mackenzie King) was too close to industrial relations; and so on. There is no sign of enthusiasm for the various reconstruction projects fashionable outside the RF during 1919–1920. When such subjects were raised, the discussion turned to ways of limiting commitments by farming them out to other Rockefeller boards or by token grants to voluntary agencies. King, Thomas, Salmon, and

[39]Embree to Vincent, 24 Sept. 1917, RF 900 21.164.

[40]Vincent, "Memorandum of expected income and expenditures," 4 Dec. 1918, RF 900 21.164.

Raymond Fosdick spoke for a broader program, but it was Wickliffe Rose's plans for a massive program in public health education that carried the day.[41]

In subsequent conferences in 1919 and 1920, doors were closed to new programs, one by one. A conference on industrial hygiene was organized in November 1919; two weeks later the board ruled it "out of program." (No reasons were given, but the outcome did not surprise one observer, who had heard too many people remark on the obvious conflict inherent in a Rockefeller organization investigating labor questions.)[42] The officers themselves decided not to create a division of mental hygiene. Thomas Salmon was still on the RF payroll and was aggressive in advocating a separate division—perhaps in a too aggressive and self-interested manner. He had allies in Embree, who administered Salmon's projects, and in Starr Murphy. The advocates of public health, however, were more numerous and of larger caliber. Victor Heiser of the IHB, William Welch, and Simon Flexner all opposed a separate division. Heiser thought mental hygiene less productive than campaigns against infectious diseases. Flexner favored a more scientific approach, building departments of neuropsychiatry in medical schools, and Welch agreed. (As head of the Johns Hopkins School of Hygiene, Welch had every reason to oppose the development of mental hygiene separately from public health.)[43]

Vincent was preoccupied—obsessed in fact—with limiting future claims on RF resources. Although he hesitated to get rid of mental hygiene altogether, he also knew that a new division would mean a new baron pressing him for more money. Seeking the middle ground as always, he proposed to distribute Salmon's work among Rockefeller agencies: aiding departments of neuropsychiatry through the GEB, research through the institute, and small grants for educational projects to the National Committee for Mental Hygiene. That left the RF with surveys, which could be easily terminated. Deeply disappointed, Salmon soon resigned.[44]

Embree later recalled this period as one of failed hopes. He blamed Vincent for not curbing the IHB and CMB by creating programs to compete with them:

[41]"Conference," 11–12 Jan. 1919, RF 900 21.164.

[42]"Conference on industrial hygiene," 14 Nov. 1919, RF 200 24.274. W. G. Thompson to Vincent, 13 Dec. 1919, RF 200 24.273.

[43]"Conference of officers and advisers," 17–18 Jan. 1920; T. Salmon, "Memorandum," 18 Jan. 1920; both in RF 900 22.165. Murphy to Vincent, 8 Mar. 1921, RF 906 2.17.

[44]Correspondence in RF 200 32.363–374 and 33.375–377. Murphy to Vincent, 8 Mar. 1921; Embree to Murphy, 12 Mar. 1921; both in RF 906 2.17.

> We missed as completely as possible, probably, the greatest chance for constructive work that foundation officers will ever have. My contention, of course, is that Vincent was responsible for policy and that I was more or less helpless as junior officer; that while I might have helped him to do great things, I could not prevent him . . . from shilly-shally indecision and lack of progress. Wherever the blame should fall, certainly a great chance was allowed to go by without a struggle, and practically without anybody realizing what was happening. . . . [Vincent] and the central administration busied themselves with war relief and other aspects of the war work which filled satisfactorily the first two years of the administration. By the end of that time, after the war work disappeared, the central administration found itself with practically no responsibilities and with the other departments thoroughly organized and entrenched. The remaining eleven years of Vincent's administration were devoted to futile and half-hearted attempts to get some control over existing departments . . . and equally half-hearted projects for the development of new lines of activity, each of which was abandoned at the first opposition.[45]

Embree's rueful reflections make one think of Woodward's swan song. Unlike the latter, however, Embree put too much weight on personality and too little on institutional structures and history. The grand but indefinite mission of the new general purpose foundations made it very difficult for anyone to define definite programs quickly enough to get them past entrenched interests like the IHB. The uncertain line of authority between board and managers impeded large new initiatives by making technical decisions into political ones. No doubt Rockefeller, Jr., placed too much faith in Vincent's social skills and too little in Greene's managerial abilities. But one wonders if Greene could have succeeded better than Vincent in containing the public health barons and getting new programs started.

Carnegie's Corporation and Carnegie's Court

It was a similar story with the early history of the Carnegie Corporation, though the balance of power there was tilted far more in favor of the board. There was no one comparable to Jerome Greene; the executive committee made policy and ruled on applications. Carnegie regarded the trustees as working executives, and to make sure they *did* work he

[45]Embree, untitled recollections, pp. 2–3, ERE box 1.

paid them a salary of $5,000 per year. Routine administration of grants was handled by James Bertram, who had started out as Carnegie's clerk and became his personal secretary for philanthropic affairs. Carnegie dominated the board, backed up by his wife and his three personal secretaries: John Poynton (personal affairs), Robert A. Frank (finance), and Bertram. All resisted shifting control to a staff of professional managers. Their own authority was a reflection of their special personal relation to Carnegie, and they did not welcome competition from outsiders. They were often joined in opposition to change by the presidents of Carnegie's other institutions, most notably Samuel H. Church, president of the Carnegie Institute of Pittsburgh. Church's motives were clear and simple: he opposed anything that might compete with his institute for CC funds.

A minority of the board believed that the corporation would be better served by professional managers and a definite program. The key members of this reform faction were Elihu Root, Henry S. Pritchett, and Robert S. Woodward. They sat on the board as heads of the Carnegie Endowment for International Peace, the Carnegie Foundation for the Advancement of Teaching, and the Carnegie Institution of Washington. Unlike Church, however, they also represented broader political, educational, and scientific interests. As Pritchett expressed it: "[W]e shall do our greatest service in the Corporation (1) by making Mr. Carnegie's institutions successes and (2) by aiding certain great causes, carefully sought out."[46] Like Woodward, he was a scientist turned administrator—their careers were variations on a theme. As a young astronomer, Pritchett took part with Woodward in the Transit of Venus commission; he was a professor (at Washington University) in the decades that Woodward was at the U.S. Geological Survey, joining the Cosmos Club circle as superintendent of the Coast and Geodetic Survey (1897–1900). While Woodward was dean at Columbia, Pritchett was president of MIT. Pritchett declined the presidency of the CIW before Woodward was offered it. Two years later (1906) he became head of the Carnegie Foundation.[47]

As the premier authority on colleges and universities (he organized the Carnegie Foundation's surveys and its college pension fund), Pritchett took on much of the burden of interviewing petitioners and administering grants. His authority in the CC rested on this practical administrative experience. It also brought him into direct and constant

[46]Pritchett to Root, 26 Oct. 1917, CC.

[47]Flexner, *Henry S. Pritchett.* Ellen Lagemann, *Private Power for the Public Good: A History of the Carnegie Foundation for the Advancement of Teaching* (Middletown, Conn.: Wesleyan University Press, 1983).

conflict with Bertram, whom he despised. He could usually count on Carnegie's goodwill (the founder was impressed by a professor with business abilities), though he did occasionally run afoul of Carnegie's dislike of being crossed or contradicted (unwelcome reminders that he was Carnegie's employee not his equal).[48] Pritchett's appoint to the executive committee in 1913 put him in a position to lead the growing movement to shift control from Carnegie's court to the professional staff. He was the activist and point man of the reform circle.

Elihu Root's authority was more political and social. A successful corporation lawyer, Root served in the U.S. Senate and as Theodore Roosevelt's secretary of state. (Had he not retreated from Roosevelt's liberal progressivism he might have been his political heir apparent.) Reputed to be the smartest man in government (though he was no deep scholar), Root liked to read geography and natural science and became a patron of Washington scientists like Hale and Walcott. He was Carnegie's social equal, and the deference he enjoyed from his friend was indispensable to the reform group.[49] Root took little part in the day-to-day operations of the CC, but as chairman of the board and its most eminent personage he held the balance of power. "You are so much larger a figure than the rest of us," Pritchett wrote, "that any policy you propose can be easily put into operation." Pritchett always informed Root of his plans in advance of board meetings; without Root he felt reform would have been impossible.[50]

Woodward was not an activist but a fairly reliable ally. Despite his ingrained mistrust of academic scientists, he could usually be counted on to vote right and do his bit in educating Carnegie's faithful secretaries in the basics of professional philanthropy. Woodward's role as head of the CIW gave him authority in scientific projects and set an example of disinterested behavior to people like Church.[51]

The reform group had their work cut out for them. The Carnegie Corporation in its early years was more like an old-fashioned family charity than a modern foundation, much more so than the Rockefeller Foundation. Carnegie created the CC because he was fed up with the bother of managing his benefactions himself, though that did not prevent him from meddling incessantly in its affairs. (He did the same in

[48]Pritchett to Carnegie, 13 Jan. 1911; Pritchett memo, 18 Jan. 1911; both in CC.

[49]Richard W. Leopold, *Elihu Root and the Conservative Tradition* (Boston: Little Brown, 1954), pp. 72–76, 176–179.

[50]Pritchett to Root, 19 Nov. 1921, 19 Mar. 1918, CC.

[51]Woodward to Poynton, 29 Jan. 1917, CC Woodward. Woodward to Bertram, 13 Nov. 1917, CC Bertram. Church to Woodward, 25 Feb., 12 Mar. 1919; Woodward to Church, 28 Feb. 1919; all in CC NAS-NRC.

his business relations, bombarding his partners from afar with memos and commands.)[52] Carnegie had an incurable habit of making promises to petitioners who caught his fancy. He could not be persuaded to stick to a specific program, and every promise to match funds raised by local groups was a lien on the CC's future income. Nothing caused Pritchett and his friends more anxiety than gaining control over spending. The trustees, Woodward expostulated, "may be making themselves liable to impeachment if not imprisonment in further mortgaging the income of the Corporation."[53] After joining the executive committee, Pritchett did manage to liquidate some of the obligations that Carnegie had piled up, but only because Carnegie was then too old and sick to contract new ones.[54]

Expenditures grew rapidly as the CC took over Carnegie's personal benefactions. Grants for public libraries and church organs peaked in 1912–1914, and a very large grant was made to the CIW.[55] What most troubled Pritchett, however, was the large and increasing number of gifts to small colleges and universities ($19 million since 1900). It was a classic example of Frederick Gates's "scatteration" or Woodward's "Havana lottery." Demand was insatiable, and there was no programmatic touchstone for distinguishing one request from another. The board could only sit passively and decide among the applications more or less on personal whim.[56]

Pritchett's and Woodward's experiences had made them deeply skeptical that colleges and universities were still a force for social and intellectual improvement. Pritchett had seen how financial dependence on alumni and local constituencies had inhibited university professors and administrators from taking unpopular stands on controversial issues. Woodward was equally certain that universities were not hospitable to modern modes of organized research. His brief against the CC's college program echoed his broadsides against the CIW's grants program.[57] The college program was neither the most expensive nor the most antiquated of Carnegie's pet projects, but it symbolized for Pritchett and Woodward everything that was wrong

[52]Wall, *Andrew Carnegie*, pp. 882–883. It was Root who suggested that Carnegie create the CC.

[53]Woodward to Pritchett, 4 Feb. 1915, CC.

[54]Pritchett to Woodward, 3 June 1915, CC program and policy (hereinafter noted as P&P).

[55]"Classification of appropriations 1911–1922," CC P&P. See also CC Grants by Categories.

[56]Bertram, "Colleges," n.d. [Dec. 1913], CC P&P.

[57]Woodward, "Observations on the duties and the responsibilities of the Carnegie Corporation," 13 Nov. 1916; Pritchett, "Fields of activity open to the Carnegie Corporation," 15 Apr. 1916; both in CC P&P.

with the CC's unprofessional governance. It became the strategic center in their battle with Carnegie's court.

Their strategy was to promote new programs more in keeping with modern philanthropic ideals than church organs and library buildings. In April 1916 Pritchett presented four options to the board: Americanization of immigrants (including information on jobs and legal rights), public education in international relations, public education in economics (to counter the antibusiness propaganda of the Walsh commission), and medical education.[58] Pritchett appealed openly to Carnegie's interest in public education and to the board's dislike of the populist side of Progressive reform movements. His personal favorite, however, was medical education, in which the Carnegie Foundation had already laid the foundation with Abraham Flexner's famous reports of 1910 and 1912. The General Education Board was moving into the field, and Pritchett clearly wanted the Carnegie group not to be left behind the Rockefellers in this great movement.[59] So eager was he, in fact, that in 1915 he pressed the executive committee to make an ad hoc grant to get reform started at Columbia. Woodward had to remind his friend that he should practice what he preached about special pleading and scatteration.[60] In early 1917 the board approved a program in medical education, and Pritchett threw himself into it with an abandon worthy of Andrew Carnegie in his philanthropic prime.[61]

World War I was far less disrupting to the Carnegie Corporation than it was to the RF, partly because its U.S. charter prevented it from engaging in European relief. But Pritchett's prophecy that the war would bring radical change to the CC was not fulfilled either. The Americanization project was a washout, and in medical education there was no competing with Flexner and the GEB. Most alarming, grants to colleges and miscellaneous grants rose sharply in 1917.[62] A flawed institutional structure encouraged old habits. The administrative burden of this retail philanthropy fell heavily on Pritchett. Academic grant-seekers preferred to work through Pritchett rather than through Bertram, partly because they confused the Carnegie Foundation and Corporation, and partly because Bertram was brusque and peremptory in his dealings with academics. Bertram resented Pritchett's intrusion on his turf; Pritchett resented having to take on what he felt Bertram

[58]Pritchett, "Fields of activity."

[59]Ibid., pp. 10–13.

[60]Pritchett to Woodward, 8 Oct. 1915; Woodward to Pritchett, 10 Oct. 1915; both in CC Woodward.

[61]Pritchett, "Report on applications for aid to medical education," 28 Nov. 1919, CC Pritchett.

[62]"Classification of appropriations," CC P&P.

was doing incompetently. The simmering conflict came to a boil repeatedly (often when Bertram treated one of Pritchett's pet clients roughly).[63] It was obvious that there could be no improvement until the secretary was replaced by a professional chief executive and program managers.

Pritchett himself was the obvious choice for president, since he had been performing the duties of a president without the title. In 1917 John Poynton urged him to take the title too: "Mr. Root feels most strongly that you can get [the CC] into shape and out of the present chaotic state."[64] But Pritchett wanted only to have the burden of routine administration lifted from his shoulders.[65]

Angell and His Programs

A few months after Andrew Carnegie's death in August 1919, Pritchett proposed a thorough reorganization: ending the practice of giving members salaries, downgrading the position of secretary, expanding the board (to dilute the power of Carnegie's "former clerks"), and creating the office of president."[66] The Rockefeller Foundation's recent reorganization was a model and a warning, and Pritchett was determined to get someone as imposing as George Vincent to head the RF's only real rival. His favored candidate, Henry Suzzallo, was strikingly like Vincent: president of the University of Washington, "perhaps the most eloquent speaker among college presidents," and adept in getting what he wanted from a populist state legislature.[67] Although the CC was less bruised by the Walsh Commission than the RF was, Pritchett worried about antifoundation legislation and wanted a leader with a national reputation and a knack for public relations. Also, he used the threat of public scrutiny and regulation to force the board to adopt a more open and modern mode of governance.[68] However, Suzzallo declined, and Root was obliged to serve as temporary president for 1919–1920. In 1921 James R. Angell became the CC's first president.

Though Angell and Vincent were unlike in character, their careers were almost carbon copies. Trained in psychology, Angell had been at the University of Chicago since 1894, rising to professor and depart-

[63]Pritchett to Bertram, 15 Feb. 1921, CC.

[64]Poynton to Pritchett, 31 July 1917, CC Pritchett.

[65]Pritchett to Root, 8 June 1921, CC.

[66]Pritchett, "Memorandum addressed to the trustees," 28 Nov. 1919, CC Pritchett.

[67]Pritchett to Suzzallo, 25 Nov. 1919, Woodward, list of candidates, CC President Selection.

[68]Pritchett, "The administration of the Carnegie Corporation," Mar. 1918, p. 8, CC P&P.

ment head, then drawn into administration, from deanships (1908) to acting president (1918–1919).[69] Angell was a more active and distinguished scholar than Vincent but less noted as an orator. His strongest professional connection was with the National Academy of Sciences. As a result of his wartime service, Angell had become an important member of Hale's circle. He served on the army's Committee on Education and Special Training and on the Committee on Classification of Personnel, remaining in Washington after the armistice to succeed John C. Merriam as secretary of the NRC. He was offered the presidency of the Carnegie Institution but chose the Carnegie Corporation instead.

It is not clear exactly how Pritchett, Root, and Woodward engineered Angell's election, but engineered it was: when Angell left after only a year, Pritchett worried that they might not be able to pull the same trick twice. He wrote to Root:

> Some of our colleagues are none too happy over the present regime and would much prefer the good old days. I am inclined to think that in the endeavor to bring in an effective regime, I have had to encounter sufficient odium to bring down more or less opposition on what I might propose. Doubtless it will be wiser if the suggestions for the future come through you. You carry enough guns to be independent of any such feeling.[70]

Or as he put it more colorfully to Hale, he would have to "lie down on Uncle Elihu and see to it that the suggestions all come from him, but it will be necessary to furnish him with the proper ammunition first."[71] This was how the reform clique got around Carnegie's courtiers: Pritchett had the ideas and planned the campaign; Root bulldozed the board. The method worked but not without generating suspicion and resentment. (Samuel Church, convinced that Woodward and Angell were conspiring to axe CC subsidies to the Carnegie Institute, retracted his vote for Angell and demanded that the election be run again.)[72] The misgivings of the old guard were not ill-founded.

Angell's plans for change went beyond even what Pritchett and Woodward had imagined. He proposed to terminate capital grants to the other Carnegie organizations "save in so far as they find a natural

[69]W. S. Hunter, "James Rowland Angell 1869–1949," *Biog. Mem. Nat. Acad. Sci.* 26 (1951): 191–208.

[70]Pritchett to Root, 21 Feb. 1921, CC.

[71]Pritchett to Hale, 21 Feb. 1921, GEH reel 29.

[72]Church to Root, 9 Apr., 1920; Pritchett to Church, 10 Apr. 1920; Church to Pritchett, 14 Apr. 1920; all in CC Church. Woodward to Pritchett, 12 Apr. 1920, CC Woodward.

place in any general program" (as, e.g., the CIW did in administering the corporation's grants for research).[73] He boldly asserted that the CC's charter was a mandate for promoting higher education exclusively—blithely ignoring Carnegie's many other benefactions. While aid to medical schools would continue, Angell proposed to concentrate on grants for graduate education in the arts and sciences. That, he felt, was the cause most worthy of the nation's richest foundation:

> The graduate school, so called, in our great American Universities has not as yet fully found itself, and its attempt on the one hand to train teachers and on the other hand to produce creative scientists and scholars has landed it in certain difficulties. . . . There is at this point a real possibility for an agency like the Corporation to do a constructive and significant piece of work. Were I to remain as president [he already knew he was leaving] . . . I should strongly urge a study of this entire problem.[74]

Angell exhorted the trustees that they could do nothing better than to help with the training of scientific researchers. He pointed to the importance of research in industry and to the war-born shortage of researchers, and urged the board to commit itself "as a matter of permanent policy to the furtherance of research."[75] He did not mean to end grants to the Food Research Institute or the Brookings Institute (both established with CC funds). But the great opportunity, he argued, was in grants to individual and cooperative projects in universities.

The model for Angell's program was certainly not anything in the CC's constitution or history (his saying so was pure rhetoric). Rather, Angell was drawing on his own experience as an academic entrepreneur and dean and as an activist in Hale's circle during the war. Their agenda was his, and it anticipated in a striking way the rationale of foundation programs in science in the 1920s. It could have served as a blueprint for the programs of the four Rockefeller boards and did serve, for a few brief years, for the Carnegie Corporation.

The most striking thing about Angell's brief tenure, however, is the gap between reform ideals and actual practice. Angell spent most of his time trying to cope with the burden of routine administration. It was a frustrating experience:

> I came out of my year as President of the Corporation with a renewed sense of the extraordinary difficulty of dealing consider-

[73]Angell, "Proposals with reference to general policy," 9 May 1921, pp. 2–3, CC P&P.
[74]Ibid., pp. 4–5.
[75]Ibid., p. 5.

> ately and intelligently with the thousand and one small requests and with determining, as a matter of principle, the relation which such requests ought to sustain to the larger and more inclusive program. . . . I was particularly impressed by the extraordinary number of local requests which, by various skillful devices, were given a personal flavor.[76]

Though he "struggled manfully," Angell failed to install more efficient methods of dealing with the deadening routine of retail business.[77] No wonder it took him almost his whole year to formulate a program.

Angell, however, was not exactly a passive victim of inertia. The record reveals that miscellaneous expenditures *increased* between 1918 and 1922, from 2.4 to 9.2 percent of total outlays. So, too, did expenditures on colleges, public libraries, and the Carnegie Institute of Pittsburgh.[78] Angell's fatal inability to resist "emergency" appeals was partly his inexperience but also partly principle. Angell believed that the CC should be eclectic and experiment with all options. Specialization was a necessity for small foundations, he felt, but for the great general purpose foundations it seemed an evasion of their special responsibility. He looked askance at the RF's self-imposed restriction to medicine and public health.[79] But the disadvantages of not concentrating were no less real. Leaving every option open only gave the tactical advantage to ongoing programs over new initiatives. The result was that Angell specialized not in medicine but in miscellany. It was a dilemma peculiar to the general purpose foundations, and neither Angell nor Vincent managed to resolve it.

Perhaps if Angell had stayed longer he would have succeeded in diverting the corporation's vast resources to graduate training and research in universities. Perhaps the CC, rather than the various Rockefeller boards, would have become the key patron of science. Root, Pritchett, and Woodward would have supported him, if for no other reason than that they believed in a strong president. They were devastated when Angell announced his imminent departure for Yale, seeing the wreck of their hopes for reform. (Pritchett was especially distressed, since the news of Angell's leaving immediately diverted the flood of academic petitioners back to his office.)[80] But without Angell to back it, his program had little chance to survive. Pritchett and Woodward hardly shared Angell's enthusiasm for university scientists, and Pritchett had not lost his enthusiasm for medical education. It was Pritchett himself

[76]Angell to Keppel, 25 Apr. 1927, CC P&P.
[77]Pritchett to Root, 21 Feb., 8 June 1921, CC.
[78]"Classification of appropriations," CC P&P.
[79]Angell, "Proposals with reference to general policy," 9 May 1921, CC P&P.

who, as acting president in 1921–1922, presided over the retreat from Angell's program.

Pritchett, Then Keppel

Pritchett's first priority after Angell left was to complete the reform of the CC's administration, and in this he succeeded. He persuaded Bertram and Robert Frank to give up their salaries ("a real accomplishment" he thought).[81] The Carnegie Corporation finally got a modern structure, with clear separation between officers and board. Pritchett also persuaded the trustees to terminate the college program. With it, however, died Angell's plan for a system of research grants: Pritchett chose to regard that as an extension of gifts to colleges. He also reaffirmed the CC's responsibility to the five Carnegie organizations—"Carnegie's children." The new programs that Pritchett favored were quite different from Angell's and much closer to Carnegie's old favorites: college libraries and library schools, art education, and educational research.[82] A few projects in medical education and scientific research would be continued but only as token demonstrations.

Every detail of Pritchett's program reveals how decisively he turned back to tradition. He went beyond routine obeisance to Carnegie in citing Turgot's text about trusts losing their vitality when their founders die. Whereas in 1916 he had thought Carnegie's grants for public libraries old-fashioned, now he proclaimed them the height of wisdom, having done more good and less harm than any other efforts to promote knowledge. From that base rose Pritchett's new programs to bolster librarians' professional status and study adults' reading habits. Pritchett's new emphasis on the CC's responsibility for sustaining "Carnegie's children" was in the same vein.[83] Most of the CC's income for the next twenty years went to shoring up Carnegie organizations, expecially his own Carnegie Foundation (whose pension fund was heading toward actuarial disaster and had to be bailed out).[84]

Pritchett's disenchantment with Angell's vision was not so much with science as such as with mass higher education. Long experience with academic petitioners had engendered in Pritchett, as it had in Wood-

[80] Pritchett to Root, 21 Feb., 8 June 1921, CC.

[81] Pritchett to Root, 19 Nov. 1921, CC.

[82] Pritchett, "A policy for the Carnegie Corporation," 16 May 1922; Pritchett, "Memorandum prepared in connection with work of Committee on Policy and President," 15 May 1922; both in CC P&P.

[83] Pritchett, "A policy," pp. 1, 5–6.

[84] Pritchett, "Memorandum," 15 May 1922.

ward, a conviction that any relationship with universities was bound to end up by making foundations into their helpless servants.

> Any sum given by [foundations] . . . is a bagatelle in comparison with the sums asked. Furthermore, American . . . colleges are today embarked upon a program which is educationally superficial, unduly expensive, and in which the granting of moderate sums of money by the Corporation has no other effect than to increase in small measure the present-day tendencies in education. . . . Our gifts . . . merely add small sums to a great stream of money, largely spent in promoting a program of education which results in a vast, sprawling, superficial regime, needing most of all serious scrutiny rather than assistance in its present tendencies.[85]

Unlike Angell or the Hale group, Pritchett was unable to perceive a mechanism for the Carnegie Corporation to participate in developing communities of science without being co-opted. Like Woodward, Pritchett never saw beyond the prewar system of science patronage. Angell brought from his experience with the NRC circle a vision of a new system of institutional patronage. But he did not stay long enough at the CC to change its older perceptions and ways of doing things.

Some of the persons mentioned as possible successors to Angell might have continued his policy: Edwin G. Conklin, professor of biology at Princeton and an activist in the NRC; NRC secretary Vernon Kellogg (Angell's suggestion), or John C. Merriam (who unfortunately had already been appointed to succeed Woodward at the CIW).[86] The majority of candidates, however, were university "handymen" like Jerome Greene or Embree.[87] (Carnegie's courtiers would have left the presidency vacant or appointed one of their own—Angell was "a good deal shocked.")[88] The new president, Frederick Keppel, was also of the handyman type. As secretary of Columbia University, he was not a high-caliber personage in the academic world. However, he had also served as undersecretary of war (1918), director of the foreign operations of the American Red Cross (1919), and director of the American Chamber of Commerce in Paris; probably it was his war record that brought him to the attention of the CC board. He was an efficient manager familiar with a range of nonprofit organizations. His strong point was a capacity for hard work.

As president of the corporation, Keppel proved a capable admin-

[85]Pritchett, "A policy," pp. 3–4.

[86]Root to Pritchett, 13 Aug. 1921; Pritchett to Root, 1 Apr. 1921; both in CC.

[87]"Report of the committee appointed to propose names," 6 June 1921, CC P&P.

[88]Angell to Root, 8 Aug. 1921, Elihu Root Papers, box 138, LC.

istrator but not an independent or imaginative leader. He had a knack for picking up and articulating trendy issues in the foundation world, but he invented none. (He later took credit for originating the CC's program in adult education, but it was Pritchett's idea.)[89] It did not take long for Keppel to settle down to the task of terminating old operating programs, liquidating obligations piled up by Angell and Pritchett (including those in science), and bailing out Carnegie organizations. Public libraries were phased out by 1925, medical projects by 1929, scientific research by about 1931.[90] Keppel had no program because he remained a hostage and wet-nurse to "Carnegie's children." Keppel was for all practical purposes the president of a small foundation. He was a one-man show: everything came across his desk; what did not appeal to him went off the desk into the waste basket, what did appeal was approved.[91] In contrast, the Rockefeller boards had become elaborate organizations staffed by professional managers.

Managers and Policies

The major accomplishment of the Rockefeller Foundation and Carnegie Corporation in their first decade was to make philanthropy a business. Many vestiges of individual and family philanthropy were removed; full-time managers took over from part-time trustees; the business of philanthropy was professionalized. Pritchett coined the term "giving corporations" and told Root that "the conduct of these great endowments has become almost a profession."[92] "Almost," nothing: managers, by virtue of their knowledge and control of day-to-day operations, had a de facto authority in matters of policy that was at least a match for the de jure authority of their trustees. These new professionals were not a homogeneous group. There was conflict between presidents like Vincent, who tried to control policy from the center, and field operators like Roger Greene or Wickliffe Rose, who had the practical power to decide what was done. There was conflict between the new generation of handymen and the slightly older generation of generalists like Abraham Flexner and Rose.

These conflicts within the ranks of professional philanthropists, to-

[90]Keppel, "Summary of unpaid obligations," 30 Sept. 1928; "Memorandum for the executive committee upon future financial obligations of the corporation," 15 Apr., 1929; chart of expenditures 1924–1946 attached to "Classification of appropriations 1911–1922," 15 Nov. 1922; all in CC P&P.

[91]Warren Weaver, oral history, pp. 401–403, CUL.

[92]Pritchett to Root, 1 Apr. 1921. Pritchett, "The function of a giving corporation," n.d. [1916 or 1921], CC.

gether with the conflict between officers and board, produced the pattern of initiative and retreat that characterized the first decade of the RF and CC. The loose structure of the large foundations, resembling that of holding companies, encouraged indecision and made it hard to start new programs, as Jerome Greene discovered when he tried to create departments for his "university of human need." The fact that decisions were made in the glare of hostile public scrutiny compounded the difficulties of taking decisive action. What got done was what there was already precedent for doing: public health and medical education in the RF, and education in the CC. In the corporation, the reform group was in control for only a few brief years and themselves led the retreat to more traditional programs. In the foundation, Vincent settled in to defending the budget against raids by the health barons or by anyone else with big ideas for new programs.

There was talk in the 1920s of a "crisis" in foundation work, of a brain drain of talent. Abraham Flexner ran down the personnel of the eight or ten "so-called foundations" in New York City in 1924 and found "very few men of really pregnant intelligence." With such mediocre leadership, he warned, "money is apt to be a source of embarrassment rather than otherwise."[93] Flexner blamed bureaucratic routine: "Instead of having time to read and think and grow, we are overwhelmed with engagements, interviews, telephone calls—all that hodge-podge of feverish and indiscriminate activity, which under the alluring title of executive work, tends to injure the better American minds."[94] Raymond Fosdick was less querulous but no less critical. "The idea of systematic philanthropy is no longer a novelty," he wrote to Rockefeller, Jr., "and men are not as easily attracted as they were, perhaps, a dozen years ago. Consequently, the Carnegie Corporation has to appoint a Fred Keppel as president, and other foundations are wobbling along under even more mediocre leadership."[95] The Rockefeller boards, he warned, were in danger of "dying from the neck up." Fosdick blamed competition from universities: "In some cases the universities outbid us in salary; in other cases . . . in living conditions," he wrote. "[They] provide a community atmosphere and intellectual companionship of a kind we cannot easily create here in New York."[96] The one-way flow of "university handymen" into the sister profession of foundation management became an ebb and flow.

[93] A. Flexner, "Foundations—ours and others," 18–19 Jan. 1924, RF 900 22.165.
[94] Ibid., pp. 14–15.
[95] Fosdick to Rockefeller, Jr., 6 Oct. 1927, RF 900 17.123.
[96] Ibid.

It was in this period of initiative and retreat that a new system of science patronage was created: the CC's endowment of the National Research Council and the RF's grants to the council for research fellowships (1919), and the programs for building centers of science of the General and International Education Boards (1923). In the next chapter we will see how these decisions were made and why. It is important to keep in mind that the new system of science patronage was created at a time when the large foundations were struggling to modernize their governance and discover their distinctive mission; when a new generation of middle managers was trying to work out their relations, not only with potential clients but among themselves and with their boards of trustees. Thus decisions about specific programs—like research grants, fellowships, and endowments for scientific institutions—were also decisions about the authority and careers of the managers of science.

Chapter Four

The National Research Fellowships

It is a commonplace that World War I was a watershed in the social relations of science. Certainly it was in the relations between scientists and their patrons. In Britain and Continental Europe, the war gave a strong boost to state support for scientific research and development. Across the Atlantic, war service gave natural scientists a visibility in the public eye that they had not enjoyed since the great Western surveys of the 1870s. It is no accident that the Carnegie and Rockefeller boards made their first major benefactions to science within months of the Armistice. The war made possible what hitherto had not been. It is no less true, however, that the war interrupted changes in the patronage system that had been stirring since the turn of the century. Much of what came to fruition in 1919 had been sown and cultivated in the prewar years. The war did little to alter the hopes and dreams of scientists, though it did make them more adept and flexible tacticians. The war was a catalyst of change, but catalysts, chemists will tell you, only accelerate, they do not alter the direction of change.

To understand how a new system of patronage was created we need to look not just at the crucial decisions of 1919 but at the extended process in which these decisions were a pivotal episode. The Carnegie Corporation's gift of $5 million to the National Academy and National Research Council, and the Rockefeller Foundation's program of NRC fellowships were the culmination of almost a decade of foray and reconnoitering. To scientists, though not to the leaders of the CC and RF, they were the first installment of a larger system of patronage. Thus, we need to examine prewar initiatives, which failed but whose failure illuminates what the fundamental issues were and what changed between 1916 and 1919. And we need to examine subsequent efforts in 1919–1923 to complete the system of patronage, to see how the process bogged down, and how temporary negotiating expedients became a system of philanthropic practice that lasted to the end of the 1920s.

The historical process out of which a new system of patronage emerged was like an ocean wave, gaining impetus slowly and unseen and finally washing up on a shore, but only up to a point. So, too, social movements acquire momentum and lose it in the process of creating change. It is not enough to look just at the key events; we need to look at the process as a whole. The interesting question is why a wave of social innovation goes so far and no further. That cycle of change is the subject of this chapter and the next.

The argument in brief is that a new rationale for the patronage of science was created in the changed context of postwar reconstruction. Unlike the nineteenth-century system of aiding deserving but underprivileged academics, the new system focused on training, on developing the community of science as a whole. Concerned as everyone was in the postwar years with manpower shortages and reconstruction, foundation leaders could be persuaded that training a new generation of academic researchers was a legitimate and affordable undertaking. Unlike the nineteenth-century system of individual grants-in-aid, the new system of community development limited potential commitments to particular individuals and institutions—that is what made it workable for foundations. The National Research Council was also crucial to the new system of patronage (another side effect of the war). By providing an administrative buffer between the foundations and individual institutions, the NRC lessened the risk that patrons would be deluged with claims by individuals and institutions—till then the chief deterrent to a real partnership.

This new rationale was a trimmed down alternative to much more elaborate proposals for institution building—especially schemes for endowed regional centers of research and training. The new rationale was improvised quickly when the NRC negotiators, realizing that the RF would not buy institutional endowment, decided to do what was doable in the short term. The key to understanding this pivotal episode lies in the local agendas of university scientists and presidents, who aspired to make their institutions into national centers. For the scientists negotiating with the CC and RF, local and national ambitions were inextricably entwined. Finally, in the next chapter, the largely unsuccessful efforts by the NRC elite to complete the grand institutional plan postponed in 1919 will be studied. These efforts reveal the limits of a system of patronage mediated by national research councils. Grants to individuals, projects, and institutions remained largely out of bounds, because foundation leaders like George Vincent could not yet conceive a rationale and a financial vehicle that would limit their obligation to science.

First Initiatives

It was the scientists, not the foundation managers, who made the first moves toward getting science on the agendas of the CC and RF. Disappointed by their experience with the CIW grants program, but with their appetities whetted, scientists hoped that the new general purpose foundations might prove more amenable patrons. They approached Greene and Pritchett not as individuals seeking grants, however, but as representatives of national scientific societies. That was something new. George Ellery Hale came on behalf of the National Academy of Sciences; Edward Pickering and James McKeen Cattell, of the American Association for the Advancement of Science. Hale wanted a building for the National Academy and salaries for full-time research staff. Cattell had in mind a program of individual grants-in-aid, like the CIW programs but administered by committees of the AAAS. They were institutional proposals, and how they appeared to Jerome Greene and Henry Pritchett depended on the record of the academy and AAAS as national institutions.

Both, in fact, had poor records of participation in national affairs. The National Academy had been established (in 1863) to put scientific expertise at the service of the government, but for most of its fifty years it had been a blue-ribbon club, restricted in membership and dominated by a gerontocracy who shunned practical service as dangerous invasions of scientific laissez-faire. The AAAS, with its open membership and interest in popular science, was broader based than the NAS, but was constrained by its poverty and parliamentary politics from launching effective reform movements. Both societies depended on volunteer management; in both, prevailing laissez-faire ideals and disciplinary rivalries stood in the way of anyone who tried to introduce new ideas of organization.[1]

Since 1900, however, reform circles in both organizations had begun to make some changes. In the academy, Hale, Walcott, and others (Cattell, before he broke with Hale) had expanded the membership, improved the quality of scientific meetings, published scientific proceedings, endowed a lecture series, and tried to organize cooperative research programs. These measures were still a long way from Hale's grand plan, which he laid out in a series of articles in *Science* in 1913–1915. Hale's agenda included a new building, a "temple of science," with laboratories and full-time research staff; an endowed program of

[1] Rexmond C. Cochrane, *The National Academy of Sciences* (Washington, D.C.: National Academy of Sciences, 1978). Sally G. Kohlstedt, *The Formation of the American Scientific Community* (Urbana: University of Illinois Press, 1976).

research grants; and endowment for projects to demonstrate the value of organization and cooperation in science.[2] Cattell pursued similar ends in different ways. Membership in the AAAS had increased enormously after he became the editor of *Science* (its official journal), and Cattell saw this national constituency as a "powerful machine" for raising standards of scientific practice and increasing public support for research. Like many reformers in the Progressive years, Cattell believed that change was a matter of communication: he put his faith in surveys, publicity, and consensus building. Despite Cattell's antipathy to the exclusiveness of the National Academy, he and Hale were closely related variants of a Progressive reform type.[3] They shared a belief in large-scale organization and, despite Cattell's populist self-image, both worked through small circles of influentials to set the direction of change at the grass roots.

They faced different problems in reconstructing their organizations: Hale had to expand the narrow and exclusive base of academicians; Cattell had to discover a way for a small group to lead a large and unwieldy rank and file. Their solutions, however, were quite similar: both devised an institution within an institution, manned by an activist elite. A system of committees ensured a fairly broad participation in the activists' organizing efforts, but without diluting their leaders' ability to set agendas. Cattell organized the AAAS's Committee of 100 on Research in December 1913.[4] Hale's comparable creation, the National Research Council, followed in April 1916. Not limited to academy members, the NRC gave Hale the broad base he needed to spread his ideals. It took the American declaration of war in April 1916 to overcome academicians' distrust of such a change, and Hale exploited the crisis to the full. Obviously designed to overlap as much as possible with the Committee of 100, Hale's organization quickly overshadowed Cattell's.[5] Like the general purpose foundations, these new scientific organizations were late manifestations of the fashion for large-scale organization set by corporations and trusts around 1900.

Money was crucial to the reform cliques in both the National Acade-

[2]George E. Hale, *National Academies and the Progress of Research* (Easton, Pa.: New Era Printing Co., 1915). Nathan Reingold, "The case of the disappearing laboratory," *Amer. Quar.* 29 (1977): 79–101. Reingold and Reingold, *Science in America*, ch. 9.

[3]Nathan Reingold, "National aspirations and local purposes," *Trans. Kansas Acad. Sci.* 71 (1968): 235–246. Michael M. Sokal, "*Science* and James McKeen Cattell, 1894–1945," *Science* 209 (1980): 43–52.

[4]"The committee of One Hundred on Scientific Research of the AAAS," *Science* 41 (1915): 317. "Grants for scientific research," *Science* 43 (1916): 680–681; 44 (1916): 50–51, 229–232. Cattell to Greene, 25 Mar. 1916, ECP II.

[5]Daniel J. Kevles, "George Ellery Hale, the First World War, and the advancement of science in America," *Isis* 59 (1968): 427–437.

my and AAAS: organizing science required capital; the efforts of volunteers were not enough. And no source seemed more promising than the new general purpose foundations, with their congenial reform ideals. Nothing would have done more to entrain the conservative rank and file in the cause of reform than support from these foundations. From the foundations' view, however, the academy and AAAS, in their conservative scientific parochialism, seemed hardly worthy of support. Without reform there could be no money; lack of money stood in the way of reform. But how to approach these new organizations? Through Carnegie and Rockefeller themselves, or through friends on the boards of trustees, or foundation officers? The activist scientists and foundation managers alike were uncertain of their agendas and authority and had no experience in working together. A relation had to be worked out by trying everything and seeing what worked.

As head of the CIW solar observatory, Hale naturally gravitated to the CC. (He professed not to want the academy connected in any way with the "Rockefeller interests.")[6] He had drafted an appeal to Carnegie about 1910 but had been persuaded (probably by Elihu Root) that the time was not propitious.[7] In 1913 the academy's approaching sesquicentennial seemed a good occasion to renew his appeal.[8] Hale worked through his friends on the CC board, Root and Pritchett, carefully framing his argument to appeal to Root's interest in international relations and Pritchett's in higher education.[9] The key to his plan was the academy building, which Hale hoped would inspire the academicians with a larger sense of public service. He had already arranged for a group of Boston architects to draw up a plan, to be ready should a benefactor appear.[10]

Pritchett did not share Hale's optimism about the academy: a grand building, he thought, was unlikely to make academicians less indifferent to a larger vision of science in national life.

> [The academy] has been a little inclined to take the position that congress and the country should accept it and defer to it as the scientific center of influence, without at the same time giving to

[6]Hale to Carnegie, 3 May 1914, GEH reel 9. Hale worried that Congress might balk at providing funds for a building site if the academy were associated with the name of Rockefeller; but the more compelling reason, he assured Carnegie, was that he preferred to deal with the CC's trustees.

[7]Hale to Root, 29 Dec. 1913, CC NAS-NRC.

[8]Hale to Carnegie, 2 May 1914, GEH reel 9; and in Reingold and Reingold, *Science in America*, pp. 213–215. Hale to Walcott, 2 Jan. 1913, GEH reel 36.

[9]Hale to Pritchett, 4 Jan., 12 Feb. 1913; Hale to Root, 3 and 10 Mar., 29 Dec. 1913; all in CC NAS-NRC.

[10]Hale, *National Academies*, pp. 138–149.

> Congress and the country any real leadership outside of the work which men do in their own . . . laboratories. . . . In fact, many individuals in the country have far more influence both upon public opinion and upon Congress than has the National Academy.[11]

The first step, Pritchett thought, was to install full-time executive officers in Washington. Hale was not discouraged. He had been quietly lobbying to give more power to resident Washingtonians, and was ready to make his move in the coming election of the academy.[12] Besides, Root was enthusiastic for the project and promised to do his best if Pritchett would go to see Carnegie. Pritchett thought it would be safer to wait and apply to the CC, but he promised Hale that he would "tackle our beneficent friend" if Hale could give him a concrete plan. Hale immediately suggested a figure of $800,000 for a building and $600,000 for endowment.[13]

Pritchett and Root presented Hale's scheme to Carnegie in March 1913. Carnegie seemed to react favorably, but a year went by with no action.[14] In March 1914 Hale renewed his suit with Carnegie directly. Carnegie obviously had forgotten all about it and was not interested, dismissing the National Academy as "just one of those fancy societies." Returning with Pritchett the next day, Hale found their "beneficent friend" positively hostile to the plan. (Hale thought James Bertram had set his mind against it.) Carnegie angrily accused Hale of abusing his hospitality, and a stormy argument ensued, which was calmed only when Hale produced a letter of support from Root. A few weeks later, Carnegie informed Hale that he could not take an interest in the building project.[15]

Such scenes were not unfamiliar to those who courted Carnegie. His philanthropic mood was notoriously volatile, and he coped with unfamiliar problems by tuning them out. Although Hale had heard that Carnegie's wife had brought him around on the academy building, Pritchett advised Hale to lie low. Carnegie was in a happier mood, he

[11]Pritchett to Hale, 3 Feb. 1913, p. 3, CC NAS-NRC, and in Reingold and Reingold, *Science in America,* pp. 197–199.

[12]Hale to Pritchett, 12 Feb. 1913, CC NAS-NRC. Cochrane, *National Academy,* pp. 200–202.

[13]Pritchett to Hale, 25 Feb. 1913; Hale to Pritchett, 3 Mar. 1913. Hale to Walcott, 24 Jan. 1913, GEH reels 29 and 36.

[14]Pritchett to Hale, 27 Mar. 1913; Walcott to Hale, 24 Mar. 1913; both on GEH reels 29 and 36.

[15]Helen Wright, *Explorer of the Universe: A biography of George Ellery Hale* (New York: Dutton, 1966), pp. 308–311. Hale to Evelina Hale, 18 Apr. 1914; Carnegie to Hale, 11 May 1914; both cited by Wright. Hale to Carnegie, 3, 5 May 1914; Pritchett to Hale, 7 May 1914; Hale to Pritchett, 11 May 1914; all on GEH reels 9 and 29.

reported, but only because he was avoiding business and "other perplexing things." Hale wondered if Pritchett's heart really was in the academy project, and he turned up the pressure on Root, hoping to gain access through him to the CC.[16] He had already begun, at Root's suggestion, to write "National Academies and the Progress of Research," the first installment of which had appeared in November 1913. Publicity and history, they both knew, would help legitimate Hale's cause in the eyes of people like Carnegie.[17] Hale would have taken his case to the RF, too, had he not been advised by Simon Flexner that the moment was inopportune.[18] (So much for his qualms about guilt by association with the Rockefellers.)

While Hale cultivated Carnegie, Cattell worked the Rockefeller side.[19] To head the Committee of 100's fund-raising efforts, he recruited Edward Pickering, his old friend and veteran bird dog of philanthropists. Their first priority was to secure funds for a program of small research grants for underprivileged academics. The idea was for the RF to make an annual grant of $25,000 or $50,000 to the Committee of 100, which would select the most worth recipients. It was the AAAS grants-in-aid program, scaled up to CIW size. Pickering pointed explicitly to the CIW's experience, arguing that it had demonstrated the need for a program aimed at "aiding the man of genius, especially where he is isolated and has no other means of carrying on his work" (Carnegie's exceptional man). Experience had also shown, Pickering argued, that a program of research grants was better handled by a committee of experts than by one individual. He reminded Greene of scientists' resentment of Woodward's personal management of the CIW grants and tried to pique Greene's interest with some examples of small projects that could be done on a shoestring.[20]

Greene knew all about the CIW's experience, but his sympathies were entirely with Woodward, not his academic critics. Small grants-in-aid to needy individuals just did not fit the RF's plan to concentrate on a few large and important problems. Pickering's examples of what could be done with small grants—research on the paleontology of western Virginia, the rotifera of plankton, and sexuality in fungi—might have been designed to prove Greene's case. The meager results of the Rocke-

16Wright, *Hale,* pp. 308–311. Pritchett to Hale, 11 May 1914, GEH reel 29.

17Hale to W. M. Davis, 12 Feb. 1920, GEH reel 11. Hale to Root, 29 Dec. 1913, CC NAS-NRC. Hale to Carnegie, 5 May 1914, GEH reel 9.

18Hale to S. Flexner, 1 Apr. 1915; S. Flexner to Hale, 16 Apr. 1915; both on GEH reel 14.

19Reingold, "Disappearing laboratory," pp. 88–90.

20Pickering to Greene, 24 Oct. 1913, 21 May 1915, RF 915 1.1. Cattell to Greene, 25 Mar. 1916, ECP II.

feller Institute's grants program also revealed the error of making miscellaneous grants without specific goals; as the institute's business manager, Greene must have known all about it. Greene suggested to Pickering that he might do better to organize committees to promote research on a few fundamental problems, or to think in terms of gifts of endowment, or a large central fund that would distribute funds to individuals in institutions already provided with facilities for doing research. The trick, he observed, would be to give scattered recipients a proper sense of responsibility to their patron and accept a measure of supervision and control.[21] His words might have been lifted from one of Woodward's diatribes against small grants. For both parties, the CIW's experience was the single most important thing shaping their expectations of a future patronage of science. Greene and Pickering just read different lessons from the story.

Undeterred by Greene's advice, Cattell and Pickering began to lobby their friends on the RF board. Pickering concentrated on Charles W. Eliot, while Cattell worked on William Welch and Simon Flexner. Flexner was crucial. His endorsement in any matter could be decisive. (Embree recalled that Flexner could get support for projects in scientific medicine "without objection and almost without discussion.")[22] Flexner liked the idea of a research fund and pressed Greene to meet with the scientists. He also let Cattell know that the RF had a large unspent surplus just then and urged Pickering and Cattell to strike while the iron was hot. A week later Cattell was in Greene's office with Pickering's proposal in hand.[23]

Cattell rolled out all the traditional arguments why small grants-in-aid were more cost-effective than investment in new facilities and again referred to academics' disappointment with the CIW.[24] Greene was all set, however, to dispel Cattell's "image of an enlightened money bag." He professed not to know if the discontent with the CIW grants program was the result of Woodward's autocratic management or the pique of those who had been refused. What he did know was that the experience of both the Rockefeller Institute and the CIW had led them to prefer in-house institutes. Individual grants were just not compatible with organized philanthropy. Greene did agree, however, to present Pickering's plan to the board, not as a proposal but as a topic for

[21]Greene to Pickering, 7, 27 Oct. 1913, 14 June 1915; Greene to S. Flexner, 13 Mar. 1916; both in RF 915 1.1.

[22]Embree, untitled recollections, n.d. [c. 1930], ERE box 1.

[23]Pickering to Cattell, 23 Feb., 17 Mar. 1916; Cattell to Pickering, 26 Feb., 15, 17, 27 Mar. 1916; all in ECP II. Greene to S. Flexner, 13 Mar. 1916; Cattell to Greene, 25 Mar. 1916; both in RF 915 1.1. Reingold and Reingold, *Science in America,* pp. 244–245.

[24]Cattell to Greene, 25 Mar. 1916. Pickering to Cattell, 23 Feb. 1916, ECP II.

discussion: Should the RF adopt scientific research as one of its fields of interest? In April 1916, the executive committee directed Greene to prepare a memorandum laying out the pros and cons.[25] This was the customary procedure for screening new departments for the university of human need. In the process, Greene's views on science seem to have changed fundamentally. The evidence is slender but suggestive.

Greene's views are laid out in two documents: a formal memorandum to the board dated 24 May 1916, and an unsigned "dear sir" letter dated 18 April (no year), probably an earlier draft.[26] Greene did not doubt that scientific research was an important enterprise: it was, he wrote, impossible *not* to consider it as a field for the RF, but that was not the issue. The problem was that scientists would become dependent on foundation handouts.[27] Surprisingly, though, Greene exempted science from the law of pauperization, on the grounds that all new knowledge must ultimately be fruitful. Both the 18 April and 24 May drafts begin with that premise; however, they come to different conclusions. In the earlier draft, Greene concluded that the RF's programs in public health were free of the risk of dependency, because they applied and diffused existing knowledge. It was a case for steering clear of research and sticking to "propaganda of existing knowledge." In the 24 May document, Greene takes the same argument to the opposite conclusion:

> There are two ways in which the Rockefeller Foundation might look upon the proposition to make possible numerous grants in aid of research. It may be regarded like any other proposition which would involve Mr. Rockefeller or any of his Boards in carrying part of the ordinary load of education and charity which each community ought to carry for itself. There is much to be said for this point of view. . . . On the other hand, the proposition may be regarded as offering the opportunity to do a service to science and to human welfare which, to a large extent, at least, will be seriously deferred, if it does not actually fail of accomplishment, but for such outside aid. The great argument for aiding *research* [my emphasis] is that knowledge breeds knowledge, it might almost be said in geometrical proportion, and the reward of prompt aid where it is really needed is to be found in the enormous and far-reaching fecundity of the ensuing benefit.[28]

[25]Greene to W. Cannon, 18 April 1916, RF 915 1.1.

[26][Greene] to "Dear sir," 18 April, 1916, RF 915 1.1. Greene, "Application from the Committee of One Hundred," 24 May 1916, RF 915 1.6. Reingold and Reingold, *Science in America,* pp. 245–246, 249–252.

[27][Greene] to "Dear sir," 18 Apr. 1916.

[28]Greene, "Application," 24 May 1916, pp. 3–4.

Now it was not just "propaganda of existing knowledge" that was exempted from the logic of dependency but the creation of new knowledge as well. Greene seems to be constructing a case for including scientific research in the university of human need.

Why the change of heart? No doubt Greene's feelings reflected the rapidly changing political climate following the torpedoing of the *Sussex* on 24 March and President Wilson's ultimatum to Germany on 18 April 1916. The new context of preparedness and mobilization gave research a national, patriotic dimension that it did not have a few months or even weeks earlier. "In all that is said about the importance of military preparedness," Greene wrote, "no single measure advocated begins to compare in importance with the proper mobilization of the resources of this country for research, and . . . without exception all the benefits accruing from research will be no less valuable for peace than for war."[29] At Greene's suggestion, the board appointed a special committee to inquire into the needs of scientific research.[30]

Pickering, unaware of how the RF worked, saw the board's action as a rejection of their proposal. Cattell surmised that the board was temporarily distracted by war relief. Meanwhile, they set up an elaborate machinery to administer a grants fund, in the somewhat naive hope that a demonstration of need and organizational ability would encourage the RF to act. They were disappointed, of course, as were the many hopeful scientists who sent in plans for projects.[31] The RF never did organize a committee.[32] The moment was lost. It is unlikely in any case that it would have supported a program of small grants in the nineteenth-century manner. Greene's vision of research as a national service looked forward to a quite different kind of relationship.

What that relationship would be was foreshadowed by an event that occurred in the summer of 1916. Visiting the Marine Biological Laboratory at Woods Hole, Abraham Flexner told T. H. Morgan about a scheme for encouraging research by newly fledged Ph.D. scientists. As director of the GEB's programs in higher education, Flexner was concerned about the supply of top-quality university teachers, that is, those who combined teaching with research. Most scientists were obliged to go directly from graduate school to full-time teaching and thus had no chance to establish habits and lines of research. The shortage could be-

[29]Greene, "Program," p. 4.

[30]RF board minutes 11 Apr., 24 May 1916, RF. Pickering to Cattell, 3 June 1916, ECP II.

[31]Pickering to Cattell, 2, 3 June, 24, 26 Aug., 3 Oct. 1916; Cattell to Pickering, 3, 5 June, 23, 26 Aug., 7, 18, 23 Oct. 1916. Cattell to N. L. Britton, 6 Jan. 1917, ECP II.

[32]RF board minutes, 25 Oct. 1916, RF.

come acute, Flexner thought, if the war cut off the supply of talent just when the GEB's program in medical education was creating the greatest demand for teachers of basic science. Suddenly, aid to promising young academic researchers became an appropriate objective for the GEB. (In the International Health Commission, Wickliffe Rose was having the same thoughts.) Flexner asked Morgan to supply him with a list of existing sources of funds for research, and wheels began to turn. Morgan told Cattell, who organized a subcommittee of the Committee of 100 to gather information and lay plans. Morgan also informed Hale of the GEB's interest and warned Flexner that Hale was likely to be knocking on his door in the near future.[33]

Nothing came of Flexner's initiative, no doubt owing to the more pressing need to organize war relief programs. But it heralded in a striking way the new kind of science patronage that emerged immediately after the war. Flexner's rationale for supporting research did not make research an end in itself but rather a means to a greater end, namely, training academic researchers. It was not aiding needy individuals but developing a community of academic scientists whose skills were deemed vital for the national good. This rationale made research an unambiguously legitimate field for support by educational foundations. It was an idea that, in a few short years, would be the basis of a new patronage of science, quite unlike the prewar style. T. H. Morgan most clearly perceived the potential of a rationale that aimed at producing researchers ("fitting them, as it were, for the market") rather than producing research—that aided careers, not projects.[34] Morgan was the harbinger of a generation of academics like Robert Millikan and Frank Lillie, who in the 1920s forged a partnership with foundations of which Pickering's generation could hardly dream.

Flexner and Morgan's scheme was a radical departure from the model of science patronage that had prevailed since the 1880s: just how radical is apparent from Pickering's vigorous reaction against it. Pickering saw the idea of grants to fledgling teachers as a substitute for his scheme of research grants, and a distinctly less attractive one. Training researchers was all very well but, to his mind, was much less important than aiding mature and productive scientists. Cattell and Morgan tried to make Pickering understand that the RF and GEB were different agencies and that a proposal to the one did not preclude a proposal to

[33]A. Flexner to Morgan, 25 Sept. 1916; Morgan to A. Flexner 17, 27 Sept., 15 Nov. 1916; all in GEB 1.5 702.7225. Cattell to Pickering, 7, 28 Oct. 1916, ECP II. There was no mention at this stage of postdoctoral fellowships. Cattell dreamed of a degree above the Ph.D., like the French Doctorat d'Etat.

[34]Morgan to Pickering, 23 Oct. 1916, ECP II.

the other. Pickering was not mollified: research could only be promoted by direct grants-in-aid, and that was that.[35] Steeped in the experience of the Elizabeth Thompson Science Fund and the CIW grants program, Pickering was unable to grasp that foundations were more likely to accept research as an appropriate objective if research were connected to education.

Thus the habits and expectations embedded in the nineteenth-century system of individual grants-in-aid were beginning to come unstuck as Europe went to war. It would be going too far to say that anyone in 1916 saw very clearly what a new system of patronage would look like. But academic entrepreneurs were learning from their mistakes, and their various gambits were still in play when the United States also went to war. The outcome of those gambits was profoundly altered by the experience of war and its effects on the institutions and public perceptions of science.

A Trade Association of Science: The NRC

American participation in World War I was brief, from April 1917 to November 1918, but it catalyzed rapid changes in the social system of science. Most important, it transformed the National Research Council: in 1917 the NRC was an elaborate system of paper committees; in 1919 it could boast a record of solid achievements in mobilizing and deploying scientists for war projects.[36] The war gave the NRC leaders experience and a taste for large-scale organization. It demonstrated that academics could deliver useful scientific services in a business-like way. Nothing was more important in reversing foundation stereotypes of ivory-tower academics than the war record of the NRC veterans.

A side effect of the NRC's rise was the rapid decline of the rival Committee of 100. Its leaders systematically co-opted by Hale's parallel committees, the Committee of 100 remained little more than a paper organization. Cattell was no match for Hale in organization, and his quixotic pacifism did the rest (he was accused of being pro-German and dismissed from his professorship). His organization survived into the 1920s but was limited to doing surveys of scientific manpower. The NRC was the center of organizing activity in the 1920s, with its millions of endowment from the CC and its practical responsibility for admin-

[35]Cattell to Pickering, 28 Oct. 1916, 27 Jan. 1917; Pickering to Morgan 18, 25 Oct. 1916; Morgan to Pickering, 23 Oct. 1916; Pickering to Cattell, 1 Nov. 1916, all in ECP II.

[36]Daniel J. Kevles, *The Physicists* (New York: Knopf, 1978), ch.s 8–9. Robert M. Yerkes, *The New world of Science: Its Development during the War* (New York: Century, 1920). Dupree, *Science in the Federal Government,* ch. 16. Cochrane, *National Academy,* chs. 8–9. Hale, "The purpose and needs of the National Research Council," 26 Mar. 1919, CC NAS.

istering the national research fellowships. Hale's circle of activists in the NRC were at the center of every postwar effort to organize science.

The institutional character of the NRC was vital to Hale's strategy for getting foundation funds. The NRC had a strong family resemblance to voluntary agencies in public health and welfare; also, to trade associations, which multiplied rapidly after the war to become a characteristic institution of what Ellis Hawley has called "the associational state."[37] Indeed, the NRC may best be understood as a trade association for science. Like its bigger cousins, the NRC was designed to bend individualistic, competitive behavior to the common good without intruding upon individual and local prerogatives. Lacking bureaucratic authority, such organizations emphasized education, communication, and consciousness raising, and worked through networks of influence and emulation—"cooperation" was their hallmark. The NRC's aim was systematically to develop the community of science by providing information about important research trends, new methods of organizing research, and sources of funds for fellowships and sponsored research.[38]

The NRC's concern with community and infrastructure appealed to the large foundations in a way that the National Academy or AAAS never had. Designed not to confer honor but to provide services, broadly representative of the natural sciences but not hobbled by disciplinary logrolling, the NRC was an ideal intermediary between foundations and their scattered scientific constituents. The Rockefeller Foundation was accustomed to working through voluntary health agencies like the Red Cross or the National Committee for Mental Hygiene and found it easy to work in the same way with the NRC. It had already made substantial grants to the NRC during the war, virtually underwriting its medical sciences division.[39] Training and community development were congruent with the ideals of organized, large-scale philanthropy. Foundations and the NRC were both designed to inform, connect, and demonstrate the efficacy of organization and management. With the NRC to mediate, it is not surprising that a partnership between foundations and academic science emerged so readily in the postwar years, after nearly a decade of false starts.

Despite the general expectation that foundations would become patrons of science, it was unclear in 1918–1919 exactly what the terms of the relation would be. These terms were invented in the process of ne-

[37]Ellis W. Hawley, *The Great War and the Search for a Modern Order* (New York: St. Martin's Press, 1979).

[38]Kohler, "Science, foundations, and American universities in the 1920s," *Osiris* 3 (1987): 135–164, see pp. 140–147.

[39]Ibid., p. 143. RF *Annual Reports*, 1917–1920.

gotiating the Carnegie Corporation's gift of $5 million to the academy and NRC and the Rockefeller Foundation's program of postdoctoral fellowships. These key negotiations focused on institutions and infrastructure. There was no serious discussion of individual research grants—a decisive break from prewar initiatives. Fellowships were granted to individuals, of course, but their aim was in fact institutional and collective: to train a new elite of teacher-researchers and, indirectly, to build up national centers of training and research. This conception emerged as the two parties sought to discover a machinery that would satisfy both foundation and academic goals.

A Home and Endowment for the NRC

Hale had been waiting for an excuse to renew his appeal to the corporation, and in August 1917 the occasion presented itself. A few months earlier, he learned, Pritchett had proposed that the CC create "a great laboratory" for research in industrial chemistry and physics. Pritchett envisioned an organization like the Mellon Institute, that would do contract research for industrial firms for their exclusive use but with publication permitted after a period of years. This scheme was obviously a response to the American declaration of war in April 1917. There were many such initiatives to mobilize science for war production. Hale's creation of the NRC, pushed through a reluctant academy, Woodward's efforts to put the CIW laboratories at the disposal of the War Department, like Pritchett's scheme for an industrial research institute, were cases of the 1917 epidemic of mobilization fever, which suddenly dispelled inhibitions and opposition to doing new things. Pritchett set up a planning committee, and in May he wired Woodward to assemble an advisory group of leading chemists and physicists. Woodward, however, seems not to have shared Pritchett's urgency. Ever pessimistic, Woodward was sure that Pritchett's scheme would be opposed by academic scientists and by the CC board.[40] Pritchett's idea of an industrial research institute stalled.

Hale recognized a golden opportunity to make use of Pritchett's newfound interest in research to revive his own scheme for national laboratories of physics and chemistry within the National Academy.[41] The idea was to co-opt Pritchett's scheme into his own, and in August 1917 Hale went to Pritchett's summer home in Santa Barbara to win

[40]Woodward to Pritchett, 23 May 1917, CIW.

[41]Hale to Root, 3, 10 Mar., 29 Dec. 1917, CC NAS-NRC. Hale's 1913 plan for the academy building projected $50,000 for laboratory equipment and $25,000 a year for a director and research expenses.

him over. Basic research openly published, he argued, would have the most far-reaching benefits for industrial development. Why not revive Hale's 1913 scheme, for which architectural plans were already drawn up? Pritchett was "partially converted" (as Hale put it). If Woodward and Root got behind Hale's scheme, he thought, there was an excellent chance that the corporation would approve a gift of both building and endowment.[42]

Hale lost no time. Within two weeks he had secured assurances of support from Root and Woodward, alerted Millikan that he would be counting on him to design the new laboratory, and got AT&T's J. J. Carty to say that basic research was crucial for industrial progress. (Industries would do applied research themselves was the argument; basic research needed the stimulus of an extramural organization like the CC.)[43] Carty's testimonial was crucial for Hale's strategem: Who should know better than Carty what industry really needed? A detailed proposal was hastily prepared for the fall 1917 meeting of the board.[44]

It took another year and a half for the machinery to grind to a conclusion—not a surprising delay given the frantic pace of war work and the grandiose size of Hale's proposal. In May 1918 Hale, Carty, Millikan, and Walcott appeared in person before the corporation's board to argue their case for an endowment of $3–$4 million and an equal sum for a building. The board stalled, agreeing only to pay for the NRC's emergency war projects.[45] Finally, in March 1919, they approved a trimmed but still very grand scheme: $1 million for a building and $4 million for endowment. The National Academy agreed to provide a building site, and the CC agreed to pay interest on the $4 million until such time as the board felt it convenient to pay the capital. The grant was earmarked explicitly for the NRC and its activities, not the academy—thus heading off criticism of the sort Hale had heard from Pritchett and Carnegie. (Some academicians grumbled that the NRC tail was wagging the academy dog.) The NRC's role in promoting in-

[42]Hale to J. J. Carty, 27 Aug. 1917; Hale diary, 16 Aug. 1917; Hale to Millikan, 16 Aug. 1917; all on GEH reels 47, 76, 25.

[43]Hale to Carty, 27 Aug., 20 Sept. 1917; Carty to Hale, 15 Sept. 1917; Hale to Pritchett, 15 Sept. 1917; all on GEH reels 47 and 29. Hale to Millikan, 30 Aug. 1917; Hale to Woodward, 27 Aug. 1917; Woodward to Hale, 7 Sept. 1917; all on GEH reels 25 and 39.

[44]Hale, "Plan for the promotion of scientific and industrial research by the National Academy of Sciences and the National Research Council," hand dated 7 Nov. 1917, CC NAS-NRC and GEH reel 51. Hale to Root, 20 Oct. 1917; Hale to Woodward, 23 Oct. 1917; Hale to Pritchett, 23 Oct. 1917; all on GEH reel 47.

[45]Hale, "Outline of statement to be presented to the Carnegie Corporation," hand dated 7 Nov. [*sic* May] 1918, GEH reel 47. Bertram to Hale, 7 June 1918; and CC board minutes, 20 May 1918; both in NAS Grants Gen.

dustrial science and connecting universities and industries appears to have been crucial in carrying the board.[46]

Ironically, the new plans for the academy building did not include the laboratories for physics and chemistry. It is not clear why: there is no evidence of a fight, they just quietly disappeared. Probably the rise of the NRC simply made the idea obsolete. Hale's vision of a showcase of pure science was appropriate for an institution whose authority, such as it was, depended on what it symbolized, not what it did for the public. It was a nineteenth-century vision. The NRC, in contrast, was organized to bring science into closer relations with other social institutions. It was a vision that was better suited to the postwar world.

It appears that the NRC faction on the CC board engineered the decision with minimal fuss. Root and Pritchett were crucial. Pritchett told Hale that "Root's eloquence did it."[47] But Pritchett's legwork was no less essential than Root's oratory and gravitas. NRC chairman John C. Merriam met Pritchett in New York before the crucial meeting to make sure that Pritchett thoroughly understood the rather complicated financial arrangements, any item of which might draw fire from Carnegie's courtiers. Merriam meant to "leave no loophole for misunderstanding," and Pritchett was the link between the NRC circle and the CC.[48] It was Root (a top corporate lawyer) who pleaded the case before a skeptical court, but Pritchett made sure that Root had a winning case to plead.[49] Woodward, too, did his bit, mollifying S. H. Church when he objected (wrongly) that the NRC would simply duplicate the efforts of the CIW.[50] Since the CC did not have real policies, even a decision to spend $5 million was simply a matter of choosing the right moment, lining up the progressive faction, and letting Root steamroller the rear guard. A new system of patronage was thus created in the smoke-filled room, without any new principles being explicitly set forth.

Obviously, the war was crucial for this turning point in the patronage of science. It made Pritchett an advocate of industrial research and gave extra weight to the views of people like J. J. Carty. (Carty was also instrumental in getting the AT&T president, Theodore Vail, to give the NRC $25,000 a year for five years.) Most important, however, the war was

[46]Walcott and Hale to J. Bertram, 1 Feb. 1919; Hale, "The purpose and needs of the NRC," 26 Mar. 1919, CC NAS-NRC. CC board resolution, 28 Mar. 1919; Walcott to Root, 13, 31 Jan. 1920; Root to Walcott, 29 Jan. 1920; all in NAS Grants CC.

[47]Pritchett to Hale, 28 Mar. 1919, GEH reel 29.

[48]Merriam to Hale, 6 June 1919; Hale to Merriam, 12 May 1919; both on GEH reel 24.

[49]Pritchett to Root, 6 Mar. 1919, CC.

[50]Church to Woodward, 25 Feb., 12 Mar. 1919; Woodward to Church, 28 Feb. 1919; both in CC NAS-NRC. Hale to W. S. Adams, 8 Feb. 1919; Woodward to Adams, 20 Mar. 1919, and to Hale, 21 Mar. 1919; all on GEH reels 2 and 51.

the opportunity that Hale and his allies had long sought—to get the National Academy into real public service. In 1913 Hale had little to point to in the academy's record; his application to the Carnegie Corporation in 1918 was a lengthy account of the NRC's wartime accomplishments, many of them paid for by the RF and CC. In 1913 Hale had asked Carnegie to take his vision of an active academy on trust; in 1918 he was asking that the NRC's practical achievements not be cut short for lack of funds.[51] It was a winning argument. The corporation's gift came none too soon; without it the NRC probably would not have survived demobilization. The NRC had been operating on funds from President Wilson's emergency fund ($122,000 in 1918–1919), which would end for good in June 1919, along with contracts from the War Department. The war made the NRC, but only foundation patronage could keep it alive in peacetime.[52]

NRC Fellowships: Beginnings

The war was also instrumental in opening up the Rockefeller Foundation to the NRC activists. There are some striking parallels: Hale's negotiation with Vincent also began with a proposal for a research institute for physics and chemistry, not industrial but medical. Again Hale managed to co-opt an internal initiative to his own ends, though in this case he was obliged to make real concessions. Unlike the CC, the RF had programs, or at least it had officers who believed in definite programs, and it was not simply a matter of lining up friends on the board. An active process of negotiation resulted in programs in support of basic research that bore little resemblance to what either party had initially foreseen. A new system of patronage was invented in the process.

What emerged as common ground was the result of a tactical retreat from a much more ambitious vision. Nathan Reingold, who has written a thorough account of this episode, sees the question of an independent research institute as the crucial one. He places the episode in the context of Hale's quest for a national laboratory. Academics were worried that wealthy philanthropists might again bypass university research for separate institutes, that is true; and Hale did have to be weaned away from his obsession with his "temple of science."[53] Nevertheless, I believe that the issue of a central institute was not the crucial one and that Hale was not the only key player.

[51] Hale, "The Purpose and needs of the National Research Council," 26 Mar. 1919, CC NAS-NRC.

[52] Hale to Cleveland Dodge, 17 Feb. 1919, NAS Funds Gen. Angell to Root, 20 Dec. 1919, NAS Grants CC.

[53] Reingold, "Disappearing laboratory," pp. 94–97.

The sticking point of the negotiation was not the question of an institute—that was never a real option—but the scientists' desire for direct gifts of endowment to particular universities. To that Vincent was adamantly opposed, and compromise was reached only when the NRC group realized that at least they would have to postpone any such hope. The crucial context was not what was happening in Washington but what was going on at the local level as the war drew to a close, and as universities, anticipating an intense competition for graduate students and research funds, rushed to gain the advantage. Hale and his friends were negotiating not just on two fronts (the CC and RF) but on many—Caltech, MIT, and the University of Chicago. These local initiatives are crucial to understanding how a new system of patronage was constructed and how limits to that system were set.

It was Simon Flexner who put the idea of aiding chemistry and physics in George Vincent's mind, with a remark that he "would rather have on the staff of the Rockefeller Medical Institute men well educated in these subjects than men trained in biology or medicine, a knowledge of which they can readily obtain later."[54] It could not have been a casual remark, since it was made in late 1917, just when the foundation's officers and trustees began seriously to consider the transition from emergency relief to postwar reconstruction.[55] Flexner was worrying, as everyone was at the time, about the critical shortage of trained manpower in science and medicine—a result of the demographics of mobilization and the draft.

Flexner envisioned an endowed research institute for physics and chemistry comparable to the Rockefeller Institute. It is less clear what Vincent had in mind. His views on science seem to have been entirely conventional. He took for granted (as many did) that the outbreak of peace would trigger cutthroat industrial competition among European nations, the United State, and Japan. He also assumed that success in this competition would depend on basic research in physics and chemistry. This drumbeat of prewar science lobbies acquired the force almost of prophecy from Germany's wartime feats in chemical and military technology. Vincent seems to imply that the RF would sponsor some program in support of the physical sciences, but it is unlikely that he shared Flexner's enthusiasm for a large research institute. Vincent never liked expensive projects, and a second Rockefeller Institute would be very expensive indeed.

In February 1918 Vincent sought the opinions of five leading scientists: two chemists, Alexander Smith and Julius Stieglitz; and three

[54]Hale to Pritchett, 28 Oct. 1921, CC Caltech.
[55]Board minutes, 5 Dec. 1917, RF 200 37.417.

physicists, Albert Michelson, John Zeleny, and Robert Millikan. (It was not a random selection: Vincent had known all five from his years as dean and president at Chicago and Minnesota.) Vincent asked his academic advisors if a national institution was needed. Would it simply duplicate the efforts of federal science bureaus and industrial R&D laboratories? Was the NRC likely to survive the transition from emergency to permanent status as an agency for "stimulating, coordinating, and, to a degree, directing research in the sciences"?[56]

On most points, the experts were in agreement. Yes, there was a great need to support pure science to ensure national competitiveness; no, it was highly unlikely that industries and federal agencies would do it right. Only Alexander Smith thought the NRC would not survive demobilization, but that may have been more a hope (he had no use for organizers and disdained the NRC activists as "card cataloguers"). Real differences of opinion surfaced only on the question of a central institute. Michelson, Smith, and Stieglitz argued that a separate research institute would in the long run draw more talent into academic chemistry and physics. Millikan and Zeleny thought it better to invest directly in universities, where research could be integrated with graduate training. They worried that a large institute would damage universities by drawing off their top research talent at a moment when they were most vulnerable owing to the acute shortage of trained researchers.[57]

Vincent picked up on the lack of unanimity among advocates of a central institution. "The more I look into the whole question," he told Flexner, "the more complicated it becomes."[58] He acknowledged that something needed to be done to better organize research, especially in chemistry and physics, but he clearly did not think that an institute was that something. Flexner remained a staunch supporter of a research institute. He dismissed Millikan and Zeleny's arguments for keeping research and teaching together—he had heard all that before; but the experience of the Rockefeller Institute and CIW, he thought, had demonstrated the advantages of separating teaching and research.[59] Vincent was not so sure. A few weeks later he sat down with Hale and Millikan to dine and talk about ways of organizing research for national service.

In fact, the whole issue of a separate institute was a red herring. Conditioned by their experience with the CIW, scientists, given the chance

[56]Vincent to Stieglitz and others, 5 Feb. 1918, RF 200 37.417.

[57]Michelson to Vincent, 8 Feb. 1918; Smith to Vincent, 6 Feb. 1918; Stieglitz to Vincent, 15 Feb. 1918; Millikan to Vincent, 18 Feb. 1918; Zeleny to Vincent, 20 Feb. 1918; all in RF 200 37.417.

[58]Vincent to S. Flexner, 14 Mar. 1918, RF 200 36.415.

[59]S. Flexner to Vincent, 19 Feb., 13 March 1918, RF 200 37.417.

to advise, reflexively took up positions on the sensitive question of an institute. But Vincent's academic advisors were worrying about the wrong thing. An institute was not in the cards, and their obsession with it only prevented them from having an effect on the decision-making process. The most important result of Vincent's opinion poll, probably, was that it put him in touch with the NRC inner circle. From their first tête-à-tête with Vincent, Hale and Millikan were the sole connection between the RF and the scientific community. The negotiations were carried out within a very small circle: Millikan, Hale, and Noyes, Vincent, and (behind the scenes) Simon Flexner. That narrow context was crucial: unlike most academics, Hale and Millikan were acutely sensitive to the point of view of their potential patrons. They were pragmatists, always willing to put results, even imperfect results, before abstract principles. A wider and more public negotiation would undoubtedly have provoked more principled defense of individualism and laissez-faire, of the sort that Vincent got from Michelson and Alexander Smith, and that would have made compromise much more difficult.

The NCR Circle: National and Local Agendas

Hale, Noyes, and Millikan were by no means agreed among themselves at first, and as they negotiated with Vincent they also engaged in a parallel negotiation among themselves. Hale liked the idea of a central research institute. He had never been a professor, and as director of the CIW's Solar Observatory he was simply not as sensitive as academics were to the potential for competition. Besides, Flexner's idea resembled his own cherished dream of a research laboratory in the National Academy. In Hale's mind the Carnegie and Rockefeller grants, and the grant from AT&T as well, were interlocking parts of a single plan, which included a national research institute.[60] Hale, however, was not a dogmatic advocate of a research institute. At Caltech he also wore the hat of an academic entrepreneur and was actively wooing Millikan and Noyes and raising funds to create a great scientific university in Pasadena. Hale would drop the academy laboratories from the Carnegie scheme, and he would do so again to get what could be got from the Rockefeller Foundation. He was a consumate pragmatist, always willing to compromise. "I am not in the least concerned about many of the details of this scheme," he told Millikan, "provided only that some such plan can be carried out."[61]

[60] Reingold, "Disappearing laboratory," pp. 91–95.
[61] Hale to Millikan, 10 July 1918, GEH reel 23.

Millikan was the most steadfastly opposed to a research institute. He had made his reputation as an educator and was head of a growing school of research and graduate training when called to take charge of the NRC. An academic entrepreneur par excellence, he had every reason to want foundation funds channeled to research and graduate teaching in universities. Investment in researchers, in his view, was a greater need than investment in the production of research.[62] Arthur Noyes was of like mind. He, too, had made his reputation by combining research and teaching. Although his laboratory at MIT was like an institute in its extramural funding (from the CIW and Noyes's patent royalties), Noyes recruited many of his disciples as undergraduates and took a deep interest in undergraduate engineering education, both at MIT and later at Caltech.[63]

At their dinner with Vincent, Millikan and Hale proposed an alternative to Flexner's central institute. Their plan was to develop three universities into regional centers of research and graduate training: one in California, one in the Middle West, and one in the Northeast. These universities would serve as magnets for outstanding research professors and graduate students, and as sources of trained researchers for other universities—leavening for the whole system. The three host institutions would be selected for their ability to provide excellent research facilities and matching funds. The RF, through the NRC, would provide salaries for full-time research staff and fellowships for graduate students and postdoctoral researchers. In this way a generation of scientists would be trained, and other universities would have a model of the best practice to emulate. The NRC activists were jubilant when Vincent invited them to prepare a formal proposal for the RF's board meeting in May 1918.[64]

Hale, Noyes, and Millikan were not thinking of regional centers in the abstract, of course: the obvious choices for the three centers were their own institutions. Caltech and MIT were mounting large campaigns at the time to become national centers of graduate training and research, and the university of Chicago was worrying about retaining its preeminence. Competition among leading universities was unusually intense in the postwar years. The war had demonstrated the advantages of organized, resource-intensive science, and universities

[62]Millikan to Vincent, 18 Feb. 1918, NAS Plan RF. (The full title of this file is "Plan for Promotion of Research in Physics and Chemistry by RF 1918,") Robert H. Kargon, *The Rise of Robert Millikan* (Ithaca: Cornell University Press, 1982).

[63]John Servos, "The industrial relations of science: Chemical engineering at MIT, 1900–1939," *Isis* 71 (1980): 531–549.

[64]Hale to A. B. Macallum, 15 Mar. 1918; Hale to James Scherer, 14 Mar. 1918, both in NAS Plan RF.

positioned themselves to compete for anticipated new research funds from foundation and corporate sponsors. Likewise, they vied to expand and improve graduate programs to take advantage of the expected boom in the demand for Ph.D. scientists. All this was very much in the minds of the NRC activists. Forced to respond quickly to the opportunity and to the perceived threat of Flexner's plan, Hale and Millikan naturally drew upon their local agendas.

Since building regional centers entailed fund raising, university presidents became, indirectly, parties to the negotiations with the Rockefeller Foundation. Since his inauguration as president of MIT in 1909, Richard Maclaurin had been trying to transform MIT from an excellent engineering college into a first-rank scientific and engineering university. It had proved frustrating. Reform of the departments of physics and chemistry had hardly begun when World War I put it on hold. Maclaurin's hopes for amalgamation with Harvard (and access to the large McKay bequest for applied science) were being slowly dimmed by legal problems and resistence from Harvard scientists.[65] The prospect of getting access to Rockefeller money through the NRC revived Maclaurin's flagging hopes. The NRC's vision of industrial development through regional centers of research and training seemed tailor-made for MIT.

In Pasadena, meanwhile, Hale too was engaged in transforming a small technical college into a regional center of scientific and industrial development. Lacking MIT's numerous and well-connected alumni, Caltech president James Scherer relied on Hale's connections with the large Eastern foundations.[66] Scherer hoped that Rockefeller funds would be the lever to pry matching funds out of industrial and civic patrons—or vice versa. The "nubbin of the thing," he wrote to Hale, was not to wait for the RF to act but to get lumber baron Arthur Fleming to guarantee $2 million as an enticement for the foundation. Noyes, however, thought it best to wait for the RF to take the first step, so as not to make it appear as if they had been using their inside track to get an unfair advantage over other universities: "[I]t might prejudice the scheme seriously if these two institutions [Caltech and MIT] in which you and Millikan and I are interested should be brought forward at the start," he wrote. "It would almost inevitably look as though we had axes to grind."[67]

[65]Maclaurin to Hale, 5 April 1918, NAS Plan RF. The Harvard-MIT saga is richly documented in the presidental papers of the two institutions.

[66]Kargon, "Temple to science: Cooperative research and the birth of the California Institute of Technology," *Hist. Stud. Phys. Sci.* 8 (1977): 3–31.

[67]Noyes to Hale, 13 June 1918; Hale to Scherer, 14 Mar. 1918; both in NAS Plan RF. Scherer to Hale, 2 Apr. 1918; Hale to Scherer, 5 July 1918; both on GEH reel 31.

Maclaurin had similar designs on his chief patrons, the chemical industrialist Pierre Du Pont and Kodak's George Eastman. Laying on the rhetoric of science mobilized and ready for national service, Maclaurin wrote Eastman of the RF's interest. Although Vincent could do nothing until the war was over, now was the time to plan for large benefactions for research in chemistry and physics. There was no greater need, he urged the industrialist, than training large numbers of top-grade researchers for industrial careers. Maclaurin and Noyes were already reckoning how the new funds would be used.[68]

University of Chicago President Harry Pratt Judson was also informed early on that a large endowment for graduate training and research might become available.[69] Judson was not an avid fund raiser (no major fund drive was actually undertaken in his administration). However, Millikan's ideas and lobbying, plus the prospect of seed money from the RF, seem to have stimulated Judson to plan an ambitious ($10 million) development campaign. The plan, approved by the university's trustees in May 1919, revolved around applied science and technology—obviously picking up on Millikan's propaganda and on public concern about postwar commercial warfare. The idea was to create a series of "institutes" in various applied sciences: industrial chemistry, agricultural botany, mining and geology, and so on. Exclusively for graduate training and research, these "institutes" would be the core of a Graduate School of Applied Science (or Technology).[70] Millikan's handywork is obvious, as is the implicit competition with MIT and Caltech. As it turned out, Judson abandoned the campaign almost as soon as it started (he was discouraged, he said, by the postwar recession) and reverted to his usual methods of cost cutting and makeshift.[71]

Thus Millikan, Hale, and Noyes were each negotiating on two fronts at the same time: with Vincent for endowment of regional centers, and locally with their own universities to make sure that they were prepared to take advantage of the expected windfall. Local agendas naturally influenced the evolving vision of a new system of patronage. Hale did not want NRC committees to decide what universities would be designated

[68]Maclaurin to Eastman, 31 May and 5 June 1918, GEH reel 23. Noyes to Hale, 13 June 1918, NAS Plan RF.

[69]Hale to Noyes, 5, 21 June 1918, NAS Plan RF.

[70]University of Chicago, *President's Report,* 1917/1919; pp. 5–6; 1919/1920: pp. 6–9. Judson to Michelson, 14 May 1919, UC Presidents' Papers 48.18. Millikan to Judson, 19 Apr. 1919 and "The necessity for the creation of new conditions in America in the field of research in pure physics and chemistry," n.d. (unsigned, but almost certainly by Millikan); Millikan to Judson, 19 June 1920. [Judson?], "The development of research in pure and applied science," n.d., [1919 or 1920]; both in UC 40.3.

[71]University of Chicago *President's Report* 1920/1921: p. 6; 1921/1922: xii–xiii.

national centers and receive Rockefeller funds, for example, because that would put new institutions like Caltech at a disadvantage. Hale's problem was to prevent a free-for-all, but without giving the impression that he had designed the plan to benefit Caltech. His strategy was to ensure that centers were selected in open competition, but by rules that would favor institutions that could raise funds locally. Hale's trump card over rivals like Stanford and Berkeley was his knack of squeezing money out of wealthy Southern Californians.[72] Confidant of his ability to win such a competition, Hale had to find others to agree to the rules. He let the University of California geologist John C. Merriam in on his plans, to avoid giving the impression of a Caltech-MIT conspiracy.[73] In June 1918 Hale also discussed the plan with Stanford President Ray Lyman Wilbur, and he inserted a clause into the draft proposal providing that no institution should be accepted before the claims of other institutions in the region had been considered.[74]

The NRC trio envisioned a kind of scientific oligopoly, a combination of decentralized competition and central leadership. The requirement of matching funds was critical to maintaining such a system. Many institutions might lay claim to doing excellent research and teaching; far fewer had the capacity to raise large funds, and a policy of matching grants gave the advantage to those few. The NRC trio were not moved by a cynically disguised self-interest so much as by an organizational vision of a society in which activism for the general good was expected to bloom from the seeds of self-interest. It was a quintessentially American mode: neither complete laissez-faire, which in the American experience had meant low standards; nor was it centralized authority, which ran against the grain of American belief in voluntarism and local initiative. American institutions tended toward oligopoly, and scientific institutions were no exception.

Hale and Millikan's approach rested on the assumption that the Rockefeller Foundation was interested in developing institutions. This was not an unreasonable assumption, given the experience of academics with the GEB's system of matching endowment. But it was a mistaken assumption nonetheless. Vincent's conception of the RF as a nursery of philanthropic ideas required that he not get involved with university endowment and fund raising. The NRC trio had intimations that Vincent preferred a grant to the NRC to promote research how-

72Hale to Scherer, 14 Mar. 1918, NAS Plan RF.

73Hale to Scherer, 5 July 1918, GEH reel 31. Harvard chemists Arthur B. Lamb and T. W. Richards also knew of the plan, but like Merriam they had no role in the negotiations. Richards to Hale, 11 Mar. 1918, NAS Plan RF.

74Hale to Scherer, 5 July 1918, GEH reel 31. Hale to Noyes, 21 June 1918, NAS Plan RF.

ever it saw fit.[75] Despite these early warning signals, however, Hale and Millikan persisted in planning for matching endowments to university centers.

Round One: Regional University Centers

The proposal sent to Vincent in June 1918 was drafted by Noyes with help from Millikan, Scherer, and Maclaurin.[76] Essentially, it was Hale's original plan for "institutes" in three to five regional centers. Universities participating in the plan would have to raise $2 million for laboratories and research professorships. The RF would supply $100,000 per year (the interest on $2 million) for ten years to each university for postdoctoral fellows and distinguished European visitors.[77] Training and research were equally essential to the plan. Noyes emphasized that the proposed centers would not be glorified graduate schools but postgraduate institutions—"the climax or culmination of the educational system."[78]

Noyes envisioned active roles for both the RF and the NRC. Universities would apply to the foundation, which would select those best able to raise matching funds (just like the GEB's college endowment program).[79] Similarly, the NRC would administer the annual competition for fellowships, selecting fellows and assigning them to one or another of the participating centers. Noyes even proposed that the NRC "suggest and approve" the appointments of research professors, recommend promising lines of research to be undertaken, and see to it that the research in all centers was done in a coordinated way.[80] These were radical steps, and largely Noyes's own ideas. There was no precedent for giving an external agency like the NRC such powers over the internal affairs of university departments.

The Hale group were themselves divided on that point. Millikan was the most inclined to laissez-faire and local control. He did not think it proper for the NRC to appoint research professors and wondered even if it should select fellows. (He preferred to give each participating uni-

[75]Hale to Scherer, 14 Mar. 1918, NAS Plan RF.

[76]Millikan to Hale, 15 June 1918; Noyes to Hale 13 June 1918; Noyes to Maclaurin, 14 June 1918; Noyes to Vincent, 13 June 1918. Four versions of the proposal exist, the first marked "provisional," the second marked "first draft," and the last two dated 11 and 13 June, all in NRC Plan RF.

[77]"Proposal for the endowment of research in physics and chemistry," 13 June 1918, pp. 2–6, NRC Plan RF. Noyes to Vincent 13 June 1918, RF 200 36.415. Reingold, "Disappearing laboratory," pp. 95–96.

[78]"Proposal," 13 June 1918, p. 9.

[79]Ibid., pp. 4, 12.

[80]Ibid., pp. 4, 7–8.

versity a quota of fellowships to assign as they saw fit.) Hale took the middle ground, arguing that the NRC should allocate fellows and have some say in the appointment of research professors but no formal control. However, he deferred to Millikan's and Noyes's greater experience in university affairs, avowing that results were what mattered, not keeping the NRC in the limelight.[81] Noyes was the strongest advocate of activist management, and as chief draftsman he had the initiative at first.

Vincent turned down the NRC's proposal in November 1918, but probably not for reasons the NRC group would have expected. They anticipated that the decisive issue would be the choice between a detached research institute and universities. Noyes rolled out all the usual arguments in favor of universities, hoping that their monopoly in training researchers would be a trump card in a time of great concern about manpower shortages.[82] Hale worked hard to bring Simon Flexner around when the two of them spent a week together on a sea voyage to Europe. (He was partly successful: Flexner still preferred a central institute but agreed to support the NRC's plan or some variant of it.)[83] In fact, it is unlikely that the RF declined the NRC's plan because they preferred a central institute. An institute was never in the cards. It was just too expensive. Reviewing his budget in December 1918, Vincent came to the sobering conclusion that of a total income of $6.5 million, only $1.5 million was uncommitted and available for new projects. At a time when national and international reconstruction would offer many opportunities for service, Vincent did not want to put all his eggs in one big basket: "While this sum is ample for many kinds of inquiry, stimulation, demonstration and co-operation," he told the board, "it seems to preclude large gifts for buildings, equipment, maintenance or endowment, such, for example, as would be required for complete support of an Institute for Physical and Chemical Research."[84] In February 1919 Vincent estimated that such an institute would cost up to $14 million and added—a damning addition—that it would duplicate existing facilities.[85] An institute was never a practical alternative, despite Flexner's hopes and Millikan's fears.

The fatal flaw in the NRC's plan was its fundamental premise: that

[81]Millikan to Hale, 15 June 1918; Hale to Millikan, 21 June 1918; Hale to Noyes, 21 June 1918; all in NAS Plan RF.

[82]"Proposal," 13 June 1918, pp. 8–11.

[83]Noyes to Hale, 29 June 1918; Noyes to S. Flexner, 20 June 1918; S. Flexner to Hale, 7, 8 July, 12, 14 Aug. 1918; all on GEH reels 28 and 14. Hale to Millikan, 10 July, 14 Nov. 1918, GEH reel 23 and NAS Plan RF.

[84]Vincent, "Memorandum of expected income and expenditures 1920–1925," 4 Dec. 1918, pp. 1–2, RF 900 21.164.

[85]Executive committee docket, 26 Feb. 1919, RF 200 37.417.

the Rockefeller Foundation would want to aid particular institutions with grants of equipment or endowment. The prospect of universities "present[ing] their claims for consideration," intended by Noyes to whet the appetites of the RF's trustees, could only have frightened them off. As would also Noyes's earnest forecast of a long period of joint planning of laboratories and research programs, or Hale's enthusiastic picture of local benefactors inspiring the foundation with feats of largesse.[86] In their enthusiasm for their local agendas, the NRC leaders were slow to perceive that they were dealing with an institution whose mission was not the same as that of the universities or the GEB.

Round Two: National Fellowships

The process of redrafting began in January 1919, with the appointment of a committee chaired by Noyes and including Simon Flexner and Millikan. With Noyes still in charge, the early drafts of a revised proposal differed only in details from the one rejected. Participating universities would still be required to raise new funds (now $1 million) and would be selected by the RF for a trial period of five years. The NRC too retained its active role, assigning fellows to participating institutions—to give the program a "national" character and prevent "unseemly competition."[87] Noyes still presumed that the foundation would agree to a scheme of matching grants.

So, too, did Hale, who began to draw other university leaders discretely into the inner circle. The most promising new entry was Yale University, whose Sheffield Scientific School was being reorganized and integrated into the college and graduate school. Hale got wind of this impending change from a friend and perceived an opportunity for developing physics and chemistry in connection with an old and distinguished school of engineering and science. It was the same model as Caltech and MIT, with the additional benefit of a connection to a first-rate university.[88] Yale President Arthur Hadley shared Hales's vision of regulated and coordinated leadership by an oligopoly of institutions:

[86]"Proposal," 13 June 1918, p. 12; Hale to Noyes, 21 June 1918.

[87]"Memorandum of a plan for the promotion of research in physics and chemistry in co-operation with educational institutions," n.d., pp. 1–3, in Hale to Vincent, 6 Feb. 1919; Hale to Vincent, 24 Feb. 1919; Walcott to Vincent, 12 Mar. 1919; S. Flexner to Embree, 15 Mar. 1919; all in RF 200E 169.2065. Memorandum, 14 Jan. 1919; Noyes to Hale, 26 Jan. 1919; both in NAS Fellowships Gen. (The full title of this file is "Research Fellowship Board in Physics and Chemistry General 1919".)

[88]Hale to A. T. Hadley, 6 Feb. 1919; Hadley, "Memorandum of conversation with Dr. George E. Hale, April 22, 1919;" both in NAS Fellowships Gen. Russell H. Chittenden, *History of the Sheffield Scientific School of Yale University 1846–1922* (New Haven: Yale University Press, 1928), ch. 19.

> As long as our universities are appealing independently for private endowment the appeal will be competitive, and we shall have the same sort of duplication which has resulted so disastrously in the cases of railroads or steamships. I have seen this danger and have tried to avoid it, and so have some other college presidents. But the presidents of one or two universities cannot by their unaided efforts stop laboratory duplication, any more than the presidents of one or two railroad companies could stop track duplication.[89]

Only extramural agencies like the NRC and RF, Hadley believed, had the perspective and financial leverage to make a few universities into regional centers of graduate training and research. Matching funds seemed no problem: Hadley hoped to use the recent large bequest by John W. Sterling to endow research professorships and build research laboratories. (Sterling's trustees preferred traditional collegiate projects, but Hadley hoped that the prospect of Rockefeller funds would change their minds.)[90]

As word of the Rockefeller project spread through the NRC divisions, Hale began to worry about setting off a gold rush. Noyes urged the chairman of the chemistry division, Edward W. Washburn, not to raise the hopes of the University of Illinois president, Edmund James, since state universities might not legally be able to receive funds from private foundations. Caution was especially needed, Noyes noted, because James was the "fighting type" of land-grant president.[91]

Both hopes and fears were premature. The final proposal that was sent to Vincent in March 1919 rested on an entirely different premise. The fellowship program was detached from any plan to build up regional centers. Universities were no longer expected to raise funds as a condition for receiving fellows, and would receive no funds directly from the RF.[92] The NRC scientists' views changed as the result of discussions with Vincent and others between January and March 1919. Gradually it became clear that they had underestimated the practical risks of combining a fellowship program with institutional fund raising.

Millikan was the first to grasp this reality, as the result of a conversation with Martin A. Ryerson, an influential trustee of both the RF and the University of Chicago. Ryerson liked the NRC plan but doubted that a five-year commitment would be a sufficient incentive to univer-

[89]Hadley to Hale, 20 Mar. 1919, NAS Fellowships Gen.

[90]Ibid.

[91]Noyes to Washburn, 4 Feb. 1919, NRC Fellowships Phys. Chem.

[92]Walcott and Hale to Vincent, 22 Mar. 1919, RF 200E 169.2065. A draft dated 18 Mar. is in NRC Fellowships Phys. Chem.

sities to make major investments in science. He felt that the foundation alone should select participating universities, after doing a national survey. Also, he believed that the NRC should leave the control of the fellowship programs strictly to the universities.[93] Ryerson's comments revealed to Millikan that university administrators had their own agendas and that a meeting of minds might not be easy to achieve. It seemed unlikely that the NRC would be able to control the process. A national survey was bound to be time consuming and political. Moreover, delays in raising funds locally could very well throw a monkey wrench into the fellowships also.

Millikan concluded that the safest course was to do first what could be done most easily, leaving the more controversial, institutional part of the agenda for another day:

> There are two somewhat distinct things which we have attempted thus far to fuse into a single plan: (1) the creation of a limited number of research institutes in Physics and Chemistry in connection with American universities and (2) the diverting of able men into Physics and Chemistry, and the training of these men through research . . . fellowships. I hope that both of these results may be brought about, but it may be wise to *separate them.* The Fellowship matter may be handled easily as a national undertaking through a Board [of] . . . the National Research Council.[94]

Millikan had no intention of giving up the original plan; he went on to sketch out for Hale a new scheme for three regional centers. A separate fellowship program was a tactical retreat only, reflecting Millikan's new understanding of the political realities. The real danger, he perceived, was not a separate research institute but the multiple pitfalls of cooperative institution building.

Although Hale and Flexner brushed aside Ryerson's specific criticisms, Millikan's idea of separating fellowships from institution building gradually prevailed. Hale was getting signals from Flexner that suggested Millikan was right about conflicting agendas and the dangers of the NRC being co-opted. Flexner objected to Noyes's idea that research professors would instruct postdoctoral and even doctoral students. Teaching would interfere with their research, he argued, and competition from research staff could demoralize regular faculty.[95] Indeed, Flexner thought that Noyes had been taking too much upon himself. "Someone else and quite independently of Noyes, a physicist

[93] Millikan to Hale, 2 Feb. 1919, NAS Fellowships Gen. and RAM III 8.5.
[94] Ibid.
[95] S. Flexner to Hale, 6 Feb. 1919, NAS Fellowships Gen.

preferably, should elaborate a plan," he suggested. "What is best for chemistry may not be just right for physics."[96] Hale rushed a copy of Flexner's note to Millikan and asked him to draw up a new plan.[97] Noyes continued to participate, but his distinctive ideals of central planning of research had less and less influence. Millikan's pragmatic and opportunistic style set the course of the final negotiations.

Within two weeks, Hale had conferred with Vincent and hurriedly drafted a revised plan, this one for fellowships exclusively. Only remnants remained of the elaborate apparatus for institutional development. Any university could participate, so long as it met the fellowship board's minimum requirements. (Hale hoped that these requirements would be set high enough so that even the best institutions would be obliged to make new capital investment.) Provision was also made for one-third of the fellows to work in institutions that could not meet the NRC's minimum requirements (but which might, Hale hoped, get grants for additional equipment from the RF.)[98] Even these weak linkages between fellowships and institutional development disappeared in the final proposal. No conditions at all were put on fellows' selection of institutions in which to work. The only vestige of the original scheme was a vague hope that universities would be inspired to raise funds for research.[99] Within a week, the plan was approved by the RF's executive committee, assuring approval by the board in May 1919.[100]

A New System of Patronage

Thus, in only two years the National Academy and the NRC received a fine new home, an ample endowment, and an important and visible new responsibility, administering the national research fellowships. Permanent connections had been made between the large foundations and American science.

The ease with which negotiations were concluded is testimony to the crucial importance of the new idea of a limited patron-client relation.

[96]Ibid.

[97]Hale to Millikan, 7 Feb. 1919, NAS Fellowships Gen. and RAM III 8.5.

[98]Hale to Vincent, 24 Feb. 1919, NAS Fellowships Gen. and RAM III 8.5. A draft dated 19 Feb. was not sent.

[99]Walcott and Hale to Vincent, 22 Mar. 1919, RF 200E 169.2065.

[100]Executive committee docket, 26 Feb. 1919, RF 200 37.417. Embree to Noyes, 27 Mar. 1919; Embree to Hale, 10 April 1919; both in RF 200E 169.2065. Reingold, "Disappearing laboratory," pp. 97–98. The executive committee authorized Vincent to choose among five alternatives: a five-year program of 50 fellowships, grants to five universities for research professors, research endowment funds in five universities ($500,000 toward $1.5 million each), a combination of fellowships and endowment, and an independent research institute.

Vincent could live with a form of patronage that was clearly educational (not for research) and that was clearly disassociated from any particular institutions. Education and fellowships were proven kinds of patronage, whereas grants for particular lines of research had a chequered history and a worrisome potential for getting out of hand. All the disabilities of schemes for individual grants-in-aid or matching endowments, which had stymied every effort to enlarge the nineteenth-century system of patronage, became of no account. With a system of fellowships administered by the NRC there could be no fear of recapitulating Woodward's troubles with his academic constituents. Thus, a rationale was invented for a limited partnership in developing the community of science as well as a workable social machinery—endowment for the NRC and a national system of research fellowships. The potential that T. H. Morgan had glimpsed in Abraham Flexner's initiative in 1916 was realized in a way that could hardly have been anticipated. First, the war changed both scientists and the large foundations in ways that made a partnership more likely. Then a new rationale was improvised in a process that, but for the tactical insight and nimble footwork of Millikan and his friends, might well have failed as it had before.

The existence of the NRC was also crucial: without it a new system of patronage would probably not have been possible in 1919. The general purpose foundation was a new and controversial institution in 1918. A new generation of managers were still feeling their way into new roles and seeing how far they could go in changing traditional philanthropic agendas. They had no experience working with scientists and every reason to be suspicious of them. Delegating authority to the NRC was a way of avoiding difficult ideological and administrative problems. It was a way for foundations to get involved with academic scientists but not more involved than they liked. It was the NRC, just as much as the war and Millikan's entrepreneurial skills, that made the difference between the failed efforts of 1913–1916 and the swift changes of 1918–1919.

If the NRC was a welcome buffer for foundations, the foundations were absolutely crucial for the survival of the NRC in the postwar demobilization. Without the Carnegie Corporation's annual checks, the NRC's numerous committees would have been paper organizations (many were that anyway). Administering fellowships was the NRC's most important (some would have said only) real job. Financially the NRC was heavily dependent on foundations, since industrial support never lived up to Hale and Millikan's high hopes. AT&T's 1919 grant of $125,000 remained the only sizable grant from industry; in general, businessmen wanted specific services for their money. (The same pat-

TABLE 4-1

FOUNDATION CONTRIBUTIONS TO THE NATIONAL RESEARCH COUNCIL (IN $1,000S)

	Carnegie Corp.		*Rockefeller Boards*							
Year	*Gen.*	*Spec.*	*Gen.*	*Fellowships*	*Small Grants*	*Publication*	*Res. Com.*	*Misc.*	*Other Found.*	*Total Income*
1916–18	100		74						9	270
1919	100		18	29				27		236
1920	170		13	29				11	12	368
1921(1/2)	85		8	27				2		254
1921–22	185		7	76		30	1	3	8	538
1922–23	182			150		20	25	20	4	615
1923–24	183	2		202		20	42	45	11	703
1924–25	178			253		16	45	51	8	1,077
1925–26	159			324		34	52	51	12	1,067

1926–27	121			309		40	55	58	8	998
1927–28	127			287		56	99	28	10	968
1928–29	115			315	25	60	115	42	17	1,025
1929–30	140			291	25	75	135	11	25	970
1930–31	127			348	75	81	142	11	9	1,032
1931–32	128			426	100	85	148	11	5	1,023
1932–33	90	10		307	55	100	155		4	823
1933–34	91		32	351	50	79	167	6	1	851
1934–35	80		38	219	55	94	114	6		672
1935–36	73		45	126	42	11	123	3	2	541
1936–37	75	10	15	84	12	24	122	6		466
1937–38	65	98	35	111	10	15	115	5	40	589
1938–39	65	127	35	89			115	2	58	616

tern obtained in industries' support of research in their own trade associations.)

The NRC became, in effect, a channel for allocating foundation funds. Between 1916 and 1940 nearly $12 million flowed from foundations to the NRC, most of it (98 percent) from the Carnegie and Rockefeller boards (table 4.1).[101] The Carnegie endowment supported the work of the general divisions—research information, industrial and government relations, and so on. Rockefeller Foundation grants went to the fellowship boards and to special cooperative research projects organized by the scientific divisions. (A similar relation existed between the Laura Spelman Rockefeller Memorial and the Social Sciences Research Council.)[102] The Hale circle and satellite NRC elites, like Frank R. Lillie's in biology, owed their remarkable ascendency to their role in building and tending the channels through which resources flowed from the large foundations to scientific fields. For a few years, until the large educational foundations got into the act, the NRC was the center of the new system of patronage.

But there were limits to this system of financial irrigation of science (it had been possible to build that relationship precisely and only because it *was* limited). Both the Carnegie Corporation and the Rockefeller Foundation remained wary of direct endowment of university programs or grants for individual research projects. The NRC, like all trade associations, was incapable of resolving institutional rivalries and so could not act as intermediary for institutional grants (it did all right, however, with systems of grants-in-aid). That was the structural reality to which Millikan and Hale accommodated when they separated fellowships from institutional projects. It took a while, however, for that reality to sink in.

[101]NRC *Annual Reports:* 1918–1939. Foundations accounted for about 60 percent on average of NRC receipts in the 1920s and about 80 percent in the 1930s. Industrial funds fluctuated with the business cycle.

[102]Social Sciences Research council, *Decennial Report 1923–1933:* pp. 104–105. Kohler, "Science, foundations, and American universities in the 1920," pp. 135–164.

Chapter Five

A Limited Partnership

The two grants to the NRC in 1919 were a new kind of patronage, in which individual grants were superceded by a new relationship with whole communities of scientists. At first the system was not perfectly understood by either party. Foundation leaders did not see (or did not care to see) the possibilities for expanding into institutional grants. Scientific leaders did not perceive the limitations of the NRC-mediated relation. Science was not really integrated into the philanthropic agendas of the Rockefeller Foundation and Carnegie Corporation. Their grants to the NRC were an improvised response to the postwar enthusiasm for science and not, as scientists hoped, the first steps toward comprehensive programs of patronage. The limits of the patronage relation were gradually revealed as practical efforts were made to expand it. The social process did not cease in 1919 but continued into the mid-1920s, with generally disappointing results.

I will deal in this chapter with three episodes. First, the extension of the NRC fellowships to the biological and other sciences: this episode reinforces the point that fellowships were an ideal way for the RF to aid science without being sucked in further than it wanted. What worked in 1918–1919 continued to work even as the postwar mood of reconstruction receded in the mid-1920s. Second, I will recount how Millikan, Hale, and Noyes kept trying to sell the RF on the idea of research endowments to a few universities—the part of their vision set aside in 1919 to secure the fellowships. These efforts failed, as did also plans for RF support of NRC cooperative research projects. Millikan's efforts to get research endowment for Caltech from the CC, though they succeeded in the short term, failed to establish a precedent for programs of institutional support. Institutional grants were not compatible with the internal politics of the general purpose foundations. The same lesson is drawn from a third episode, Edwin Embree's short-lived program to give institutional structure to a new discipline of "human biology."

In the 1920s, the RF and CC would not or could not expand and diversify the system of patronage that they had created in 1919. That may seem a strange thing to say when they gave so many millions to the natural sciences in the 1920s. A closer look, however, reveals that both boards were for the most part just making good on commitments made in 1919. The real concern in the 1920s was to hold the line. This pattern is apparent in the Rockefeller Foundation's expenditures (table 5.1).[1] The NRC fellowships remained the single most important program, extending to medicine and biology and growing to a steady state of about $500,000 per year by 1925. But all expenditures on the natural sciences constituted less than one-tenth of the foundation's total budget, and its only new initiative was Edwin Embree's short-lived program in "human biology." Significantly, Embree's largest appropriation was to the Marine Biology Laboratory at Woods Hole, which, like the NRC, channeled aid to a large scientific community while protecting the foundation from claims from individuals and institutions.[2] The Carnegie Corporation spent far more on science than the RF did, but most of it went to the CIW, NRC, and Caltech as a result of initiatives made in 1919.

Extending the Fellowships: Biology

Fellowships were the most important area for expansion of the new patronage system in 1920–1923—indeed, just about the only one. The Rockefeller Foundation was not just willing but eager to extend the fellowship program to other sciences. Abraham Flexner and Vincent took the lead in organizing a system of medical fellowships in 1922.[3] (They differed from those in the natural sciences in being a more integral part of RF and GEB programs in medical education.) The biology fellowships followed along in 1923, and in the same year the physics and chemistry program was extended to mathematics, astrophysics, and geophysics, all with an ease that astonished some NRC regulars.[4] The biology fellowships were more troublesome and took three years to negotiate, but the problems were on the scientists' side, not the foundation's. The limiting factor was entrepreneurial experience and know-how: the biologists had trouble seeing how the world looked from the

[1]RF *Annual Reports,* 1913–1928.

[2]RF 200 147.1812–1814.

[3]Relevant materials are found in NRC files "Medical Fellowships" and "Medical Meetings," and in RF 200E 169.2066 and GEB 1.5 711.7289.

[4]Kellogg to Millikan, 25 Oct. 1923; Kellogg to Vincent, 31 Oct. 1923; Millikan to Kellogg, 31 Oct., 17 Nov. 1923; Hale to Kellogg, 19 Nov. 1923; Kellogg to Hale, 24 Nov. 1923; all in NRC Fellowships Gen.

TABLE 5-1
ROCKEFELLER FOUNDATION EXPENDITURES IN NATURAL SCIENCES (IN $1,000s)

		NRC		NRC Fellowships			Div. Studies			
Year	Rockefeller Inst.	Med. Com.	Biol. Abstr.	Phys., Chem.	Med.	Biol.	Res. Proj.	Marine Biol.	Other	Total[a]
1915	1,000									
1916	2,002									
1917	3,128									
1918	576	14								14
1919	101	6		14						20
1920	5			50	4					54
1921	6		15	61	18					96
1922	3,422			82	51					133
1923			20	84	84	14				202
1924			15	86	226	41	2	500		870
1925			28	105	346	55	47	50	22	653
1926			69	118	283	64	86	50	19	689
1927			48	129	255	79	202		39	752
1928			57	105	289	67	215		12	745
Total	10,240	20	252	834	1,556	320	552	600	92	4,226

[a]Excluding Rockefeller Institute.

patron's point of view. They were obsessed with getting individual research grants and only slowly learned that they would do better to approach the RF on terms it could accept—a lesson that the physicists and physical chemists had grasped more quickly. But Vincent and his fellows were patient teachers. For them fellowships were an ideal arrangement: not too expensive, low risk, and universally popular.

The three-year negotiation over the biology fellowships reveals how important the structure and mores of disciplinary communities were in shaping patronage relations. It also underlines how much depended on a small and experienced leadership circle. Hale, Millikan, and Noyes were able to act as if on behalf of all chemists and all physicists, even though their views may not have been representative of all. There was no such group among the biologists, at least not at first. The members of the NRC's Division of Biology and Agriculture were more inclined to a parliamentary style of leadership in which all interests were represented, and that made it much more difficult for the negotiators to adapt and improvise. Eventually a group did emerge, led by Frank R. Lillie and Vernon Kellogg, who were much more like the Hale group in their leadership style. Taking their cues from the experience of the physicists and chemists, they soon had a fellowship program in place. Thus, disciplinary differences reveal fundamental features of the process by which a new social relation was created—especially the importance of entrepreneurial experience and styles of leadership.

It is no accident that relations between foundations and science were created by physicists and physical chemists. These disciplines were unusually homogeneous communities; or, at least, they had an unusually authoritative core elite and a clear-cut hierarchy among their varied subfields. Not so biology, which was a congeries of competing and contentious subspecialties or subcultures, with varied relations with medicine, agriculture, psychology, and natural resource management, all of which offered attractive but competing opportunities for discipline building. Hale, when he was organizing the various disciplinary committees of the NRC in 1917, had had particular trouble with the biologists. They were reflexively suspicious of organizers and of anyone who tried to define a central core of problems. The decentralized structure of the biological sciences, and their multiple boundaries with other disciplines, encouraged individualism and laissez-faire. Resentment of the physicists' dominant role in the NRC's wartime projects engendered more mistrust. In 1919, for example, Robert Yerkes warned Hale to expect trouble from the biologists: "[S]erious misconceptions still persist [about the NRC], conspicuous among which are first the idea that this is your attempt to centralize and control scientific research in America, and second that the only possible role of such an organization is the destruction of valuable research men."[5] Perhaps Simon Flexner was not just giving in to academic snobbery when he sensed that a physicist should be the one to design the relation between science and foundations. Building a new social relationship was more easily done by an elite group that did not have to worry too much about internal disciplinary politics.

Envious of the physicists' and chemists' good fortune, the biologists of the NRC began at once to lobby for a fellowship program of their own. In December 1920 they took the usual first step, gathering ex-

[5]Yerkes to Hale, 31 July 1919, GEH reel 40.

pressions of support from eminent biologists. Michael Guyer, who chaired the fellowship committee, naively suggested that a few fellowships might be paid for out of unused funds of the chemistry and physics program. A similar proposal was made by the psychologist Carl Seashore, who hoped that psychophysics might qualify under the physics program. Both suggestions were quickly rebuffed. Guyer's group seemed not to know what to do next, and for over a year nothing more was done.[6] The announcement of the medical fellowships in April 1922 heightened the biologists' frustration but left them still "at a loss to know what to do." They did not dare to do another survey of opinion, for fear of once again arousing expectations that might again be dashed. So many applications for fellowships had come in after the first survey that Guyer had been forced to make a public disclaimer.[7] It was not a good start.

Obviously, Guyer and his group had nothing like the entrepreneurial know-how and connections of the Hale circle. The members of their committee were a diverse group, obviously chosen because they represented important biological interests, not because they knew how to raise funds from Eastern philanthropists. Guyer was a professor of zoology at the University of Wisconsin, who worked in experimental evolution and eugenics and was best known as a teacher and author. Wilfred H. Osgood was a mammologist from the Field Museum of Natural History in Chicago, a stronghold of traditional systematics. Lewis R. Jones, a professor of plant pathology at Wisconsin, represented the connection with agriculture in the land-grant schools. So, too, did the plant physiologist Albert F. Woods, who had made his career in federal agricultural bureaus and agricultural colleges. All successful in their different ways, none had the connections with medicine or private universities that might have given them insights into how the Rockefeller staff expected things to be done.

The exception was Frank R. Lillie, who had proved himself as an entrepreneur in his fifteen years as chairman of zoology at the University of Chicago, as director of the Marine Biology Laboratory (MBL) at Woods Hole, and as leader of a large research school in experimental developmental physiology. Most important, Lillie already had connections to the philanthropic world. As director of the MBL, he had led a successful campaign to get grants of endowment from the Carnegie and Rockefeller boards. As a member of the NRC's Committee for Research in Problems of Sex, he had been instrumental in arranging a

[6]Guyer to members of the biology division, 13 Apr. 1921, 23 Apr. 1922, NRC Fellowships Biol. The full name of this file is "Board of National Research Fellowships in Biological Sciences." Seashore to H. G. Gale, 28 Oct. 1921, NRC Fellowships Gen.

[7]Guyer to fellowship committee, 8 Dec. 1922. NRC Fellowships Biol.

program of grants-in-aid, supported by a personal gift from John D. Rockefeller, Jr.[8] As a member of the biologists' fellowship committee, however, Lillie represented just one biological subculture among many. Factional jealousies and suspicion of strong-minded organizers made it difficult for any one person to conceive a winning strategy and carry the rest along with him.

Guyer's committee were passive agents of their various constituencies. They had no clear vision of their role as decision makers. They did not try to discover what, practically, foundations might be prepared to pay for, but were guided solely by what their constituents wanted. Their survey had revealed that biologists wanted support for researches by experienced investigators, not fellowships. Indeed, many biologists believed that fellowships would go begging, owing to the temporary shortage of newly fledged Ph.D.s. (Similarly, a group of physiologists led by Joseph Erlanger had argued against medical fellowships, for fear that fellowships would divert attention and resources from the more fundamental problems of more jobs and better pay for biomedical scientists.)[9] Guyer saw that the alternatives were to fund more production of research in the short term or to increase the supply of researchers for the long term; he himself preferred the long-term strategy. But obedient to grass-roots opinion, he proposed a hybrid scheme of six fellowships: two each for new Ph.D.s, mid-career academics, and established professors.[10] While that might make professors of zoology happy, it was unlikely to appeal to foundation officers already sensitized to the dangers of scientists' desire for individual research grants.

The fact that the NRC's biologists shared a division with agriculture was also a problem. The vast investment in agricultural science by the federal and state governments made it an unappealing area for private foundations. Certainly, Vincent showed no interest. It was becoming clear, at least to people like Frank Lillie, that biologists would do better with foundations if they emphasized the connection of biology to medicine (as he himself did). They might, for example, team up with the NRC's behavioral scientists. Psychologist Raymond Dodge, chairman of

[8]B. H. Willier, "Frank Rattray Lillie 1870–1947," *Biog. Mem. Nat. Acad. Sci.* 30 (1957): 179–236. Sophie D. Aberle and George W. Corner, *Twenty-five Years of Sex Research* (Philadelphia: Saunders, 1953). Lillie to CC, 2 Dec. 1921; Vincent to Pritchett, 23 Jan. 1922; Pritchett to McClung, 14 Feb. 1922; all in CC Marine Biology Laboratory. Lillie to Vincent, 5 Aug. 1921 and "Statement concerning the Marine Biological Laboratory," RF 200 147.1812.

[9]"Exhibit A," minutes of the Medical Sciences Division, 27 Feb. 1920, and much correspondence, NAS.

[10]Guyer circular letter, 13 Apr. 1921, NRC Fellowships Biol.

the Division of Anthropology and Psychology, was as eager as Lillie was to get a piece of the foundation pie. But he had the same problem with his constituents as did the biologists. An opinion poll, which Dodge hoped would show support for a joint program of fellowships, revealed instead an overwhelming preference for a separate fellowship program. Psychologists wanted nothing to do with biologists, and they wanted support for their own research, not fellowships, which they felt would only produce more competitors for limited research funds.[11] Psychologists, like biologists, were a highly factionalized community; that, and the contested turf between the two disciplines, made cooperation a highly charged political issue.

The most important practical result of the psychology gambit was that it brought Frank Lillie closer to center stage. Lillie's research on reproductive biology straddled physiology, organic and biochemistry, and psychology. The medical connection, plus his experience in fund raising, put him in a strategic position to take the lead in the matter of fellowships. In January 1923, Dodge and Lillie drafted a proposal that was tailored to what they believed the RF would like to hear. They dropped the idea of grants to established researchers and called for a program of postdoctoral fellowships on the model of physics and chemistry. They also dropped agriculture, focusing on biology and psychology as the crucial intermediate links between the physical sciences and the social sciences and medicine. A fellowship program in biology and psychology, they argued, would unite and strengthen the existing fellowship programs.[12] These were winning arguments. Dodge and Lillie were guided not by rank-and-file biologists' notions of patronage but by what experience had shown was doable.

Lillie was guided in part by his own professional ambitions. The NRC's proposal and rationale bear the unmistakable stamp of Lillie's own research program in what some would call human biology:

> The realization is . . . spreading that outside of the field usually regarded as belonging to medicine lie great problems of racial and social health, for example problems of motivation, problems of sex research, human migrations, eugenics and the like. As population tends to approach its maximum the above problems must become the problems of the day, because they are essential to na-

[11] Dodge circular letter, 24 Dec. 1922; C. H. Judd to Dodge, 18 Dec. 1922; E. Boring to Dodge, 18 April 1922; Dodge to Ruml, 22 Dec. 1922; Ruml to Dodge, 26 Dec. 1922; all in NRC Fellowships Biol.

[12] "The desirability of post-doctorate fellowships in the biological sciences," 18 Jan. 1923, pp. 1–2, RF 200E 169.2065. Kellogg to Merriam, 18 Jan. 1923, NRC Fellowships Biol.

> tional well-being. These problems lie in the fields of biology and of psychology and anthropology.[13]

This was precisely the vision that Lillie was pushing as chairman of the NRC's committee on sex research. It was the program he would soon be taking to Wickliffe Rose, and later to the RF, hopefully to inspire them to endow a research institute for his group at Chicago.[14] The point is not just that Lillie was pursuing his self-interest; it is, rather, that Lillie's scientific interests gave him a better perspective than most biologists had on what philanthropists would accept.

We cannot assume that Vincent was inspired by the specifics of Lillie's program. The resemblance to the physics and chemistry fellowships was undoubtedly more important. Dodge and Lillie's proposal revealed that the new leaders among the NRC biologists were not driven by disciplinary logrolling and were willing to make a deal. They were representative of experimental biology and medicine, not traditional zoology and botany. Their political experience was gained in private universities, not land grant. The leadership style of the NRC biologists was becoming more like that of physicists and physical chemists, and it served them well with the Rockefeller Foundation.

Lillie's new role as chief entrepreneur was greatly aided by Vernon Kellogg, the NRC's first permanent secretary. Like Michael Guyer, Kellogg came from the more conservative side of biology (he was an entomologist and evolutionist) and was best known as a teacher and popular science writer. He also had a knack for organization, however, which became apparent during World War I, when he served as director of the American Committee for Relief of the Belgians and as Herbert Hoover's assistant in the U.S. Food Administration.[15] He worked closely with Hale and Millikan in NRC politics and shared their entrepreneurial style. Kellogg's work in food distribution and relief in Europe also brought him into close contact with the RF, and in February 1922 he became a member of its board of trustees. (In 1920 he declined an offer to become an officer.)[16] An insider in foundation circles and the Washington science scene, Kellogg was strategically placed to assist biologists in their efforts to acquire a patron.

Kellogg helped Dodge and Lillie tailor their proposal to RF standards and shepherded it through the RF's administrative machinery. In February 1923, the board approved an appropriation of $75,000 a year

[13]"Desirability of postdoctoral fellowships," p. 3.

[14]Lillie to Rose, 17 June 1924; Lillie to Mason, 5 June 1931; both in RF 216D 8.103.

[15]C. E. McClung, "Vernon Lyman Kellogg 1867–1937," *Biog. Mem. Nat. Acad. Sci.* 20 (1939): 245–257.

[16]Embree diary, 12 Mar., 4 May 1920, RF.

for five years.[17] Millikan congratulated Kellogg on his achievement, as one old hand to another: "It is my conviction that we are doing more for research in this country by these fellowships than by any other activity which we have undertaken since the organization of the National Research Council."[18]

It is a little surprising that the biologists were not quicker to emulate the visibly successful model of the physical science fellowships. (Clearly, one should never underestimate the differences between disciplinary subcultures.) It is not at all surprising, however, that the model was eventually extended to all the major groups of sciences. It was natural, from the foundation side, to be systematic in funding fellowships. Fellowships helped the various communities of scientists without committing the RF to support of particular individuals, lines of research, or institutions. It addressed the shortages of expert manpower, which seemed a problem of national importance at the time. Fellowships were an ideal vehicle for the new partnership between foundations and science that had been worked out in 1919. But the limits of that relationship were becoming apparent at the very time that the fellowship programs were expanding. We have examined what was easy to do; no less revealing of the patronage system are the things that proved difficult or impossible to do, like grants for individual research projects and for institution building.

The RF and NRC: A Limited Partnership

Despite the great success of the NRC fellowships (everyone thought they were one of the best things the RF had done), they never were the nucleus of a program or division in the natural sciences. Vincent steadfastly resisted efforts to expand support of scientific research. The foundation in his view was a philanthropic laboratory, and the fellowships were an experimental demonstration, and that was all. Vincent liked to say that the fellowships promoted international cultural exchange and goodwill, a belief that fit science to the foundation's mission without committing it to specific projects. In practice, the fellowships were a sideshow to the RF's large operating programs in medical education and public health. Fearful of creating another expensive operating division like the IHB and CMB, Vincent leaned over backward to prevent projects like the NRC fellowships from developing into permanent commitments to institutions.

[17]Kellogg to Vincent, 23, 31 Jan., 1 Feb. 1923; Vincent to Kellogg 29 Jan. 1923; all in RF 200E 169.2066. Kellogg to Merriam, 18 Jan. 1923; Kellogg to McClung, 24 Feb. 1923; Kellogg to Gano Dunn, 6 Mar. 1923; all in NRC Fellowships Biol.

[18]Millikan to Kellogg, 8 Mar. 1923, NRC Fellowships Biol.

For the NRC inner circle, though, the fellowship program was a foot in the RF's door, the first step in their grand plan for regional centers and university "institutes." Responsibility for the next step fell naturally to the physicists and chemists of the NRC's fellowship board, who, in addition to administering fellowships, also served as an ad hoc planning committee. They occupied a strategic position. A small group of people who knew each other well and met regularly, the fellowship board could maneuver more adroitly than the more cumbersome and dispersed NRC divisions. Simon Flexner was chairman, and Millikan, Hale, and Noyes were all founding members. The Caltech troika intended that the board would actively raise funds for regional university centers and use fellowships as an incentive to universities to improve research facilities and staff. This had to be done discretely, of course: even the appearance of personal deals with university presidents would be fatal to the fellowship program. To avoid this problem, Millikan proposed that appeals to university presidents be made by the chairman of the NRC, who did not have fellowships to hand out and thus could not be suspected of logrolling.[19]

Simon Flexner was no less an activist, but he had a somewhat different agenda: namely, to pursue his cherished idea of a research institute. (No doubt this is why he agreed to chair a committee of chemists and physicists.) He saw the fellowship board as analogous to the Rockefeller Institute's first board of scientific directors, who at first had allocated extramural grants but who had become a planning group for inhouse programs—"the natural instrument to launch the central research laboratory." Flexner anticipated that the NRC's fellowship board would develop in the same way. The fellowships were a temporary expedient, in his view: a first step toward building new research institutions outside the university system.[20]

Millikan's and Flexner's visions were combined, somewhat uneasily, in an application made by the NRC's fellowship board to the Rockefeller Foundation in May 1920. The first part called for a research institute in physics and chemistry, with permanent laboratories ($1 million) and a permanent research staff ($250,000 per year). The second part proposed a decentralized system of graduate training and research in universities. Both parts emphasized the need for scientific manpower: "[A]ny plan . . . must involve as its primary feature, not the output of research, but the development of research and the stimulation of research throughout the universities of the country; for the universities are and must remain not only the most important centers of

[19]Millikan to Hale, 17 Mar. 1919, NAS Fellowships Gen.

[20]S. Flexner to Hale, 18 Sept., 5 Nov. 1919, GEH reel 14.

research activity, but the main agencies for training research men."[21]

It was Noyes and Hale's 1918 plan, altered a bit to make it politically more palatable. It called for smaller grants to more (ten to twelve) institutions: $20,000 per year for ten years from the RF and matching $10,000 from the universities. (The requirement of matching funds would restrict participation without giving other institutions grounds for feeling unjustly excluded.) Second, grants would be made to departments of physics and chemistry to aid faculty research and graduate teaching. Finally, central coordination by the NRC was limited to conferences; no formal controls would be imposed.[22]

This trend toward broader participation and decentralized decision making reflected the rapid decay of the wartime enthusiasm for planning and organization. Academic scientists were fed up with the bureaucratic controls of government projects, and, as they returned home, local ambitions and rivalries inevitably seemed more important than new national institutions. The temporary but acute shortage of scientific manpower (caused in part by a vastly expanded industrial market for Ph.D. physicists and chemists) intensified the normal competition among universities for top research talent.[23] Everyone sought to gain an advantage by exploiting special local resources or special connections with industrial or foundation patrons. Hale and Millikan were acutely aware of the potential of any system of grants to stir up local jealousies. One of the advantages of Flexner's central research institute, they noted, was that it would be less divisive.

The central institute scheme also differed from what Simon Flexner had initially envisioned. It was less like the Rockefeller Institute and more like, say, Caltech or the University of California—not surprisingly, since the plan seems to have been worked up by Hale and Gilbert N. Lewis, the dean of the School of Chemistry at Berkeley.[24] They put the emphasis on training of researchers and on cooperation with universities, and characterized the institute as "the climax of the university system of the country." Elaborate provisions were proposed for postdoctoral fellows and for visiting academic researchers, who would come for the summer months or on sabbattical leave and would take back home enthusiasms and habits acquired in working with the per-

[21]"Plans for the promotion of research in physics and chemistry," May 1920, p. 1, NRC Fellowships Gen. Enclosed in Angell to Embree, 8 May 1920, RF 200 37.417.

[22]"Plans," pp. 7–8.

[23]Geiger, *To Advance Knowledge*, ch. 5.

[24]G. N. Lewis to Hale, 12 Sept., 14 Nov. 1919, 8 Mar. 1920; Hale to Lewis, 6 Jan. 1920; [Lewis], "Plans for the promotion of research in physics and chemistry," May 1920; [Lewis?], "Government of the research institute of physics and chemistry," n.d.; all on GEH reel 52.

manent institute staff.[25] It was Flexner's ideal of an institute, modified to suit the largely academic and local ambitions of scientist entrepreneurs like Millikan, Noyes, and Lewis. Both options, however, called for large capital expenditures by the Rockefeller Foundation, and that spelled their doom.

Vincent was no more eager for a large institute in 1920 than in 1918, and he certainly was not interested in the fellowship board's plan for cooperative fund raising: the beauty of the fellowship board for Vincent was that it buffered the foundation from further claims. Neither the institute nor the joint fund-raising schemes got to first base. In fact, experience was already revealing that the NRC circle's hopes for local fund raising had been overly optimistic. Even such an experienced entrepreneur as Robert Millikan was frustrated. At the University of Chicago, President Judson abandoned the campaign for a graduate institute of technology and science only a year after it was announced. He blamed the postwar recession, but it seems clear that his heart was just not in it.[26] Millikan pressed him at regular intervals to revive their grand scheme but finally gave up and accepted Hale's long-standing invitation to join him at Caltech, where the idea of an institute for graduate study and research was fast becoming a reality.[27] The NRC's fellowship board gave up trying to be a planning group and settled down to the routine of processing applications and monitoring the progress of NRC fellows.

A similar fate met efforts to get the RF to support the NRC's various cooperative research programs. In June 1920, NRC chairman James Angell tried to persuade Vincent to support at least one or two of the most important research projects in each division. It was vital, he argued, to counter the growing criticism from rank-and-file scientists that the NRC was better at producing committees and reports than it was at pushing specific projects to completion.[28] Scientists' expectations were aroused by the publicity surrounding the gifts from foundations, and for them patronage meant grants for their own research projects: "[T]heir confidence in us is, as we have reason to know, going to be seriously shaken if we are not rather promptly able to make some tangible beginnings, however modest, in the way of direct assistance to research in pure science, especially in the groups other than physics and chemistry, and more particularly in biology."[29] The biologists' disaffection, Angell warned, was approaching a crisis.

[25]"Plans," May 1920, pp. 2–6.

[26]University of Chicago, *President's Report* 1920/1921: p. 6; 1921/1922: pp. xii–xiii.

[27]Millikan to Judson, 19 June 1920; 11 May and 2 June 1921, UC 40.3 and 17.12.

[28]Angell to Vincent, 3 June 1920, RF 200 36.416.

[29]Angell to Vincent, 7 June 1920, RF 200 36.416.

Nevertheless, the foundation's executive committee discouraged a formal application.[30] It is not hard to guess why. The last thing Vincent wanted was to get involved in supporting academic disciplines. The NRC's numerous scientific divisions were designed to spawn projects for cooperative research, and nothing was easier to do. The orgy of projecting in the postwar years would have scared off any foundation, since support of one project was bound to invite a flood of applications from every division. Without a programmatic touchstone of its own in science, the RF would have no rationale for limiting its commitment. Subsequent requests for medical research projects likewise were rejected out of hand.[31]

So sensitive were Vincent and Flexner to the dangers of individual research grants that they automatically brushed aside any proposal having to do with research. In April 1919, for example, Millikan and Hale asked the RF to support subcommittees of the NRC's physics division that were planning conferences and compiling bibliographies in new and productive research specialties. Vincent and Simon Flexner would have none of it, until Millikan craftily explained that the proposed bibliographies would make the fellowship programs more effective (presumably by directing fellows to new lines of research). A small grant ($20,000) was then made, on the understanding that it was experimental and not a commitment to support the physics division. The NRC executive committee had to guarantee that the RF would not be deluged with requests from other divisions.[32] Ten years later that fear was still very much alive.[33]

Fellowships, which everyone except Vincent expected to be a temporary first step toward institutional programs, thus remained the one permanent link between the Rockefeller Foundation and scientists in universities. It was lucky for them that Millikan thought to do first what could be done most easily: had he stuck to his principles, it is possible that nothing would have been done at all. But Millikan was mistaken in thinking that a first step would make further steps easier. Vincent's practical wariness of open-ended commitments to individuals and institutions outweighed any enthusiasm he had for science in the abstract.

[30] Vincent to Angell, 11 June 1920; Angell to Vincent, 12 June 1920; both in RF 200 36.416.

[31] Kellogg to Vincent, 7 Nov. 1922; Vincent to Kellogg, 21 Nov. 1922; Embree to Kellogg, 19 Dec. 1922; all in RF 200 36.416.

[32] Millikan to Vincent, 9 May 1919; Merriam to Vincent, 29 May 1919; Embree to Merriam, 4 June 1919; Millikan to Embree, 15, 21 June 1919; Embree to Merriam, 20 June 1919; all in RF 200 36.415. Embree to S. Flexner, 4 June 1919; Flexner to Embree, 13 June 1919; Embree to Angell, 4 Dec. 1919, RF 200E 169.2065.

[33] Officers conference, 19 Oct. 1928, RF 904 2.14.

Carnegie Programs in the 1920s

The Carnegie Corporation's partnership with science was also a limited one in the 1920s, notwithstanding the $17 million plus it spent on the natural sciences between 1915 and 1933 (table 5.2).[34] Nearly half of that total went to the CIW, which testifies more to the holding company philosophy than to any enthusiasm on Keppel's part for natural science. (The analogous grants by the RF to the Rockefeller Institute ceased in 1919.) Most of the other half of the corporation's outlay went to only two institutions, the NRC and Caltech, and though the rate of spending peaked in the late 1920s, this is not an indicator of increasing interest in science. Quite the contrary, they represent the liquidation of commitments made in 1919 and 1921.

Thus, the CC's vast expenditures were the result of momentum created by one sharp impulse just after World War I, when Angell and the reform clique gained control briefly at a moment when science seemed the key to national progress. The large expenditures of the late 1920s hide the fact that there were no longer any scientific projects to which the corporation had a long-term, programmatic obligation. For Keppel and his board the CIW and NRC were convenient buffers against further claims by university grant-seekers. Caltech was exceptional only because of the special personal connection between the Hale circle and their allies on the corporation board. That connection eroded steadily through the 1920s, the Caltech troika improvising appeals that they hoped Keppel could not refuse, and Keppel parrying every move. Money changed hands, but the unpleasantness of the experience made it a pyrrhic victory. No story better reveals the limits of the partnership between general purpose foundations and university science.

The Caltech troika's original application in 1921 was full of hope for a close institutional relation. An annual grant of $40,000 for three years would support Millikan's and Noyes's research on radiation and the constitution of matter. A capital gift of $500,000 would upgrade the undergraduate engineering degree course from four to five years and introduce more research into the training of an engineering-science elite.[35] The Caltech group did not request capital endowment of departments, but the annual grant to Millikan and Noyes was in effect the annual interest on anticipated endowments of their departments.

The combination of research and training had been a winning one in

[34]R. M. Lester, *Summary of Grants Primarily for Research in Biological and Physical Sciences* (New York: Carnegie Corporation, 1932). Carnegie Corporation, *Annual Reports* 1921–1931. CIW *Yearbooks*, 1916–1931. There are minor discrepancies in the figures from these sources.

[35]Fleming and Hale to CC, and application, 17 Sept. 1921, CC Caltech.

the fellowship negotiations. In the context of an appeal for a particular institution, however, it just set off hair-trigger alarms. Pritchett told Hale he could not support such a "mingled plan of research and administration." In language that could have come from the lips of Robert Woodward, Pritchett argued that research and education were two different things and that mixing them would only ensure that both would suffer. The corporation meant only to aid outstanding researchers, and Pritchett urged his friends not to justify a grand program of research by its relevance to education.[36] (Hale, Millikan, and Noyes assumed that Pritchett was taking issue with the Caltech philosophy of combining research and engineering education. More likely, Pritchett wanted to send money their way, knew the CC board would not approve a grant for education, and was trying to provide Hale with a rationale for dropping the request for general endowment.)

Pritchett knew what Hale did not: namely, that the corporation meant to administer any grant through the CIW. The institution's new president, John C. Merriam, had agreed to the arrangement but only for grants specifically marked for research. It was improper, he felt, for the CIW to administer funds for educational purposes. This is undoubtedly the reason why Pritchett tried to talk Hale into soft-pedaling education. And, indeed, in November 1921 the board approved a grant of $30,000 a year for five years for the divisions of chemistry and physics, to be administered by the CIW. The request for general endowment was deferred.[37] It was almost a replay of the negotiations with Vincent in 1918–1919, only this time Millikan did not voluntarily split off the institutional part of the scheme. The Carnegie Corporation did it for him.

Ironically, the educational thrust of the Caltech application had been James Angell's doing. Millikan had first approached Angell on the assumption that the corporation would be interested only in research, but Angell had protested that he would be even more eager to assist Caltech's experiment in educating an elite corps of research engineers.[38] Unfortunately for Millikan, Angell left almost immediately after presenting the Caltech proposal to the board, and Pritchett, who inherited the nascent project, had a more limited idea of the corporation's mission.

Pritchett had neither the scientific expertise nor the expert staff to administer a program of research grants, and, like Vincent, he shrank

[36] Pritchett to Fleming and Hale, 21 Oct. 1921, CC Caltech.

[37] Minutes of executive committee of the CIW, 21 Oct. 1921; Merriam to Pritchett, 1 Nov. 1921; Pritchett to Merriam, 22 Nov. 1921; Pritchett to Fleming and Hale, 21 Oct. 1921; CC board resolution, 17 Nov. 1921; all in CC Caltech.

[38] Millikan to Pritchett, 31 Oct. 1921, CC Caltech.

TABLE 5-2

Carnegie Corporation Expenditures in Natural Sciences (in $1,000s)

Year	*Carnegie Inst.*		*NRC*		*Caltech*		*Univ. of Pa.*		*Harvard*		*Research*	
	Gen.	*Spec.*	*Gen.*	*Spec.*	*Chem., Phys.*	*Gen.*	*Phipps Inst.*	*Med.*	*Phys.*	*Marine Biol.*	*Med./Sci.*	*Misc.*
1915–16	150										8	
1916–17	150		50									
1917–18	150		100								7	
1918–19	150		100								8	
1919–20	150		1,450							25		
1920–21	150		170	35						100	16	
1921–22			185		30		25				35	5

1922–23		5	183		30		25				75	4
1923–24		10	180		30		25				40	
1924–25			178		43	100	25				66	5
1925–26			699		42		25				34	25
1926–27	1,219	60	150		30	200		250			25	40
1927–28	1,158	5	1,142		30	200		3			47	15
1928–29	1,128	46	1,072	10		200						10
1929–30	1,067	31	1,043	10						10	25	
1930–31	1,029	18		10		7	25		50	42	5	10
1931–32		66					25					
1932–33		67					25					
Total	6,501	308	6,702	65	235	707	200	253	50	142	401	139

from creating such a staff. Indeed, the relationship between the CC and CIW was very much like the one between the RF and NRC. The CIW was insulation, protecting the corporation against claims by other universities that would have seen a direct grant to Caltech as an inviting precedent. It was good camouflage, and it was in everyone's interest.[39] To conservative members of the CC board it would look like assisting one of "Carnegie's children." For Pritchett it was a convenient way of channeling money to science without having constantly to do battle with the conservative faction. Scientists on the CIW's board had their own reasons to oppose institutional grants. William Welch, for example, thought it a fine thing for the corporation to give money to academic science but feared that an institutional grant would appear to create a favored relation with Caltech. A grant to Millikan and Hale by the CIW would leave the door open for future grants to luminaries at Yale, Harvard, or—most to the point for Welch—Johns Hopkins.[40] Everything in the history and politics of the Carnegie boards conspired to favor grants that, whatever their intent, did not have the appearance of grants to institutions.

From Millikan's negotiations with the CC, as from his negotiations with the RF in 1918–1919, forms of patronage emerged—fellowships, annual subsidies to special groups—that were least likely to set a precedent for future claims by universities. In both cases, operating responsibilities were delegated to agencies—the NRC and CIW—whose expert scientific staff could insulate patrons from direct involvement with academic scientists. This limited partnership was as far as the large foundations would go to adapt their traditional missions to the postwar enthusiasm for science.

Millikan Wins a Battle and Loses a Campaign

Hale and Millikan did not abandon their campaign for endowment, however, as they had in 1919. Their response to Pritchett's well-intentioned advice was to go to the barricades for the principle of educational endowment. A barrage of large-caliber letters was fired off—Hale's ran to nine pages—defending Caltech's distinctive educational philosophy.[41] By only deferring the request for general endowment and referring it back to Pritchett, the CC board had left the door of the president's office ajar for future lobbying.[42] Each year between 1922 and

[39]CIW board minutes, 9 Dec. 1921, vol. 3, pp. 114–119, CIW.

[40]Ibid.

[41]Hale to Pritchett, 28 Oct. 1921; Noyes to Pritchett, 29 Oct. 1921; Millikan to Pritchett, 31 Oct. 1921; Hale to Root, 1 Nov. 1921; all in CC Caltech.

[42]Pritchett to Hale, 22 Nov. 1921, RAM 20.9.

1926 the Caltech group renewed their attack, and each year they were put off, first by Pritchett and then by Keppel. Keppel did not argue but simply asserted that financial obligations did not allow a capital gift just then.[43] No doubt, but the record of scattered grants, mostly small but adding up to a substantial sum, suggests that the more fundamental reason was that Keppel was as eager to avoid a precedent for institutional patronage as Hale and Millikan were to set one.

In 1923 and 1924, the Caltech group shifted their strategy somewhat, focusing specifically on endowment ($250,000 each) of a department of geology and a five-year engineering course.[44] By then, however, Keppel had become immune to the Caltech lobbyists—allergic, in fact. So he was not swept away when Millikan laid on the charm and sweet talk about Caltech's excellence and the benefits of demonstrating that the corporation stood by its friends.[45] Keppel was also lobbied by Merriam, who was eager to assist in the development of geology in California, and Keppel finally relented. Millikan got $25,000 for two years for geological and geophysical research, but no promise of future capitalization and nothing for engineering.[46]

By 1926 Millikan and Hale had abandoned any hope of getting a general endowment, and they began to lobby instead for a grant of $600,000 to capitalize the recurring grant for research in chemistry and physics. There were several reasons to change tactics once again: the 1921 grant was running out, and Keppel was not eager to renew. Also, Millikan was getting restless from the uncertainty and frustration of continual fund raising, and Hale worried that Pritchett and Root were getting old and might not be around much longer to help. It seemed to be now or never with the CC.[47]

Millikan, Hale, and Noyes had by then definitely worn out their welcome mat. "Between you and me," Keppel confided to Pritchett, "I am getting pretty sick of the attitude of our friends in Pasadena. . . . [T]hey are literally the only people of the hundreds with whom I have come in contact who act on the assumption that I am neither sincere nor truth-

[43]Noyes to Pritchett, 27 Apr. 1922; Fleming and Millikan to CC, 5 Oct. 1922; Millikan to Keppel, 29 Sept. 1923; Fleming to Keppel, 6 Feb. 1926; Keppel to Merriam, 5 Dec. 1924; all in CC Caltech.

[44]Millikan to Pritchett, 2 May 1923; Millikan to Keppel, 29 Sept. 1923; both in RAM 20.9. Merriam to Noyes, 26 May 1924, GEH reel 24.

[45]Keppel to Merriam, 5 Dec. 1924, CC Caltech.

[46]Ibid. Merriam to Keppel, 6 Dec. 1924, CC Caltech. Merriam to Noyes, 26 May 1924; Hale to Merriam, 8 June 1924, 8 July 1925; Merriam to Hale, 16 June 1924, 8 Jan. 1925; all on GEH reel 24.

[47]Hale to Merriam, 10 Mar. 1927; Hale to Pritchett, 10 Jan., 27 Oct. 1927; all on GEH reel 29. Keppel to Hale, 23 Nov. 1926, CC Caltech. Hale to Merriam, 10 Mar. 1927, GEH reel 24.

ful."[48] But Root and Pritchett, who wanted Caltech to get its $600,000, had already begun to arrange things behind the scenes in precisely the same way that they had in 1918–1919, with Merriam providing somewhat lackadaisical support.[49] "Uncle Elihu and I have had rather a hard time over the whole matter," Pritchett reported to Hale in early 1927, "but we have done our best."[50] Quashing one final effort by Keppel and his allies to derail the scheme, Pritchett and Root finally obtained a research endowment for Millikan and Noyes.[51] Pritchett's partisanship was not just loyalty to old friends: he did not like the way Keppel and the board were throwing small sums of money at minor academic projects. He was distressed, too, by the revival of a nineteenth-century style of collegiate culture in universities, as athletics and social activities eclipsed prewar scholarly ideals. Caltech, Pritchett felt, was upholding ideals of research that he held dear.[52]

The grant to Caltech was the last major grant that the Carnegie Corporation made to science. No precedents had been set, no philanthropic principles established, no extension of the patronage system negotiated. The battle was won, but the campaign was lost. In hindsight, it was lost when Angell left in 1921 and the corporation did not become a foundation for graduate education and research. Keppel liked occasionally to support the research of famous scientists. Millikan had only to ask for $20,000 for his work on cosmic rays in 1932—no fuss, no hassle—and a similar sum (both laundered by the CIW) went to Arthur Compton. But Keppel cold-shouldered Millikan's plea for $20,000 a year to make up for the collapse of Caltech's endowment in the depression, as he did anything that implied a specific commitment to an institution.[53]

A few years later Edwin Embree reflected on the importance of having a definite, philanthropic program:

> I should think the influence of the Carnegie Corporation has been far less [than the RF's], and many others have told me the

[48]Keppel to Pritchett, 24 Mar. 1927, CC Pritchett. Pritchett to Hale, 22 Nov. 1926, CC Caltech.

[49]Hale to Merriam, 10 Mar. 1927, GEH reel 24. Merriam to Keppel, 5 Dec. 1924, CC Caltech. Pritchett to Hale, 1 June 1928, GEH reel 29.

[50]Pritchett to Hale, 4 Jan. 1927; Hale to Pritchett, 10 Jan. 1927; both on GEH reel 29.

[51]Keppel to Hale, 23 Nov. 1926; Keppel memos, 29 Nov., 20 Dec. 1926; MAC to Keppel, 2 Apr. 1926; Merriam to Keppel, 2 Apr. 1926; executive committee resolution, 20 Jan. 1927; board resolution, 19 May 1927; all in CC Caltech.

[52]Pritchett to Hale, 18 Jan. [1927]; Hale to Merriam, 10 Mar. 1927; both on GEH reels 29, 24. Geiger, *To Advance Knowledge,* ch. 3.

[53]Keppel to Millikan, 20 Jan. 1932, 18 May 1933; Millikan to Keppel, 15 May 1933, RAM 20.11.

> same, and apparently because of a lack of defined and somewhat limited objectives. . . . I have had the strong impression that the R.F. became an intellectual and public force for good whereas the Carnegie Corporation seems to me in the minds of most persons I know merely one of the places where they might get some money.[54]

Embree could have gone on to say that the lack of programmatic commitments was something from which the RF also suffered—not the operating divisions but Vincent's office. Embree knew: he had tried and failed to get the RF to expand beyond fellowships in his ill-fated program in "human biology."

Squeezed Lemon: Embree and Human Biology

The vehicle of Embree's ambitions was a new Division of Studies, created in January 1924 with himself as director. (It was his idea originally.) The new division was an umbrella for a number of small projects in nursing education, dispensaries, the remnants of the stillborn mental hygiene program, and other things too insignificant to catch the eye of the powerful medical barons. For Vincent it was to be a nursery of philanthropic ideas and surveys. For Embree it was that and more: a nursery of new operating programs. Though he made light of the name ("Division of Trial and Error, Division of Lame Ducks, Division of Half-Squeezed Lemons—it's really too easy to caricature"), the Division of Studies was Embree's chance to become a divisional baron himself.[55] Unlike Vincent, who feared new operating programs as much as those that already plagued him, Embree felt that the IHB and CMB could be kept in check by competing programs controlled from the office of the president. He urged Vincent to occupy the field of humanities quickly before Flexner and Rose did so.[56] Vincent's failure to move forcefully into new fields finally drove his loyal lieutenant to try his own hand at being a philanthropic entrepreneur. Embree set out to occupy the field of "human biology," co-opting the Division of Studies in much the same way that Beardsley Ruml and Rose were co-opting the Laura Spelman Rockefeller Memorial and GEB for the social and natural sciences.

54Embree to Fosdick, 8 Jan. 1936, RF 903 1.2.

55Embree to Vincent, 4 May 1923, RF 913 1.1.

56Embree to Vincent, 4 May 1923, RF 913 1.1. "Memorandum of conference at Gedney Farms, 18–19 Jan. 1924"; Embree, "Expansion of programs of Rockefeller boards," 18–19 Jan. 1924; A. Flexner, "Foundations," p. 18; Embree to Vincent, 4 Feb. 1924; Embree to Fosdick, 7 Feb. 1924; all in RF 900 22.165.

"Human biology" was a fashionable but ill-defined rubric for a congeries of related fields: human heredity, growth and development, psychology, anthropology, constitutional medicine, and others. It is probably best understood as one of many attempts, in the wake of the medical reform movement, to capture the biological and behavioral side of medicine for university science departments. In any case, Embree probably picked up the idea from the biometrician Raymond Pearl, one of the liveliest leaders of the movement and a trustee of the RF.[57] Embree's plan was to develop the biological sciences underlying individual behavior, much as Ruml was developing the sciences underlying social behavior. (In fact, the two men discussed a joint program.)[58] It was a somewhat odd mixture of things that Embree presented to the board as a program in February 1925: racial mixing in the Pacific, sexuality and human development, human heredity and experimental evolution, anthropometry of ethnic groups, brain physiology, and mental hygiene. Amazingly, the board authorized him to survey the field and present proposals. With much hustle and bustle, and with the distinguished Princeton biologist Edwin Conklin in tow, Embree set off on a grand tour of ethnological and biological institutions in Japan, Australia, and the Pacific Islands.[59] In the next year or so, a program of fellowships was begun, and grants were made to half-a-dozen projects, most notably to Pearl, a charismatic and cosmopolitan figure, to whom Embree attached himself, as he had before to Clarence Day and Vincent.[60] For a few years Embree zestfully emulated the grand manner of Ruml and Rose.

Embree's road show was short-lived, however, in part because "human biology" lacked the coherence and scientific legitimacy of the biomedical sciences, and in part because Embree's behavior alerted Vincent and others to the fact that another little barony was in the making. Even Fosdick, who was deeply interested in mental hygiene and human behavior, worried about where it would all go and what it would cost. Simon Flexner declared bluntly that he could make no sense of it. By late 1926 Fosdick had concluded that the Division of Studies should be closed down and its more viable programs (nursing education,

[57]"Memorandum of Conference."

[58]"Conference of members and officers, Princeton," 23–24 Feb. 1925, RF 900 22.165. Embree to Fosdick, 26 Aug. 1925; Embree to Vincent, 27 Apr. 1925; all in RF 915 4.33.

[59]Conference, 23–24 Feb. 1925. Embree, "Studies in human biology," 23–24 Feb. 1925, RF 900 11.165. Embree to Fosdick, 26 Aug. 1925, RF 915 4.33.

[60]"Division of studies," Jan. 1926, RF 900 22.166. Correspondence and reports in RF 609D 1.1. Correspondence in Raymond Pearl Papers, American Philosophical Society, Philadelpha, Pa.

dispensaries, and premedical sciences in China) incorporated into Richard Pearce's Division of Medical Education.[61]

The one program that did not fit into medical education was "human biology." Pearce had long been interested in having a program in the basic biomedical sciences, but the behavioral and social sides of Embree's program had no appeal. With Vincent's approval, Fosdick concluded that they should be transferred temporarily to Pearce's division and terminated as soon as possible. "I find," he wrote, "that they do not amount to anything anyway."[62] That left Embree without a job: unsuited for promotion to vice-president and overqualified for a post as Pearce's assistant, he was clearly on the way out. In December 1926 he was sent to Europe to survey the field of nursing education and prepare for a posting to the foundation's Paris office. Ignoring injunctions to drop the subject of human biology, Embree continued to play the role of enthusiastic patron, arousing the hopes of psychologists and anthropologists at the University of London. That made Vincent and Fosdick nervous, but they found it difficult to rein Embree in.[63] A few months later Embree assured Stanford President Ray Lyman Wilbur that the RF's moratorium on human biology was only temporary.[64]

By imitating the baronial behavior of Ruml and Rose, Embree just added to the internal rivalries and conflicts from which he himself had suffered and to which Fosdick and Vincent were determined to put an end. There would be no place for Embree's scheme of human biology in the reorganized Rockefeller Foundation or for Embree himself. In 1928 he moved on to a productive and always controversial career as head of the Julius Rosenwald Fund.[65]

Embree blamed himself for the failure of the human biology program:

> If I had that job today [c. 1930] I think I could build up a program that would be not only brilliant but would command support even

[61]Conference, 23–24 Feb. 1925. S. Flexner to Vincent, 27 Jan. 1924; Embree to Fosdick, 26 Aug. 1925; both in RF 915 4.33. S. Flexner to Fosdick, 4 May, 1926; Fosdick to S. Flexner, 28 Oct. 1926; both in RF 900 17.121 and 17.122. Board minutes, 23 Feb. 1927, RF 913 1.1.

[62]Fosdick to S. Flexner, 8 Oct. 1926, RF 900 17.122. Fosdick, "Report of the Committee on Reorganization," 5 Nov. 1926, pp. 6–8, RF 900 19.136.

[63]Fosdick to Vincent, 21 Dec. 1926; Vincent to Fosdick, 23 Dec. 1926; Vincent to Embree, 19 Dec. 1926; Fosdick to A. Flexner, 12 Jan. 1927; all in RF 100 4.35.

[64]Embree to Wilbur, 11 July 1927; Wilbur to Vincent, 21 May 1926; both in RF 915 4.33.

[65]Johnson, "Phylon profile X: Edwin Rogers Embree." E. R. Embree and Julia Waxman, *Investment in People: The Story of the Julius Rosenwald Fund* (New York: Harper, 1949).

> from the conservative. But for three years I sweat blood on the job in New York. On the whole I think the results were those that would be expected from a very immature person. The things I got done were not bad, but I did not get the philosophy of the thing for years and years and was therefore unable to formulate a really statesmanlike program.[66]

Certainly there is some truth in that, but it seems unlikely, even if Embree's program had been better conceived and managed, that the RF would have been converted to supporting programmatic research projects. Vincent was determined to permit no new baronies to come into his world, and his theory of the foundation as a philanthropic laboratory made his conservatism seem a positive virtue. The same ideas and expectations blocked Millikan's efforts to build on the NRC fellowships and doomed Embree's hopes for human biology. The demise of the Division of Studies was part of the larger pattern of innovation and retrenchment.

Conclusion: A System and Its Limits

The structure, history, and internal politics of general purpose foundations made it difficult to expand the limited partnership that was created in 1919. The CC and RF originated as holding companies for quasi-independent operating agencies created previously by Carnegie and Rockefeller. Thus, history left them with central managements that were weaker than their operating parts. Central managers had carte blanche to enter new fields, but that freedom only made it harder to decide what the mission of the new foundations was and to resist financial claims by their operating divisions. Vincent and Keppel tried to solve this problem by seeing their offices as think tanks of philanthropic ideas, but that only made it difficult for good, doable ideas to become operating programs. There were differences: "Carnegie's children" were separate from the CC; the IHB and CMB were (in principle) subunits of the RF. Vincent never accepted the dominance of the medical barons, as Keppel did the dominance of the Carnegie presidents. Keppel embraced a policy of no policy; Vincent failed to have a policy, but he did try. In the end the structure and values of these philanthropic holding companies limited their ability to connect with academic scientists. They could not afford, financially or politically, to create national institutes or to risk direct support for individual scientists or their local institutions. So they eschewed institutional grants and

[66]Embree, "Rockefeller Foundation," n.d., c. 1930, ERE box 1.

kept the NRC and CIW as buffers between themselves and individual scientists. The CC and RF gave millions to science in the 1920s, but the partnership that began in 1919 remained a distinctly limited one.

Creating the new system of patronage was an extended process. It gained momentum slowly after about 1910 as entrepreneurial scientists probed the new foundations, trying to discover what kind of relationship would work. World War I set off a brief period of rapid change, giving a new public meaning to the physical sciences, a new authority to the Hale circle, and making the conditions right for the invention of new bridging institutions like the NRC. All this happened rather quickly, in part because the Hale circle knew from experience to do quickly what was doable. The process of extending the new system of patronage continued into the 1920s but with far fewer successes. The limits of the trade-association relationship were quickly reached. Enthusiasm of the scientists for national institutions waned as wartime nationalism gave way to local or disciplinary interests and aspirations. The large general purpose foundations, stymied by internal conflicts, shunned entanglement with local institutions. Only by looking at the whole twenty-year process can we understand how a new partnership was constructed, and how its limits were defined.

The general purpose foundations were not, for the natural and social sciences, where the main action was in the 1920s. By deciding not to make institutional grants in science, the RF and CC left a philanthropic void that was filled by the International and General Education Boards and the Laura Spelman Rockefeller Memorial. It was in these educational foundations, with their traditions of field operation and systematic institution building, that large-scale institutional programs in science developed in the mid-1920s. Wickliff Rose and Beardsley Ruml were no less inspired by scientists' war records than were Vincent or Flexner, and they were not impeded by internal conflicts from diverting their foundations to building scientific institutions.

Vincent, Pritchett, and Angell created new social roles and a new social space, into which others moved when they did not. They opened doors with their idea of foundations as philanthropic nurseries and of the NRC and CIW as administrative surrogates. But that vision and institutional machinery also ensured that it would be the heads of operating foundations who would walk through those open doors.

1. Robert S. Woodward, c. 1920. Courtesy of the Carnegie Institution of Washington.

2. Henry S. Pritchett, c. 1912. Courtesy of the Carnegie Foundation for the Advancement of Teaching.

3. James R. Angell, c. 1916. Courtesy of the National Academy of Sciences Archives.

4. General Education Board, 1915. From left to right and bottom row to top: Edwin A. Alderman, Frederick T. Gates, Charles W. Eliot, Harry Pratt Judson, Wallace Buttrick; Wickliffe Rose, Hollis Frissell, John D. Rockefeller, Jr., Eben Sage, Albert Shaw, Jerome D. Greene, Abraham Flexner; George E. Vincent, Anson Phelps Stokes, Starr J. Murphy. Courtesy of the Rockefeller Archive Center.

5. George E. Vincent, mid-1920s. Courtesy of the Rockefeller Archive Center.

6. Edwin Embree, mid-1920s. Courtesy of the Rockefeller Archive Center.

7. Arthur A. Noyes, George E. Hale, Robert A. Millikan, early 1920s. Courtesy of the California Institute of Technology Archives.

8. National Academy of Sciences Building under construction, 1922. Courtesy of the National Academy of Sciences Archives.

9. National Academy of Sciences Building, 1924. Courtesy of the National Academy of Sciences Archives.

10. Wickliffe Rose, mid-1920s. Courtesy of the Rockefeller Archive Center.

11. Augustus Trowbridge, early 1930s. Courtesy of Princeton University Archives.

12. Halston J. Thorkelson, mid-1920s. Courtesy of Martha Thorkelson Riddell.

Part II

Institutional Relations, the 1920s

Chapter Six

Americans in Paris: Wickliffe Rose, Augustus Trowbridge

The biggest and most active patrons of science in the 1920s—the General and International Education Boards and the Laura Spelman Rockefeller Memorial—were big and active because they were operating foundations, not holding companies and nurseries of philanthropic ideas. Wickliffe Rose and Beardsley Ruml were experienced field operators, and they inherited going concerns. The GEB had twenty years of experience in giving grants for the building and endowment of colleges and medical schools when Rose took charge of it and the IEB in 1923. He was accustomed, from his public health work, to developing national systems of institutions, and as head of the GEB he had a mandate to do equally big things in science. So, where Vincent and Keppel avoided (or tried to avoid) direct grants to universities, Rose and Ruml made institutional grants the mainstay of their programs in the natural and social sciences. Where Vincent and Keppell saw institutional grants as pitfalls of uncontrollable expenditure, Rose and Ruml developed the appropriate machinery for making institutional grants serve their programmatic goals. "Make the peaks higher" was Rose's watchword, and he and Ruml came close to putting into practice the system of national centers of research and training that Millikan had envisioned for the RF and CC in 1919. The IEB, GEB, and the memorial completed the 1920s system of science patronage by doing what the RF and CC could not or would not do.

The officers of the operating foundations were also a new and distinctive group: not generalists or university handymen like Jerome Greene, Vincent, Embree, or Keppel, but scientists turned philanthropists. Most of the people Rose got to run the programs of the IEB and GEB—people like Augustus Trowbridge or Albert R. Mann—were academic scientists of an entrepreneurial bent. Many of them moved back and forth between the academic and foundation worlds.

(Trowbridge and Mann, after leaving the IEB, became clients of the Rockefeller boards.) They were active program managers, often on the road making contacts, asking questions, getting connected into scientific communities and networks. They were active participants in making projects. Trowbridge did not delegate the running of the IEB's fellowship program to the NRC, as Vincent did but ran it himself. The architects of institutional projects intervened more boldly than their predecessors in the business of science (though not as boldly as their successors of the 1930s would do). They were second-generation managers of science.

In the history of IEB and GEB projects, policymaking will play a less dominant role than it has so far. That is not because the IEB and GEB operated without policies, but because philanthropic practices were made primarily in the field in the process of negotiating projects. "Make the peaks higher" could mean many things; what it did in fact mean depended in part on what was doable. Thus, we need to pay more attention now to the art and craft of project making, in which opportunism and improvisation played as great a part as policy. And we need to look more broadly than we have till now at international communities of scientists and national systems of universities. The IEB and GEB programs were meant to aid whole systems of institutions, and the large-scale structures of these systems largely determined what the partners in science could and could not do. So we need to take an ecological approach and focus on the interaction between foundation managers and the various scientific communities in which they worked.

In this chapter we will explore, first, the sources of Rose's vision of science patronage, both in his prior experience in public health operations and in the unfinished business of the Rockefeller Foundation. I will argue that Rose acquired his enthusiasm for science—more precisely, his conviction that science was a cause for the premier educational foundation—by watching Vincent pass up opportunities in this field. There is some evidence, in fact, that Rose was directly inspired by Millikan's conception of systems of regional centers of research and training, the scheme that Vincent rejected in 1919. We will then see how Rose and his chief lieutenant in Europe, Augustus Trowbridge, set about creating a mechanism for making large institutional grants: how Rose established connections with natural scientists, and how Trowbridge learned the ropes of field practice. To conclude this chapter, we will survey, through Trowbridge's eyes, the state of the natural sciences in Europe in the 1920s. We will see how different countries and university systems had recovered from the war, what strategies academics employed to keep up or get ahead in different levels of the university system, and where in this diverse system the opportunities lay for Rose and Trowbridge to engage in institutional development on a truly grand scale.

Sources of Rose's Vision: Unfinished Business

One of the most striking things about the history of the IEB and GEB is the almost total absence of any formal discussion about policy, even though diverting the GEB from general education to scientific research and training was a major policy change. That part of the story is starkly simple: Rockefeller, Jr., asked Wickliffe Rose if he would take Wallace Buttrick's place when the latter retired as head of the GEB in 1923. Rose said he would on condition that he could take the GEB into the natural and agricultural sciences and in Europe as well as the United States. (Having expanded the Hookworm Commission to a worldwide operation, Rose could hardly be expected to be content with an American horizon.) So as not to risk losing the GEB's charter (it was restricted to the United States, and recourse to law might have invited political opposition), Rockefeller, Jr., created a new board, the IEB, and put Rose in charge of both.[1] All this is evidence of the exalted position that Rose enjoyed in the Rockefeller circle. As Embree irreverently recalled:

> As Rose's prestige increased, he took on the outward habits of a god. In office discussions he never allowed himself to take part in the preliminary skirmishes, but toward the end of a conference, he summed up the whole matter, and his unquestioned ability in analysis, together with his solemn tones, established a precedent which grew almost into a fetish that Rose having spoken, nothing more need or should be said.[2]

No one had more influence with Rockefeller, Jr., than did Rose, except possibly Raymond Fosdick. There was no agonizing over policy: Rose wanted to be a patron of science, and what Rose wanted he got.

The question is, How did Rose know what he wanted? Where did he get an interest in postgraduate education? How did he come to believe that natural science was the most important field that the GEB could enter? Why did he rely on capital grants to scientific institutions as his vehicle of system building? To answer these questions, we need to look at who Rose was talking to and what he was doing in the years just before he left public health to become the premier patron of science.

Although Rose is known for his work in public health, his first love

[1]George W. Gray, *Education on an International Scale* (New York: Harcourt Brace, 1941).

[2]Embree, recollections, n.d. [c. 1930], p. 4, ERE box 1.

was education. Education had rescued him (in his twenties) from a life of hard-scrabble farming in Tennessee. For two decades, as professor of philosophy and education and dean of the George Peabody College for Teachers, Rose had devoted himself to education in the South. He was an active member of the Southern Education Board and the Peabody and Slater funds. When offered the presidency of Peabody College shortly after becoming head of the Hookworm Commission, he had come very close to accepting. He had an educator's omnivorous tastes and could deal as an intellectual equal with scholars in such varied fields as medicine, law, and economics. Though not trained in science, Rose had a logical turn of mind that was, to quote Simon Flexner, "essentially scientific."[3] In the long view, Rose's dozen years in public health could be seen as an interlude in a career devoted to general education.

Even in those twelve years Rose was never far from educational matters. Indeed, it was his work in public health that brought him to the specific problem of the education of scientists. As director of the IHB in the years of its greatest expansion, Rose was constantly faced with a shortage of expert manpower, a bottleneck created by his expanding networks and his insistence that public health officials have academic medical credentials. By 1916 it had become an immediate practical problem, and after talking it over with Abraham Flexner, Rose jotted down a scheme to increase the output of medical scientists:

> 1. Ascertain the three or four departments of medical science in which there seems to be the greatest need of increase of productive men. . . . 2. Find the one or two or three masters in each of these fields. 3. Arrange to increase the educational influence and productivity of each of these men . . . by supplying the scientist with all the equipment which he needs both for research and for educational purposes, and by making it possible for him to surround himself with a select group of men, as many as he could use and teach to advantage. 4. . . . Make it financially possible by means of scholarships or fellowships for men to pursue this course of training even through a long period of time. The chief inducement . . . will come in the form of opportunity for a career in this field.[4]

Flexner's role in Rose's scheme for educating medical scientists is not entirely clear. Rose hinted that he drew up his own plan when Flexner

[3]John Ettling, *The Germ of Laziness* (Cambridge, Mass.: Harvard University Press, 1981), pp. 113–117. Flexner, "Wickliffe Rose 1862–1931," *Science* 75 1932: pp. 504–506.

[4]Rose, "A plan for increasing productivity in medical science," 24 May 1916; Pearce to Rose, 24 May 1916; Rose to Embree, 13 Jan. 1919; all in RF 906 1.3.

seemed unwilling to commit the GEB to the problem.[5] On the other hand, it was only months after Rose jotted down his scheme that Flexner asked T. H. Morgan to make a case for a GEB program in medical scientific education. Whether Flexner and Rose were moving toward cooperation or competition, it is clear that Rose's educational ideas were very different from the GEB's standard practices in medical education. All the distinctive features of Rose's philanthropic style are apparent in his 1916 memorandum: the idea of making a few peaks higher, the combination of fellowships and grants for institutional development, the personalized language of mentors and acolytes. This was not Flexner but vintage Rose.

Rose's 1916 scheme bears a striking resemblance to the plan for the natural sciences that he laid before the IEB board in 1923:

> 1. Begin with physics, chemistry and biology; locate the inspiring productive men . . . ; ascertain of each of these whether he would be willing to train students from other contries; if so, ascertain how many he could take at one time; provide the equipment needed, if any, for operation on the scale desired. 2. Provide by means of fellowships for the international migration of select students to these centers of inspiration and training; students to be carefully selected, and to be trained with reference to definite service in their own countries after completion of their studies. 3. Give limited assistance as opportunity may offer, particularly in the more backward countries, to enable the returned scientists to establish themselves.[6]

The nearly identical phraseology leaves little doubt that Rose simply revived his earlier scheme when obliged, at short notice, to draw up a program for education in the natural sciences. It appears also that Rose was sending a clear signal that he meant to go his own way and not simply adopt the GEB's operating style. Rose's emphasis on individuals and mentorship, like the international scope of his scheme for science, came directly out of his experience in public health organization.

But where did Rose acquire his fervent belief in science? As he put it to the board in 1923:

> 1. This is an age of science; all important fields of activity, from the breeding of bees to the administration of an empire, call for an

[5]Rose to A. Flexner, 16 Dec. 1915; Flexner to Rose, 17 Dec. 1915; both in GEB 1.5 702.7225. Pearce to Rose, 24 May 1916, RF 906 1.3.

[6]Rose, "Scheme for promotion of science on an international scale," pp. 6–7 in docket for IEB board, 30 Apr. 1923, IEB minutes, 30 Apr. 1923, pp. 39–41. Rose diary, 20 Feb. 1923, vol. 1, pp. 27–28, RF.

> understanding of the spirit and technique of modern science. 2. The nations that do not cultivate the sciences cannot hope to hold their own; must take an increasingly subordinate place; must become more and more a drag to general progress; and must in the end be dominated by the more progressive states even though these states do not seek to dominate. 3. The nations now cultivating the sciences are but a small minority of the peoples of the world. 4. It should be feasible to extend the field for the cultivation and the service of science almost indefinitely. 5. Promotion of the development of science in a country is germinal; it affects the entire system of education and carries with it the remaking of a civilization.[7]

The problem is that such views were commonplace in the early 1920s in scientific circles; so the question is, Where exactly did Rose intersect with those circles?

One important point of intersection, obviously, was at the Rockefeller Institute, where Simon Flexner was putting into practice his conviction that progress in medicine came from physics and chemistry. Simon Flexner and Rose were quite close. As head of the IHB, Rose made a practice of delegating the IHB's research to the institute. He also relied on Flexner for technical advice, for example, in the selection of IHB fellows. This division of labor kept the two men in regular contact, and their correspondence reveals a relationship of deep mutual respect and trust.[8]

The NRC was another point of intersection with natural scientists. The war brought Rose into contact with the scientists manning the U.S. Food Administration in Europe, most notably Vernon Kellogg and the nutritional biochemist Alonzo Taylor. With their wide firsthand experience of conditions in postwar Europe, Kellogg and Taylor were useful sources of information for Vincent and Rose. In September 1919 Kellogg briefed Rose on health conditions. A month later Rose called on NRC secretary James Angell to inquire about the NRC's postwar plans and to borrow the NRC's manpower rosters for recruiting IHB fellows.[9]

Rose's postwar travels also brought him into contact with European promoters of science as an engine of social and economic reconstruction. One of the most interesting of these circles was the Spanish Junta para Ampliación de Estudios (Board for the Extension of Learning).

[7] Rose, "Scheme for the promotion."

[8] Rose to F. F. Russell, 13 Aug. 1922, RF 1.1 62.886. Rose-Flexner correspondence, SF.

[9] Rose diary, 11 Sept. 1919, RF 700 16.121. Rose to V. G. Heiser, 22 Oct. 1919, RF 1.1 39.613.

Led by José Castillejo, this group endeavored to make science the vehicle for rejuvenating Spain's antiquated culture and economy. Rose took part in discussions with Vincent, Embree, and Castillejo in December 1921 and visited Madrid in March 1922, reporting back to Vincent that Castellejo's group was "the most hopeful force" for progress in Spain.[10] Rose also came into contact with British scientific and medical activists, whose success in linking science to national reconstruction was an example and inspiration to their counterparts in the NRC and elsewhere. Alonzo Taylor introduced Rose to these circles in 1919, and, as the patron and midwife of the London School of Hygiene, Rose got to know them well and to admire their idealism and practicality. "[T]he war has shaken loose many old traditions in these Isles," he wrote Vincent in 1919. "[I]t is much easier now to do new things. In medical education the current is running with the rapidity of a mill race. The English seem to be thoroughly aroused to the importance of scientific research in all fields, and medicine is coming in for its share of this general interest."[11]

A tantalizing bit of evidence also links Rose directly with Noyes and Millikan's unsuccesful efforts to get Vincent to sponsor a system of regional university centers. In July 1919, Millikan gave an address at the University of Chicago, in which he made public the plan that had been proposed privately to Vincent. Coming only a few months after the conclusion of the fellowship negotiations, his address was clearly part of Millikan's continuing campaign to realize the rest of the original plan for a network of regional centers of research and graduate training. Millikan challenged philanthropists to help:

> How can such institutions be created? Perhaps by government initiative. But if we may argue from the past the development is likely to come about in America in another way. We have developed in the United States a highly patriotic and highly intelligent public sentiment which stimulates men of wealth and power to devote themselves and their fortunes to great public enterprises. . . . The great opportunity in science then for the man who wishes to invest his funds where they will count most for his country and his race lies in the endowment of research chairs, or better semi-research chairs, in a few suitably chosen educational institutions. . . . If some such program . . . for producing scientific men and for creating centers of research . . . can be adopted in the United States . . . then in a very few years we shall . . . see men coming from the ends of the earth to catch the inspiration of our leaders

[10] Embree diary, 2 Dec. 1921, RF. Rose to Vincent, 2 Mar. 1922, RF 1.1 62.885.

[11] Rose to Vincent, 30 Dec. 1919, SF. Rose diary 15 Nov. 1919, RF 700 16.121.

> and to share in the results which have come from our developments in science. If we fail to seize these opportunities then the scepter will pass from us and go to those who are better qualified to wield it.[12]

Millikan's sermon for science was aimed at Judson, patrons of the University of Chicago, Vincent, and leaders of all the great foundations.

Simon Flexner sent Rose an offprint of Millikan's address in November 1919. Rose replied, in a brief but eloquent note, that it had struck him as "one of the most stimulating things I have read in a long time. It sets one's mind going in a number of directions and in germinal lines."[13] Rose was not then in a position to take up Millikan's challenge, but it is tempting to think that it came into his mind two years later, when he was unexpectedly handed the chance to divert the vast resources of the GEB into new channels. The importance of the Millikan connection is that it goes beyond vague ideological affinities. Rose's faith in science was commonplace and overdetermined; his idea that science was the proper business of large foundations was not at all common, not at least in foundation circles. That idea, I believe, was inspired by Rose's knowledge of Millikan's unfinished business with the Rockefeller Foundation.

Millikan's was not the only scientific business that Vincent had left unfinished. Rose also played a minor part in abortive efforts by the RF to stake out agricultural and educational programs in postwar Europe, efforts that look very much like Rose's plans for agricultural and general education. These initiatives produced only two small programs of emergency grants to European medical scientists in 1920 and 1922. Yet in these vestiges of unfulfilled hopes are clues to what led Rose to think that science was a fitting task for the IEB.

It is a story with a familiar plot. Vincent was under pressure from all sides in 1919–1920 to do something to help Europeans recover from the devastations of war.[14] The trustees were eager to terminate emergency relief and begin constructive programs. Abraham Flexner and Raymond Fosdick badgered Vincent to consider programs in general education in Central Europe. Simon Flexner made an impassioned plea for an emergency grant to the Vienna Medical School. Embree urged his chief to be more open to opportunities to aid the new nations created out of the ruins of the Austro-Hungarian and Ottoman em-

[12] Robert A. Millikan, "The new opportunity in science," *Science* 50 (1919): 285–297, see p. 297.

[13] Rose to S. Flexner, 15 Nov. 1919, SF.

[14] Rose, "Conditions in Eastern European countries," 14 Mar. 1920, pp. 3–4; Embree diary, 27, 29 Dec. 1919; 8 Apr., 5 May 1920; all in RF 700 16.121.

pires.[15] In October 1919 the board gave Embree an informal portfolio in the Balkans and Near East and in December authorized Vincent to send a commission there to survey opportunities in general education and agriculture.[16] In March 1920 it instructed Vincent to appoint an experienced person to represent the foundation in Europe, no doubt hoping that a permanent officer would initiate permanent programs. The post was never filled.[17]

Vincent, in fact, had little enthusiasm for grand schemes of European reconstruction. Inexperienced in European matters, he was much more interested in China, and, as always, he worried that a new program would only mean another baron to plague him. So he stalled.[18] Besides, everyone who came back from Eastern Europe in 1919–1921 agreed that what was needed was relief on a vast scale. Reconstructing education, public health, and cultural institutions was impossible until economies and political systems recovered, and those were tasks quite beyond the capacity of private philanthropy.[19] Hard realities conspired with Vincent's indecision to turn large opportunities into a modest program of relief to European medical schools. Embree was sent to Central Europe to survey larger possibilities in agriculture and general education, but a few weeks in the field convinced him that public health alone was more than the foundation could handle.[20] This pattern of expansion and retreat was repeated in 1922, in connection with a survey of the medical sciences in Germany, Austria, and Poland by Richard Pearce and Alonzo Taylor. The result was a second temporary relief program, and another opportunity to expand into the basic natural sciences in Europe was allowed to slip by.[21]

As a trustee of the RF, Rose was an interested and canny observer of

[15]Embree diary, 18, 23 Dec. 1919; Vincent to Embree, 7 Feb. 1920; both in RF 700 16.121. Pearce to Taylor, 25 May, 14 June 1922, RF 700 16.122. V. Heiser to Rose, 26 Mar. 1920, RF 1.1 45.688.

[16]Embree diary, 20 Oct. 1919, RF. Board minutes, 26 Feb., 3 Dec. 1919, RF 100 56.553.

[17]Embree diary, 12 Mar., 4 May 1920, RF. Board minutes, 24 Mar. 1920, RF 100 59.553. Pearce to Vincent, 17 Sept. 1920, RF 700 16.121. Vincent to Pearce, 29 May 1922; Vincent to S. Flexner, 29 May 1922; both in RF 700 16.122.

[18]Embree diary, 19 May, 20 Oct. 1919, RF.

[19]Embree diary, June–Oct. 1921, p. 1 of Exhibits, RF. Merle Curti, *American Philanthropy Abroad* (New Brunswick, N.J.: Rutgers University Press, 1963), pp. 259–263, 268–278.

[20]Board minutes, 26 May 1920, RF 100 56.553. Embree diary, 4, 7 June 1920, RF. Embree to L. R. Williams, 27 May 1920; Embree to Vincent, 21 July 1920; Embree, "Medical education and nurse training in Central Europe," n.d. [Oct. 1920]; all in RF 700 16.121.

[21]Kohler, "Science and philanthropy: Wickliffe Rose and the International Education Board," *Minerva* 23 (1985): 75–95.

these philanthropic feints and retreats. He was not one of those who were pressing Vincent to expand the foundation's horizons. He did not oppose expansion but consistently argued for sticking to traditional programs in medicine and public health.[22] He was quite willing, however, to administer the foundation's surveys and emergency projects. In December 1919 he agreed to survey Eastern Europe (Vincent thought it would be cheaper than sending Embree and less likely to arouse expectations of grants). Rose recommended against constructive programs for the time being.[23] The two emergency medical grants programs were administered by the IHB staff headquartered in the foundation's Paris office. These routine tasks gave Rose invaluable grass-roots knowledge of science and education in Europe, opened doors to European cultural elites, and expanded his vision of philanthropic possibilities.

An instinctive entrepreneur, Rose noticed opportunities that others let slip by and felt impelled to do himself what others did not care or dare to do. For the Rockefeller Foundation, the years 1916–1921 were replete with lost opportunities, and Rose was always there, watching and thinking about what might be possible for foundations to do.

World War I made science seem a cure-all for the world's ills, and Rose shared this idealistic faith. But the practical specifics of his scheme—the emphasis on mentorship and centers of training, the combination of science and agriculture, the association of science with the reconstruction of national and international culture—these specifics were drawn from Rose's knowledge of Millikan's and Vincent's unfinished business.

A Working Network

Inventing a policy for the IEB was relatively easy; it could be done from behind a desk. Constructing the machinery of an operating foundation was a more complex practical task. Though Rose knew little of the world of natural scientists, he did know from the IHB how the machinery of an operating foundation was constructed, and the IEB was built on this experience.

Rose's first task was to build a network of expert advisors and potential mentors of IEB fellows. Naturally, he turned first to the NRC. In May 1923 he asked Vernon Kellogg to send him names of a few eminent leaders in physics, chemistry, and biology. Working from lists provided

[22]Embree diary, 4 April, 31 Oct. 1922, RF. Fosdick to Rose, 18 July 1922, IEB 1.1 2.37.

[23]Embree diary, 18, 23, 29, 30 Dec. 1919; Rose to Vincent, 1 Feb. 1920; Embree to Rose, 4 Mar. 1920; all in RF 700 16.121.

by the NRC division heads, Rose sent out a further request for names of the most eminent American scientists. By July he had a master list of fifty-four names.[24] Rose also turned for advice to individuals he knew and trusted: T. H. Morgan, chemists Phoebus Levene and Henry Dakin, T. W. Richards (signed up when Richards came to pay a begging visit).[25] Rose had the warmest personal relation with Robert Millikan and through him with Arthur Noyes. Even when his circle of scientific acquaintences was much wider, Rose still turned to a few trusted friends like Morgan and the Caltech group.

The first job Rose set his expert advisors was to get a program of fellowships underway by suggesting mentors who in turn would suggest promising candidates. "We shall not undertake to set up any definite machinery for this purpose at the present time," he wrote to Millikan. "We should prefer to begin more informally, depending upon a few individuals here and there whose judgement we trust and whose contacts are such as to bring to their attention the type of men we have in mind." Informality and discretion were essential, Rose went on, to avoid "arousing undue expectations," and he asked Millikan and Noyes to regard their advisory roles "as betwixt us."[26]

Rose's system of confidental advisors stands in sharp contrast to the elaborate public machinery of the NRC fellowship committees. Rose seems never to have considered delegating the IEB fellowships to committees. The reason, no doubt, was that the IEB fellowships had a dual purpose, which the NRC fellowships did not. In addition to recruiting talents to careers in science, the IEB fellowships were an integral part of Rose's long-term plans for developing institutional centers. Fellowships, he realized, provided occasions for discovering which institutions had the facilities, organizing talent, and resources for research training on a large scale. Vincent had separated fellowships from institution building, precisely so that the selection of fellows could be delegated to the NRC. He was determined that fellowships *not* lead to support of research in particular universities. Rose put back together what Vincent and Millikan had separated. He meant fellowships to lead to bigger things. And because fellowships had implications of institutional support, Rose was obliged to keep control of them within his own office and a small circle of trusted advisors.

Rose also used his advisors to aid his self-education in science. He became an avid reader of popular science books, and Simon Flexner

[24]Kellogg to Rose, 11, 12, 14 May 1923; Rose, "Scientists 1923"; all in IEB 1.1 10.141.

[25]Rose diary, 17 Nov. 1923; Morgan to Rose, 3 Dec. 1923; both in IEB 1.1 10.144. P. Levene to Rose, 3 Oct. 1923; Dakin to Rose, 8 Nov. 1923; Pearce memo, 15 May 1923; Richards to Rose, 30 Oct. 1923; Rose to Richards, 31 Oct. 1923; all in IEB 1.1 10.141.

[26]Rose to Millikan, 9 Oct. 1923; also, 17 Oct. 1923; both in IEB 1.1 10.144.

took pains to provide him with good ones.[27] Rose delighted in playing the role of talented amateur in discussions with his new scientific friends. For example, he wrote Millikan about an idea suggested by Millikan's popular account of his discovery of the so-called *M* rays. With a display of diffidence and apologies about laymen interfering with foolish suggestions, Rose suggested an experiment to test Millikan's theory. Tactfully, Millikan replied that Rose's suggestion, far from foolish, had also been made to him by several eminent scientists. "I was very glad to see," he wrote, "how much of a real scientist you are."[28] (In fact, he had already tried the experiment, with negative results, and *M*-rays turned out to be a will-o'-the-wisp.)

Rose expanded his survey of the scientific world in the fall of 1923, in preparation for his European grand tour. Again he turned to Kellogg, Millikan, and Morgan for lists of eminent European scientists.[29] Rose was also eager to learn more about the intellectual structure of the sciences: What were the major specialities of physics, chemistry, and biology? What lines of research were especially productive? Where were the centers of activity in each specialty? In the next few years, his geographical and intellectual mapping of the sciences became ever more detailed and systematic. The master of the public health campaign set about developing scientific communities in the same methodical way that he had set about eliminating hookworm or yellow fever from rural communities.

From time to time Rose went "afield" himself to meet people, inspect facilities, and get a firsthand sense of the morale and atmosphere of scientific centers. His most extensive trip was his tour of Europe from December 1923 to April 1924.[30] Everywhere he asked questions, checking and cross-checking his information about up-and-coming people, institutions, and research fields. The diaries and correspondence of the Paris office staff were systematically winnowed for information, which was then entered into a card file and into Rose's "little black book" or field guide, which told him at a glance whom he should see and what he should ask about wherever he happened to be.[31] By systematically comparing opinions Rose also learned to evalu-

[27]Rose to S. Flexner, 22 Sept. 1924, IEB 1.1 3.55.

[28]Rose to Millikan, 13 Jan. 1926; Millikan to Rose, 21 Jan. 1926, IEB 1.1 1.11.

[29]Rose to Kellogg, 15, 19 Sept. 1923; Kellogg to Rose, 18 Sept. 1923; "Chemists," 19 Sept. 1923; "Physicists," 22 Sept. 1923; all in IEB 1.1 10.141. Kellogg to Rose, 20 Sept., 3 and 10 Oct. 1923; J. E. Zanetti to Kellogg, 25 Sept. 1923; all in IEB 1.1 10.144.

[30]Rose diary, Dec. 1923 to Apr. 1924, IEB 1.1 3.49–50.

[31]Rose to A. R. Mann, 16 Dec. 1926, IEB 1.1 10.145. Rose's little black book, IEB 1.1 10.142.

ate individuals' judgment about people.[32] He got to know the personalities and pecking orders of half-a-dozen disciplines.

Rose's Lieutenant: Augustus Trowbridge

Although he arranged the first few projects himself (Niels Bohr's, e.g.), Rose always intended to delegate the management of fellowships and institutional projects to a professional staff. His grand vision attracted ambitious and capable people, of whom he was a seasoned picker. The Paris office was free of the malaise and sense of crisis that afflicted the RF's New York office. Rose's lieutenants were recruited from active and even distinguished scientific careers, not from university middle management. At the head of the natural sciences program in Europe, the physicist Augustus Trowbridge was the most able person in the Paris office. Wilbur E. Tisdale, an ex-physicist, was recruited in 1926 from the NRC to take charge of the IEB's fellowship program. As the executive secretary of the NRC's division of physics and chemistry, Tisdale had done the routine administration of its fellowship program. In charge of the program in agriculture were Albert R. Mann, the energetic dean of Cornell's School of Agriculture, and Claude B. Hutchison, a plant geneticist and plant breeder from Cornell and future dean of the University of California's School of Agriculture. Mostly in their early forties, they were an energetic and capable group.

Augustus Trowbridge was fifty-four years old and a distinguished professor of physics when Rose plucked him from the Princeton faculty. He came from a well-to-do New England merchant family, devoutly Christian (two of his sons went into the clergy), and deeply imbued with ideals of public service. Especially appealing to Rose was Trowbridge's deep knowledge of European languages and cultures. He had been sent off at an early age to Italy, where he was taught by a private tutor and became a confirmed Europhile. He spoke French, German, and Italian fluently.[33] Only two years Hale's junior, Trowbridge belonged to a generation for whom science and education could still be as much a mission and a calling as a career.

As a researcher, Trowbridge was accomplished but not brilliant. He acquired an interest in radiation and optics as a graduate student in Berlin, taking the infrared as his special research field. Although he

[32]For example, W. Lund, "Comparative statement embodying the views of G. N. Lewis and . . . Hugh S. Taylor . . . ," 10 Mar. 1927, IEB 1.1 10.142.

[33]Karl T. Compton, "Augustus Trowbridge 1870–1934," *Biog. Mem. Nat. Acad. Sci.* 18 (1937): 219–243, see p. 219–221.

came to Princeton at a time (1906) when the university was remaking itself into a major research center in mathematics and mathematical physics, Trowbridge was and remained an experimentalist. He was never very ambitious or prolific about publishing and liked best to design new apparatus. (He is best known for his part in perfecting the echelette diffraction grating.) His interest in instrumentation also had a practical side, and he worked on electronic amplifiers of telephone signals and other electrical equipment. This practical bent found expression during the war, when he worked on acoustical range finding and organized a field service in France to deploy the new technology for locating artillery fire. Returning to Princeton in 1919, he developed new instruments for measuring vibration of automobile crankshafts and flame speeds in internal combustion engines.[34]

Trowbridge also displayed a talent for academic administration. He had broad cultural interests, social graces, tact, and practical judgment. He was an excellent judge of human character, tolerant of human frailties, and an effective administrator. The war nourished this managerial side of his character, and after 1919 he was drawn further from research into university business. He was interested in student affairs and had a flair for managing building projects, of which there were many after the war. That experience stood him well in his later work for the IEB. He was also active in the NRC, serving on the fellowship board and as vice-chairman and chairman of the Division of Physical Sciences.[35] It was there, no doubt, that he came to Rose's attention.

Trowbridge was an ideal person for Rose: widely known as a physicist but not devoted to research, experienced in laboratory design and construction and devoted to scientific education, an ardent internationalist. Trowbridge responded instantly to Rose's temptations. He and his wife were thoroughly at home in Europe, and he saw the potential for service to science on a scale that he could not hope to have at Princeton. Although Trowbridge felt restive as an academic administrator and hoped someday to get back to research, he knew himself well enough to know that any research he did would suffer from being "continually interfered with by pressing administrative details." The choice was not a difficult one.[36]

Trowbridge and his colleagues enjoyed a good deal of independence in the Paris office—indeed, it appears they would have welcomed a little

[34]Ibid., pp. 222–226, 228–233, and erratum page. Kevles, *The Physicists,* pp. 126–136.

[35]Compton, "Trowbridge," pp. 228, 231, 234–235.

[36]Rose diary, 1, 8 Oct. 1924, IEB 1.1 24.340. Trowbridge to Rose, 2 Nov. 1923; Rose to Trowbridge, 5 Nov. 1923; both in IEB 1.1 10.141. Trowbridge to Rose, 8 Nov. 1927, IEB 1.1 24.346.

more attention from Rose. After his 1923–1924 grand tour, Rose never returned even for a brief visit to the Paris office. When Trowbridge made his periodic trips to New York, he sometimes felt that Rose had more important things to do than to talk to him. He had a free hand to establish procedures and initiate and "mature" projects, though on important issues Rose would let it be known what the policy was. Despite some tensions between Paris and New York, it was an effective working relationship. Trowbridge and his colleagues in the Paris office were the practical inventors of the role of which Warren Weaver became the most famous exemplar in the 1930s.

Fellowships and Surveys

Trowbridge, no less than Rose, needed a quick education in several unfamiliar disciplines, and he soon discovered that the fellowship program was a good way to find out who was who in European science. He spent much of his time in the first year organizing the fellowship program. It was, he reported, "the most time engrossing, but also the easiest and the surest part of my work."[37] There was no problem getting started. During his grand tour, Rose had asked many of his hosts to send him names of promising candidates for fellowships; they did, lots and lots of names, which Trowbridge found waiting to be sorted out when he arrived in Paris. On the American side, Trowbridge could draw on the experience of the NRC fellowship boards.[38] On the European side, he learned by doing. Trowbridge soon made it a rule to personally interview all candidates and pay personal visits to host institutions. Though time consuming, this method gave him valuable opportunities to learn what was really going on at the grass roots, behind the smooth facade of formal applications. Because candidates had to be nominated by their mentors, each application for a fellowship was a chance for Trowbridge to gather information about the people in charge of laboratories and institutes, and about the equipment and support available for research and training in various universities. Information could be elicited without leading informants to think that it was an invitation to ask for a large grant. Discussions of plans for expanding institutes or research programs revealed to Trowbridge opportunities for projects

[37]Trowbridge to Rose, 29 Oct. 1925, IEB 1.1 43.605.

[38]Trowbridge to Rose, 8 June 1926, pp. 2–3; Rose to Trowbridge, 8 July 1926; both in IEB 1.1 24.345. Rose to Kellogg, 21 Nov. 1923, IEB 1.1 10.144. Rose to Lillie, 18 Sept. 1924; Lillie to Rose, 23 Sept. 1924; both in IEB 1.1 9.128.

and at the same time let scientists know indirectly what the IEB was looking for.[39]

There was much paperwork, but it was not a desk job. Trowbridge was not a bureaucrat but a practical fieldworker, often on the road and always concerned with the human side of science. He worried more than Rose did about the careers of ex-IEB fellows. He organized a follow-up system, visiting IEB alumni and noting individuals who seemed likely candidates for leadership positions. (He was frequently asked for advice in appointments and promotions.) It was his idea to establish a small grants-in-aid fund to provide equipment and assistants to former fellows who found themselves temporarily in minor posts but who seemed destined for better ones, if they could just manage to keep a productive line of research going. Trowbridge also used the NRC fellowship boards and their extensive networks as informal employment agencies for IEB fellows returning from Europe.[40]

It was also Trowbridge who began using IEB traveling professors to gather information about special disciplines or fields. The idea was suggested by Harvard mathematician George Birkhoff, who in 1925 was making a tour of European institutes for his own edification and offered to share his observations with the IEB. As his knowledge became more discriminating, Trowbridge became aware of the need for expert advice in disciplines other than physics. He and Rose hoped at first to appoint full-time assistants in the various disciplines, but they quickly discovered that it was not easy to find outstanding scientists, forty to forty-five years old, with administrative flair and a willingness to interrupt research careers.[41] Such people were more than willing, however, to serve the IEB as temporary paid advisors. The idea was to provide expenses and an honorarium to eminent scientists on sabbatical, to enable them to tour the centers of their discipline and report to the IEB on research trends, facilities, and the rise and decline of scientific reputations. Because these quasi-official visitors were not seen as official representatives of the IEB, they could ask questions without causing offense or arousing undue expectations: "[T]hey would simply be eminent American scientists doing something . . . for them[selves] . . . and for American science generally." The system was also a good way to spot promising recruits for foundation work: it gave the IEB the services of the best people and gave academics a chance to see if they had a taste for foundation work. The most notable person thus recruited

[39]Trowbridge to Rose, 29 Sept. 1927, IEB 1.1 24.345.

[40]Trowbridge to Rose, 29 Oct. 1925, 29 Sept. 1927; both in IEB 1.1 43.605 and 43.607.

[41]Rose to T. Lyman, 6 Apr. 1926; Rose to S. W. Stratton, 6 Apr. 1926; both in IEB 1.1 7.109.

was Princeton chemist Lauder Jones, who succeeded Trowbridge in 1929.[42]

The surveys of the sciences grew more and more elaborate in late 1926 and 1927, as data was compiled into elaborate multicolored maps that showed at a glance the location and intensity of achievement in various fields of science across Europe[43] (see illustrations 13 and 14). Surveys had long been the customary first step in public health campaigns and social movements, and the GEB had mapped the local markets of colleges. But Rose and Trowbridge were the first to apply these methods to the natural sciences. It was a remarkable experiment in the geography of science, never tried before on such a scale and, so far as I know, never again attempted even in the era of big science.

The exemplar of the IEB's geography of science was Harlow Shapley's report on European astronomy, which included a detailed classification of research specialties with thumbnail explanations, lists of institutions for each specialty ranked by their capacity for training IEB fellows, and a map of major and minor centers.[44] George Birkhoff produced a similar analysis for mathematics.[45] Rose was enthralled and urged Trowbridge to put his data on physics in Shapley's graphic format. Trowbridge's tables and colored maps of mathematical and experimental physics was received with no less enthusiasm: it was just the systematic mapping of a discipline that Rose most needed.[46] Protozoologist Gary N. Calkins prepared similarly elaborate lists and maps for zoology.[47] Leslie C. Dunn reported on experimental biology in Russia and on genetics.[48] Lauder Jones mapped general and physical chemistry.[49] Others surveyed botany, plant physiology, plant genetics, and other subfields of the agricultural sciences.[50] In geology and paleontology and the more practical fields of forestry, fisheries, and

[42]Officers conference, 30 Nov. 1925; Trowbridge to Rose, 12 Dec. 1925, 14 June 1926; all in IEB 1.1 24.343 and 24.345.

[43]Maps in IEB 1.1 10.143.

[44]Harlow Shapley, "Classification of fields of study and research," 2 Dec. 1926, IEB 1.1 10.141. Shapley, "Brief survey of significant astronomical institutions," 6 Dec. 1926, IEB 1.1 18.265.

[45]IEB 1.1 8.110 and maps in IEB 1.1 10.143.

[46]Rose to Trowbridge, 24 Mar., 4 Apr. 1927; Trowbridge, "List of outstanding European and American workers in the field of physics," sent to Rose 3 May 1927; all in IEB 1.1 10.146. Trowbridge, "Mathematical physics and experimental physics in the United States," 7 Mar. 1927; and maps; all in IEB 1.1 10.143.

[47]G. N. Calkins, "Leading animal biologists of Europe," 7 Apr. 1927; Calkins, "Eminent zoologists of Europe," 30 Aug. 1927; all in IEB 1.1 10.146. Calkins, "Report on zoological survey," 29 Aug. 1927, IEB 1.1 11.165.

[48]L. C. Dunn to Hutchison, 2 Nov. 1927, 29 Mar. 1928, IEB 1.1 11.165–166.

[49]L. W. Jones, "General report chemical research . . . ," 29 June 1927, IEB 1.1 11.165.

[50]Material in IEB 1.1 11.163–166, 17.252, 19.285, 20.299, 23.329–337, 24.339.

oceanography (growing interests for Rose), the IEB delegated surveys to committees of the NRC.[51] By late 1927, Trowbridge and Rose were working on a master map of world centers of science in all the sciences and a comprehensive classification of primary and secondary centers.[52]

These maps were meant to be used in the IEB's now numerous projects of institution building, though it is not clear that they were much used. They seem extravagantly cartographic, done more for aesthetic than for practical reasons. Clearly, Trowbridge and Rose got caught up in the process of survey and mapping and lavished more effort on their maps than was needed to guide field practice. The maps instructed and delighted Rose and made impressive demonstrations for the trustees. But it may be that they are most significant for what they reveal about the aesthetic of organization that Rose brought with him from his public health campaigns. The classifications and maps are emblematic of the 1920s zeal for social reconstruction on the grand scale. In the 1930s the IEB veterans in the Paris office discouraged Warren Weaver from undertaking such surveys. Tisdale later recalled that their academic surveyors gave the Paris office as much embarrassment as sound advice (unfortunately, he did not elaborate). Lauder Jones found the maps of little use because they were quickly outdated by the academic game of musical chairs and by the rapid pace of change of active research fronts.[53]

In practice, the doing or undoing of projects depended on a complex set of practical circumstances: personalities, politics, and finance—especially finance. It took an up-to-date knowledge of currencies, economic conditions, and cultural and educational politics. Trowbridge had to understand half-a-dozen national university systems. He had to know where entrepreneurial know-how was concentrated, and how the competion for resources and choice research problems worked at different levels of university systems. Making projects was an art of the possible, and Trowbridge was an artist of the possible.

A View of European Science

Trowbridge made projects in a dozen of the twenty-odd countries of Europe. In that cultural patchwork there was enormous regional diversity, but it was a diversity that was structured by geography and history.

[51]David White, "Memorandum on the need for studies in paleobiology," 13 Jan. 1926; White, "Main divisions of the geological field," 13 Apr. 1926; both in IEB 1.1 21.307. Forestry in IEB 1.1 14.209–213; fisheries and oceanography in 14.216–15.219, 17.256.

[52]Trowbridge to Rose, 29 Sept. 1927, IEB 1.1 43.607. Map in IEB 1.1 10.143.

[53]Tisdale to Weaver, 25 Mar. 1934, RF 915 1.1. "Conference of trustees and officers," 29 Oct. 1930, pp. 102–106, RF 900 22.167.

A century of industrialization and cultural investment had produced gradients in academic culture, from north to south and east to west. There were aging systems, up-and-coming ones, and nonstarters. The early predominance of Germany in the natural sciences had produced marked differences in the ability of scientists elsewhere to participate in the most productive and fast-paced fields, shutting out some, showing others the way. Smaller and poorer countries, and those whose institutions were less adaptable to cultural change, had to develop strategies for coping with the scientific great powers, usually through investing in a few specialized fields. At the peripheries, lack of entrepreneurial opportunities and experience gave scientists less access to new resources like the IEB. "Make the peaks higher," Rose enjoined, but he had his eye also on undeveloped peripheries. How these conflicting imperatives were translated into practice depended on what was doable, and what Trowbridge did reflected the large-scale structures of national cultures and political economies.

By late 1925, when Trowbridge began in earnest to mature projects, most European countries had gotten over the worst effects of the war. Extremist political movements had been subdued, and liberal governments were in power in many countries, though often embattled and shakey. Most currencies were stabilized or soon would be, and some compensations had been made for the effects of inflation on the standard of living of the professional class. The worst aspects of reparations and the official hostility between France and Germany had been softened.[54] So too in science: the International Research Council was no longer being used to cordon off German scientists. Science lobbies were strengthened by the war in most countries, and funds were available for research from the Medical and Agricultural Research Councils in Britain, the *Notgemeinschaft* in Germany, and several private foundations in Scandinavia. Universities were flooded with students, and their output of research had returned to something like the prewar level.[55]

A long report by Trowbridge in June 1926 gives a fascinating bird's-

[54]Charles S. Maier, *Recasting Bourgeois Europe* (Princeton: Princeton University Press, 1975).

[55]Brigitte Schröder-Gudehus, *Deutsche Wissenschaft und internationale Zusammenarbeit 1914–1928* (D.Sc. thesis, University of Geneva, 1966). Harry W. Paul, *From Knowledge to Power: The Rise of the Science Empire in France, 1860–1939* (Cambridge: Cambridge University Press, 1985). Ludolph Brauer et al., eds., *Forschungsinstitute, ihre Geschichte, Organisation und Ziele* (Hamburg: Paul Hartung Verlag, 1930). Konrad H. Jarausch, ed., *The Transformation of Higher Learning 1860–1930* (Chicago: University of Chicago Press, 1983). Paul Forman, "Scientific internationalism and the Weimar physicists: The ideology and its manipulation in Germany after World War I," *Isis* 64 (1973): 151–180. Forman, "The financial support and political alignment of physicists in Weimar Germany," *Minerva* 12 (1974): 39–66.

eye view of European science at that critical juncture. Much of what Trowbridge saw surprised him. He had expected to find a sharp difference between the war's winners and losers, but that was not the case. He expected to find a correlation between economic prosperity and investment in science. Not so: state support of science correlated most closely with stablized currencies. Social psychology, Trowbridge surmised, was more important than economics. Countries that had not managed to stabilize their currencies were pessimistic about the future and unwilling to invest; in those that had returned to a gold standard, a feeling of confidence and prosperity resulted in larger investment in education and science. The roster of countries with inflationary currencies included all the scientifically "backward" ones in Eastern Europe and Iberia as well as France, Italy, and Belgium. The hyperinflation in Germany and Austria had been especially crippling, but by mid-1926 the worst was over, and Trowbridge sensed a growing confidence in the future. Ironically, the prospects seemed worst in France and Italy. "In spite of outward signs of a fairly general prosperity," Trowbridge observed, "there is a very universal fear of what the future may hold in store and a feeling of the futility of attempting to make long-range plans. In academic circles there is the acceptance of intolerable conditions and a rather hopeless acquiescence in the gradual deterioration of what is still a very admirable structure."[56]

The supply of talent was not the problem, surprisingly. Like most observers of the European scene, Trowbridge believed that inflation had had an especially devastating effect on professionals and civil servants, forcing them out of academia and destroying family resources that before the war would have been used to educate and support a younger generation during the low-paid early years of academic careers.[57] Yet Trowbridge reported no shortage of recruits to academic science. In Switzerland he saw an alarming surplus. The bottleneck, it seemed, was in the material resources to train a new generation of academic scientists.

As Trowbridge saw it, the war and postwar turmoil had amplified the normal cycle of investment and retrenchment. It was normal for university and government officials to put off, whenever possible, maintaining university plant and equipment; uncertainty about currency was an ironclad excuse to do what came naturally. Governments were more cautious about planning and future investment than was warranted by actual economic conditions, Trowbridge thought. Hence the

[56]Trowbridge to Rose, 8 June 1926, pp. 23–30, quoted from p. 26, IEB 1.2 24.345.

[57]Maier argues that laboring classes were the hardest hit (see his *Recasting Bourgeois Europe*, pp. 356–364).

curious gap between economic recovery and official parsimony toward academic science.[58]

Everywhere Trowbridge went he saw evidences of a cycle of growth and retrenchment in investment that waxed with the war and waned in its aftermath. He was struck by the number of new buildings. Obviously, money had been easy to get for science for a few years, and large sums had been spent, mainly on new laboratories and teaching facilities. This was especially apparent in the formerly neutral countries—Sweden, Switzerland, and the Netherlands—and Trowbridge surmised that war profits had been systematically invested in education and science. The same pattern was apparent in England, of all the belligerents the most aggressive in matching Germany's strategic investment in science and science-based industries. In Germany and France, however, Trowbridge was surprised at the relative absence of new construction. Germany had less need, he thought, since it had built extensively in the late nineteenth century (though institutes were showing signs of age). But France had visibly fallen behind in new plant and equipment.[59]

No less striking to Trowbridge were the universal complaints of the shortage of current funds for teaching and research: it was the same in the fine new laboratories of Scandinavia as in the decrepit institutes of Paris and Rome. Funds for upkeep and repair were hard to get; plans for research, expanded in the heady years of patriotic cultural rivalries, were being curtailed. Trowbridge was always being asked for grants to make up for diminished state funds, or to buy long-deferred equipment, or to do researches planned but abandoned for lack of funds. This was true not just in countries with deflated currency but was especially marked in Switzerland, Denmark, and Norway. Only Sweden seemed to have escaped the worst of the fashion for budget cutting.[60]

As Trowbridge saw it, governments had invested in new facilities in the expectation that science would be vital to international economic and cultural competition after the war. (The same belief that made Rose a patron of science.) The economic realities of postwar depression dashed grand dreams of social reconstruction. Governments retrenched, and so too did private benefactors, exhausted by innumerable fund drives during and after the war. Capital investments in facilities for research during and just after the war were not followed by comparable increases in operating funds for salaries and research ex-

[58]Trowbridge to Rose, 8 June 1926, pp. 23–30.

[59]Ibid., pp. 23–30. But see Spencer R. Weart, *Scientists in Power* (Cambridge, Mass.: Harvard University Press, 1979), chs. 1, 2.

[60]Trowbridge to Rose, 8 June 1926, p. 24.

penses.[61] The existence of new laboratories and expanded research programs only made the gap between expectation and reality in the mid-1920s more discouraging to scientists who had witnessed what seemed to be a new era in state support of science.

The IEB's methods were in fact well adapted to these circumstances. Rose never liked spending money on bricks and mortar. He liked to invest in people, and his plan to provide equipment and supplies for training and research was just what European scientists most needed. There was a problem, however: IEB grants required matching funds. Because governments were retreating from expected increases in support, it often appeared to Trowbridge and Rose that they were being asked to shoulder a responsibility that governments were trying to shrug off. It was the old problem of pauperization, and it was compounded in countries with unstable currencies and no tradition of state support for science. The IEB board was hypersensitive to anything that seemed like emergency relief or that might lead to dependency. The first principle of educational foundations was to help institutions build local constituencies. Yet in Europe those most in need were least able to meet IEB requirements.

Trowbridge took a more flexible position on this issue. He made a fundamental distinction between the Northern Protestant countries, mostly clustered around the North Sea, and the countries of the Latin South and West. On the whole, the Northern group were financially and culturally better able to engage in joint projects with the IEB. They had a strong political tradition that the state should provide higher education for all who could benefit from it; their economies were sound, their currencies stable. None of those favorable circumstances obtained in the Latin countries. Nevertheless, Trowbridge argued, the IEB's greatest opportunities lay there. France and Italy were large and populous and thus had a far greater potential than the smaller countries to sustain science and higher education. Their cultural institutions were also more venerable. As Trowbridge expressed it, Italy and France had been "the standard bearers of civilization, when the other group was only emerging from a state of barbarism."[62]

Trowbridge also observed that scientific centers in the smaller countries tended to be outstanding in one or two fields, in contrast to centers in the great capital cities, which could aspire to eminence in all disciplines. Many of the small centers had been created for outstanding individuals, often after they had won a Nobel prize and were able to shame or bully government officials into paying for new institutes.

[61] Ibid., pp. 23–30.
[62] Ibid., pp. 29–30, 45–47.

These were specialized, one-man institutes, cultural showcases for small countries that had found this an affordable way to compete with the scientific great powers on their southern borders. This strategy worked, but only for specialized fields in which small countries had some local comparative advantage, human or material. Trowbridge doubted that small countries, with their limited human and economic resources, ever would be able to develop comprehensive centers of science. Depending on single individuals, rather than a broad-based educational system, special institutes were fragile.[63]

Even such a preeminent center as Niels Bohr's Institute for Theoretical Physics seemed vulnerable. Trowbridge felt that the Danish government had about reached the limits of public tolerance of subsidies to foreign visitors, especially as times grew hard. He worried too that the whole show, despite its fame, could move overnight: "[I]f only two or three individuals were to accept calls to some larger country," he warned Rose, "the University of Copenhagen would probably revert to its former position of international insignificance in the pure sciences."[64] Trowbridge saw the same limitations in Norway, Sweden, the Netherlands, and Switzerland.[65] It was less dangerous, he thought, to invest in the larger and more populous countries, with their cultural institutions concentrated in capital cities like Paris or Rome.

Trowbridge's perceptions were obviously colored by his deep love of Latin history and culture. He admired the French system of grandes écoles, centralized in Paris, supported by the state, and based on a system of secondary schools that fed in talent and provided secure jobs for graduates. That system, he argued, had produced "a goodly number of geniuses in the pure sciences" as well as an admirably well-trained and highly motivated corps of teachers. Although advancement was achieved by seniority, rather than competition for prestigious chairs (the German method), the French style of bureaucracy had not eroded standards, Trowbridge thought. He applauded efforts by Italians to draw scientific talents from the provincial universities, with the idea of making Rome a world center of science comparable to Paris.[66]

Trowbridge hoped that the appeal of creating new centers in the small Northern countries would not cause Rose to overlook the need to preserve older centers. "[T]here will be more to develop in the Northern group and more to save in the Southern group," he wrote, "and . . . just now in the generally disorganized state of scientific education and

[63]Trowbridge to Rose, 8 June 1926, pp. 30–35.
[64]Ibid., p. 34.
[65]Ibid., pp. 31–35.
[66]Ibid., pp. 39–42, 45–51.

research in Europe the two are of somewhere nearly equal importance for world civilization."[67] Trowbridge was well aware that grants in France and Italy were likely to be emergency aid, and thus not appeal to the trustees. But he stuck to his conviction that "there is more to be saved in these countries by timely and (I hope) temporary emergency aid than there is by assistance in forward-looking plans in most of the countries of Western Europe."[68] (In 1928 he was still trying to persuade Rose of the value of emergency aid.)[69]

In fact, the process of making projects put a premium on short-term advantages over long-term promise. Since the IEB was in operation only for five years, projects that got off to a quick start were the ones that made it through the process. Local entrepreneurs had to know how to deal with an American foundation (or be quick studies), and they had to have ready access to matching funds and a program that fit IEB aims.

In Britain, cultural affinities and the experience of scientific cooperation during the war undoubtedly made it more likely that projects would come to term. Trowbridge was struck by the aggressively confident, "almost greedy" manner in which English scientists approached the IEB. (In contrast, the French maintained their pride in a studied indifference that won Trowbridge's respect.) Trowbridge thought the British saw IEB funds as a quid pro quo for honored war debts. More likely, it was confidence born of ample experience of working with the Rockefeller Foundation in medical and public health projects. (It is no accident that the most aggressive British were the London biomedical elite.) Likewise, German and French mathematicians and physicists undoubtedly benefited from the active trans-Atlantic networks of exchange in these disciplines in the 1920s, whereas the strong nineteenth-century biological networks were long gone.[70]

An Overview of IEB Projects

The distribution and timing of IEB projects reveal several distinctive patterns (table 6.1).[71] Note, for example, the number of grants (almost all early ones) for international infrastructure and communications: the International Bureau of Weights and Measures (standard constants), American Chemical Society (abstracts), the *Academia Lincei*

[67]Ibid., pp. 29–30.

[68]Ibid., p. 46.

[69]Trowbridge to Rose, 27 Mar. 1928, IEB 1.2 26.367.

[70]Hugh Hawkins, "Transatlantic discipleship: Two American biologists and their German mentor," *Isis* 71 (1980): 197–210.

[71]"Summary of appropriations by IEB as of December 31, 1930," IEB 1.1 6.92.

TABLE 6-1
INTERNATIONAL EDUCATION BOARD APPROPRIATIONS IN NATURAL SCIENCES (IN $1,000S)

Institution	*Field*	*1923*	*1924*	*1925*	*1926*	*1927*	*1928*
Copenhagen	Physics	45					
Copenhagen	Physiol.	107					
Göttingen	Physics		8				
Bureau of Weights and Measures	Physics		4				37
Naples Station	Biology		55				
Leiden	Physics		70				10
Harvard	Astron.			180			
Nat. Institute, Madrid	Physics, chem.			20	420		
Utrecht	Physics			3			
Radium Institute, Vienna	Physics, med.			9			5
Plymouth Station	Biology			5			
Edinburgh	Genetics				150		
Edinburgh	Zoology				362		
Tromsö Station	Geophys.				75		
Göttingen	Math., physics				350		
Paris	Math., physics				119		198
Paris	Chem.				10		
Utrecht	Physiol.					4	
Copenhagen	Chem.					132	
Botanical Garden, Geneva	Botany					27	
London	Zoology					586[a]	
Uppsala	Physics					12	
Stockholm	Chem.						70
Caltech Palomar	Astron.						6,000
Cambridge	Nat. Sci.						3,500

(*continued*)

TABLE 6-1 (*Continued*)

Institution	*Field*	*1923*	*1924*	*1925*	*1926*	*1927*	*1928*
Harvard	Biology						3,000
Lyons Observatory	Astron.						4
Jardin des Plantes, Paris	Botany						200
Virginia	Nat. Sci.						175
Jungfrauhoch	Geophys.						38
Uppsala	Chem.						50
Fellowships			108	175	182	170	139
Publication			35	51	15	50	5
Visiting profs.			17	11	6	2	2
Totals		152	297	454	1,689	1,053	13,363

[a]withdrawn and revised, 1931.

(publication). International centers for research were favored: the famous Naples Zoological Station and its smaller French and British imitators, geophysical stations like the one at Tromsö in Norway for studies of aurora, astronomical observatories (the great 200-inch telescope at Mt. Palomar was the IEB's most expensive project), and botanical gardens and herbaria. Such projects predominated in the first few years: they were easily identified, had no competitors, and exemplified the internationalism of science. Making grants there did not require a great deal of knowledge or experience and thus were a low-risk way to get started while Trowbridge was learning how national systems of universities worked.

Projects in universities did not begin in earnest until the May 1926 meeting of the board, three years after Rose's program was officially approved. Several patterns catch the eye: the large number of medium-sized grants in the smaller countries around the North Sea; the small number of projects in Germany (Göttingen was the exception), compared to three in Britain and a cluster in Paris. Madrid was the only major project on the undeveloped periphery of science. Finally, there is the late cluster of large planned projects (Cambridge, London, Harvard) pushed through as time was running out on the IEB.

These patterns will be analyzed in detail in the next chapter, but a few general observations may be in order here about policy and practice. Clearly, geography was an important practical element in who got what: the small Northern countries around the North Sea did much better than the large Southern countries, for example. Institutional size, too, was a vital factor: as a group, provincial universities did rather better than the large cosmopolitan universities. And timing: when in the IEB's life cycle a project was begun could spell the difference between success and failure. Early on, middling projects had the advantage; toward the end, it was the grand ones; always, delay was fatal, even for the best projects.

The distribution of projects reveals the limits of official or unofficial policy. Trowbridge's enthusiasm for the Latin cultures of Europe proved less important than the cultural and financial backing of higher education in the small Protestant countries of the North. No matter how much Rose and Trowbridge wanted to uplift science and education in the undeveloped regions of Eastern and Southern Europe, the lack of financial and cultural support for academic high culture in those regions made it difficult for them to do very much. Rose's belief in comprehensive, elaborately planned schemes proved a less practicable guide in the field than Trowbridge's practice of doing what was doable.

Given Rose's policy of making peaks higher, we might have expected more projects in Germany, especially in Berlin, the prewar Mecca of science. Trowbridge and Rose would have liked to do more in Germany, I think, but practical circumstances intervened. The University of Berlin had declined from its former preeminence; Trowbridge remarked on how different a place it was in 1926 from the place that had inspired him thirty years before. Politics, too, were an impediment, both German politics of cultural *Machtersatz* (making science and culture serve as a substitute for lost political and diplomatic power), and French politics of *revanche.* The indirect effects of political isolation made it difficult to do projects in Germany and, especially, in Berlin. Foreign visitors stayed away, and Berlin scientists seemed uninterested in luring them back. They seemed not to know about the IEB's programs; Max von Laue seemed indifferent when asked if he would take IEB fellows. Also, German universities were losing out in the competition for research funds to the growing system of Kaiser Wilhelm institutes. It was in these showcases of German scientific culture, not universities, that the most able researchers and students congregated in the 1920s, and that made it difficult for the IEB, with its explicitly educational mission, to make projects anywhere in Germany.[72] Government priorities, currency in-

[72]Paul Weindling, "The Rockefeller Foundation and German biomedical sciences, 1920–1940: From educational philanthropy to international science policy," in Nicolaas A. Rupke, ed., *Science, Politics and the Public Good* (London: Macmillan, 1988), pp. 119–

flation, and political isolation all conspired against German scientists in the competition for American patronage.

Conclusion

Before moving on to what Trowbridge actually did in Europe, let us review how he got there, and how the IEB and GEB, almost overnight, became the world's greatest private patrons of the natural sciences.

The creation of the IEB and the reconversion of the GEB from general education to science completed the new system of science patronage. Although this happened four years after the Rockefeller Foundation and Carnegie Corporation grants to the NRC (the timing was contingent upon Wallace Buttrick's retirement), these two pivotal episodes were integrally connected. Rose, no less than Vincent, Angell, and Pritchett, was responding to the warborn enthusiasm for science and to the postwar movement for institutional investment and reconstruction. More important than ideology are the specific connections between Rose's vision and the RF's and CC's unfinished business. As a ringside observer of the negotiations between Millikan and Vincent, Rose was well placed to see potential opportunities not being seized—not just vague schemes but specific, concretely realized programs of institution building. It was undoubtedly that knowledge that inspired Rose to put his practical experience as a field operator to work for science rather than for some other equally worthy educational cause, when the opportunity arose in 1923.

Rose's programs of institutional endowment complemented the CC's and RF's underwriting of research training fellowships and the communication and lobbying services of the science councils. Together, these programs constituted a distinctive system of patronage, different from the individual grants-in-aid of the prewar era and the project grants of the 1930s. It was an accident of philanthropic politics that the elements of the system were divided among half-a-dozen foundations rather than one or two. In 1919 a division of labor between the RF (general services) and the CC (institutional endowments) might have seemed most likely. But Vincent's obsession with holding the line on new programs and Angell's sudden departure from the CC altered the balance against radical change. Had Pritchett and Keppel adopted Angell's program of institution building, or had Vincent bought Millikan and Noyes's vision of a system of regional university centers, there would

140. Germany was also losing out to foreign competition in some crucial areas of science, like experimental biology and physical chemistry. Trowbridge to Rose, 8 June 1926, pp. 35–38. IEB 1.1 24.345. Trowbridge diary, 6–7 Oct. 1925, IEB 1.2 34.482.

have been little left for Rose to do for science. But had the CC and RF not already created key parts of the system of patronage, would Rose have seen the other part so clearly as an opportunity? It is unlikely.

The practical problems of a divided, cobbled-together system of patronage became clear by the end of the 1920s. In the interim, however, many millions flowed from the IEB and GEB into the natural sciences in the United States and Europe—more millions, perhaps, than would have been possible in a more rationally constructed system. We need to turn now to that side of the story and to the question of who got what, and why.

Chapter Seven

Making the Peaks Higher: European Science

WHO GOT WHAT AND WHY? To answer these questions we need to turn from foundation politics to field practice. It was practical things like university politics and finance and entrepreneurial know-how that largely determined whose proposals made it through the grant-making process and whose did not. Each project was the result of complex negotiations between local actors and the officers of the IEB. Projects were learning experiences for both parties. They were occasions to redefine the relationship between grant-giver and grant-getter, since some accommodation was always required to get the parties onto the same track. To understand how the system of institutional patronage worked, we need to take an ecological approach and examine how, in particular cases, the agendas of patron and client fit, and how each party accommodated to the other in order to get something done. We need to set projects in their particular local contexts to see what elements of scientists' local situations were most vital to the process of negotiation.

Although every project was a little different from every other one, location in the European system of universities was the key variable in most. Location is a complex variable: it has geographical, cultural, and economic elements (North South, Protestant Catholic, rich poor). Each country and region of Europe has its distinctive politics of high culture, and some were more accessible to American patrons than others. Institutional size and prestige was a second element in location, cutting across national lines. It was quite a different thing making projects in specialized institutes in provincial universities, and in the large, sprawling institutions of capital cities.

It was location, mainly, that shaped individual and institutional strategies for keeping up or getting ahead in university systems. And it was the competitive strategies of grant-seekers, more than anything else, that determined the fit with Rose and Trowbridge's strategy of developing university systems by making their peaks higher. We will see how this system of patronage worked by examining projects in three distinctive locations: the provincial universities of Scandinavia, Germany, and the Netherlands; the great cosmopolitan centers (London, Cambridge, and Paris); and Spain, on the outer margin of scientific high culture. In each case we will need first to show the connection between location and a distinctive strategy of scientific development; then to show how local strategies fit (or did not fit) the IEB's process of making projects.

In the small countries around the North Sea, academic entrepreneurs enjoyed the benefits of societies and governments that put a high value on university education and research. At the same time (as Trowbridge observed), the small size and resource base of these countries limited expansion and left universities vulnerable to raids from the scientific superpowers, most notably Germany. The problem was, How could small provincial universities compete, with their limited resources, in the newest and most productive lines of research? How could native sons who made good be kept from migrating to the more exciting cosmopolitan centers of continental science? The solution, as we have seen, was to create specialized institutes for these exceptional individuals, often outside the system of official chairs and institutes. Governments subsidized these institutes as showpieces of national scientific culture—visible displays that even small countries could participate in international science. Although provincial universities could never hope to compete with the large cosmopolitan centers in every field of science, they could compete in particular lines by means of these specialized showpiece institutes.

Trowbridge encountered such institutes in the smaller German states, Scandinavia, the Netherlands, Switzerland, and Scotland, and they proved to be attractive and practicable opportunities for making projects. Limited and provincial they might be, but these specialized institutes also had real practical advantages for the IEB: strong individual leadership, proven capacity to turn out good research, committed financial support, and a demonstrated international drawing power—an especially important thing for the IEB's system of mentorship and research training. Such practical advantages often outweighed the risks of overspecialization and vulnerability in institutes that depended ultimately on single individuals. Stockholm, Leiden, or Copenhagen could never rival Paris or London in depth and diversity, but making projects there was usually more straightforward and doable, and many got done before the IEB closed its doors in 1928.

The universities of the large capital cities adopted a different competitive strategy, which attempted to capitalize on their size and diversity. Like most institutions they had grown by seizing opportunities in fields that for one reason or another could attract funds. The result of half a century of accretion was a congeries of institutes, some new and productive, some aging and sclerotic. Faced with growing competition from smaller institutions in the early 1920s, the cosmopolitan universities sensed that the greatest advantage could be achieved at least cost by investing in large-scale organization and rationalization. They could aspire to be outstanding in many fields and had the resources to plan. In traveling the European circuit, Trowbridge found that many cosmopolitan universities were trying to unite scattered institutes and laboratories into coherent groups, which provincial rivals could not hope to emulate.

These cosmopolitan universities presented Trowbridge with a particular mix of advantages and drawbacks. They were more secure investments in the long-term than the one-man showpieces: particular institutes might wax and wane, but the reputation of the university did not stand or fall with any one. They were better places for apprentices to get a broad experience in scientific research. On the other hand, large universities had larger problems of factionalism and internal rivalries. There were more parties to accommodate at the bargaining table, and it was hard enough to get cooperation from people in a single discipline, let alone half a dozen. A more intimate knowledge of local politics was required for comprehensive projects. Usually Trowbridge had to delegate responsibility for planning to someone who had the knowledge and authority to make hard decisions about who got what. Such people were rare, and often plans matured too slowly or were spoiled by too many cooks who could not agree on a recipe.

The most serious problem with comprehensive projects, in Trowbridge's view, was the necessity of including weak departments. The policy of aiding only the already strong ("make the peaks higher") was a practical way of limiting risk. People with a track record were likely to share the IEB's standards of quality. Not so those in charge of weaker departments: for them progress meant losing authority to a younger generation or to competitors in more fashionable or productive lines of research. Self-interest lay in taking the IEB's money while trying to do as little as possible to change the status quo. Trowbridge was reluctant to risk investing in weak departments, because he never quite figured out how to deal with the uncertainties. For all these reasons, the record of big projects in cosmopolitan centers was at best mixed.

There was nothing but risk for Trowbridge in the undeveloped regions of Southern and Central Europe, where there was no tradition of state support for higher education, unstable currencies and volatile ethnic and cultural politics, and rudimentary infrastructures for recruiting and employing academic scientists. Institutions were simply not in a position to approach the IEB, and for individual scientists in a state of chronic poverty, the best strategy was to do research that did not depend on resources or access to the latest news from the centers of European science. It was a conservative strategy, avoiding competition in active fields and avoiding risky innovation. Some individuals did manage to keep productive, but their survival strategy made it almost impossible for Trowbridge to help them. Relief was what they needed, Trowbridge realized, but the IEB was not in the business of relief. Madrid was the IEB's only major experiment on the periphery.

That is how the ecology of science patronage worked in interwar Europe: different locations in the system fostered different competitive strategies; the fit of strategies to the IEB's system of making projects (or lack of fit) determined who got what. Let us turn now to some particulars.

The Northern Paradigm: Copenhagen and Göttingen

Niels Bohr's Institute of Theoretical Physics at Copenhagen exemplifies the competitive strategy of the small Northern countries. A precocious Nobelist, Bohr used his fame to get a special chair in 1916 and a new institute in 1920. The most striking thing about Bohr's institute in the 1920s was its internationalism. Located in a modest provincial university, it had been made into the center of an international network of theoretical atomic physicists, a favorite resort for foreigners who came to get the latest news, absorb the special blend of experiment and theory that was its hallmark, and bask in the warmth of Bohr's charisma. The Danish government supported Bohr's institute as a showcase of national culture, even though many of those who benefited were foreigners. As economic conditions deteriorated in the early 1920s, however, it became harder to get money from Parliament "for an institution of purely international scientific importance." National needs came first in hard times.[1]

The cosmopolitanism of Bohr's institute and its combination of re-

[1]Finn Aaserud, *Redirecting Science: Niels Bohr, Philanthropy, and the Rise of Nuclear Physics* (Cambridge: Cambridge University Press, 1989). C. Lundsgaard to A. Flexner, 6 Apr. 1923; Bohr to IEB, 27 June 1923; both in IEB 1.2 28.403. Thomas Schøtt, "Fundamental research in a small country: Mathematics in Denmark 1928–1977," *Minerva* 18 (1980): 243–283. I am grateful to Finn Aaserud for showing me chapters of his book before publication.

search and mentorship made it a nearly ideal starting point for Rose. Eager to have a part in Bohr's success, Rose gave the Danish government easy financial terms (he asked only for a modest increase in annual support), and the project sailed through the process of planning and approval.[2] Funds were provided for an addition to the institute's building, plus equipment for a new line of research on atomic spectra.

Rose was no less eager to have a hand in the mathematics and physics institutes at the University of Göttingen. Like Bohr's institute, these too stood out in postwar Europe as international centers. It was not just the roster of stars (David Hilbert, Richard Courant, and Lev Landau in mathematics, James Franck, Robert Pohl, and Max Born in physics), but the unusual ability of both groups to work together and to integrate the theoretical and experimental aspects of atomic physics. That was what set the Göttingen school apart from other German institutes and attracted droves of foreign scientists. For Rose, Göttingen was a model of the international centers he hoped to develop throughout Europe.[3] The Göttingen group had relatively little trouble getting local funds. For the Prussian government they were one of the few reminders left of Germany's once formidable cultural power. A small grant (to upgrade a building project already underway) was all that the physicists needed at the time, but Rose hoped for more from the mathematicians, who had larger plans in hand.[4]

Before World War I Felix Klein had conceived the idea of a laboratory for mathematicians, with research rooms around a central library—an idea borrowed consciously from the experimental sciences. Bringing mathematicians out of their isolated studies and into a public space, Klein hoped, would have the same vitalizing effect on mathematical practice that laboratories had had on physics a generation earlier. The new institute would be in close proximity to the physics institute, and there would be a laboratory of applied mathematics where physical models could be constructed and studied.[5] War and inflation had put Klein's plan on hold, but it was again coming to life in the mid-1920s, just in time for the IEB. It was Niels Bohr and his mathematician broth-

[2]Rose to C. Lundsgaard, 22 May 1923; Bohr to IEB, 27 June 1923 enclosing H. M. Hansen, "The Institute for Theoretical Physics," n.d.; Lundsgaard to Rose, 31 July 1923; Rose to Bohr, 21 Nov. 1923; all in IEB 1.2 28.403.

[3]Trowbridge diary, 2–3 July 1926, IEB 1.2 34.485. Lewis Pyenson, "Mathematics, education, and the Göttingen approach to physical reality, 1890–1914," *Europa* 2 (1979): 91–127. On the isolation of German institutes, see Mann to Rose, 6 Jan. 1924, IEB 1.2 34.482.

[4]Born, Franck, and Pohl to Rose, 18 Mar., 21 July 1924; Franck and Pohl to IEB, 24 Mar. 1925; Trowbridge diary, 8 Oct. 1925, p. 56, IEB 1.2 34.484. Rose to Trowbridge, 7 Apr. 1926, IEB 1.2 34.485.

[5]Trowbridge diary, 2–3 July 1926, IEB 1.2 34.485.

er, Harald, who told the Göttingen group to try the IEB when negotiations with city officials for a contested building site seemed to have stalled.[6]

According to James Franck, it was the IEB's fellowships and grant that revived the project.[7] More precisely, it was the entrepreneurial skills of the Göttingen group, and of Trowbridge, no tyro in building projects. The Göttingen project was one of the first in which Trowbridge revealed his talents as a manager of science. Richard Courant and James Franck were the chief negotiators on the university side, and they were skilled negotiators. Trowbridge characterized them as "the persuasive type, very tactful (non-argumentative)," practical and business-like, and experienced in academic politics. In contrast to Landau, who was aggressive and abrasive in his dealings with Trowbridge (Franck and Courant felt obliged to apologize for him), the two principles were "all tact."[8] They knew exactly what they wanted and what the IEB was likely to give. Always well rehearsed for Trowbridge's visits, they were adroit in setting the stage for negotiations with pleasant, informal social gatherings for Trowbridge and his wife.

Trowbridge was an old hand at that game himself, of course, and thoroughly enjoyed gambits well played. His administrative role at Princeton had given him much experience in designing laboratories, approaching patrons, and negotiating among the various interests involved in such a project. He was also thoroughly familiar and sympathetic with the Göttingen style: it had in fact been the model for the reconstruction of the mathematics and physics departments at Princeton, in which Trowbridge had taken a leading part.[9] He was no less an active participant in the Göttingen project than Courant and Franck.

Trowbridge sounded a discouraging note at first, to dispel any illusion that the IEB would be willing to give the mathematicians fine new quarters just because they were world famous. Without mentioning names, he intimated that other universities had made plans that were better prepared and financed than those of the Göttingen group. He apparently felt that the Göttingen group had been too diffident in putting their needs before university officials and knew that IEB funds were a powerful incentive for officials to loosen the purse strings. Trowbridge was assisted in this little drama by his mathematical advisor, George Birkhoff who, disclaiming any official role in the IEB, inquired with studied innocence what steps were planned to regain the

[6]Ibid., p. 4.
[7]Trowbridge diary, 8 Oct. 1925, p. 56, IEB 1.2 34.484.
[8]Trowbridge diary, 2–3 July 1926.
[9]Trowbridge diary, 8 Oct. 1925, pp. 56–57.

dominant position that the institute had enjoyed in its prewar heyday. Birkhoff perceived a certain air of complacency in the group and meant to startle them out of it.[10]

Everyone wanted a deal, of course, and one was easily reached. Although Trowbridge would have preferred not to pay for a new building, he agreed when Courant and Franck pointed out that it was almost impossible to get building funds from the state (it meant getting two rival ministries to agree; additional operating funds were easier to get because they were the sole perogative of one). The university proved more forthcoming than the scientists had expected, and the prospect of IEB money quickly dispelled conflicts among the principles that had brought earlier efforts to a halt. Construction was soon underway at a site adjacent to the physics and chemistry institutes.[11]

The Northern Group: Variations on a Theme

Where circumstances were right, the Northern strategy of specialization and international outreach was just about ideal for IEB projects, and the Copenhagen and Göttingen schools of mathematical physics were just unusually successful examples of a style widespread in Northeast Europe, especially in Scandinavia. There were limits to the strategy of specialized showpiece institutes, however, and the IEB's projects in Scandinavia and the Netherlands clearly display these limits.

There were the economic limitations of small countries that Trowbridge noted, and the limited political tolerance of subsidizing foreigners from larger, wealthier countries. But there were also problems internal to the universities and the scientific disciplines. Many local heros made their reputations by perfecting specialized laboratory instruments or by exploiting opportunities in fields that were not in the mainstream of their disciplines. A relatively narrow achievement, together with restrictions on the number of professorial chairs, made it difficult to engineer new chairs in regular university institutes. If institute heads succeeded in blocking intrusion on their disciplinary turf—as often happened— personal positions had to be improvised outside the regular academic institutes, or even outside the science faculty. Thus, it was not uncommon for Nobelists with international reputations to be found in marginal positions, working in makeshift quarters with minimal resources, and cut off from the more reliable resources that came with teaching large lecture courses. The system created such situations.

[10]Trowbridge diary, 2–3 July 1926, pp. 6–8.

[11]Ibid., pp. 11–12. Trowbridge to Rose, 19 Oct. 1926, IEB 1.2 34.485.

Trowbridge often found it difficult to know how far the IEB should get involved with these specialized laboratories. On the one hand, they were good places for IEB fellows to learn specialized instruments and techniques, though not for a rounded training. They produced exact experimental data essential for theoreticians, but often the experimentalists did not themselves do theory. Isolation tended to encourage narrow specialization. Some of these laboratories seemed to Trowbridge to have the potential of becoming "secondary centers," but their marginal positions and irregular support made it hard for him to measure that potential. Rose and Trowbridge were wary of getting involved with individuals who were too specialized or with institutions that served national rather than international constituencies.

Uppsala physicist Manne Siegbahn was such a case. Siegbahn had received the Nobel Prize in 1924 for his work on the quantitative measurement of atomic spectra, though the professional gossip was that he was not up to the caliber of other Nobelists. The spectroscopic data he produced was exact and reliable and happened to be just what theoretical atomic physicists most needed in the mid-1920s. But Siegbahn was a somewhat old-fashioned experimentalist, a perfectionist who had no interest in combining experiment and theory, as Bohr and others were. Foreign physicists came to be instructed in experimental technique, not to absorb a new style of physics. Siegbahn was interested in nothing but X-ray spectra: ". . . a one-field laboratory decidedly," Trowbridge noted. Nor did Uppsala promise to become a center of science for Scandinavia generally.[12] Without much enthusiasm, Trowbridge recommended a grant for extra equipment, to enable Siegbahn to stay competitive with the strong American schools of quantitative spectroscopy.

Similar but rather more promising was Heike Kamerlingh Onnes's laboratory for low temperature physics at the University of Leiden. It was a highly specialized and highly productive line of research. The equipment was very complex and costly. (Multiple copies of experimental setups had to be maintained to make efficient use of the large and expensive refrigerating machinery.) Since not many universities could afford to compete, low-temperature physics was a field in which a small but scientifically developed country like the Netherlands could corner the international market. Onnes himself (a Nobelist in 1913) and his specialized facility attracted a polyglot crowd of students, who were supported by ample stipends from a government anxious to have an international showcase in a prestigious and competitive field of physics.

[12]Trowbridge to Rose, 16 May 1925, 17 Mar. 1927, 13 Apr. 1928; Siegbahn to Rose, 23 Feb. 1925; Trowbridge to Rose, 20 Apr. 1925; Trowbridge diary, 2 Oct. 1925; all in IEB 1.2 41.587.

In this case, Trowbridge felt, a rather large request for additional equipment ($70,000) seemed justified.[13]

The case of Leiden also illustrates the inherent vulnerability of "one-field" experimental laboratories. Onnes died soon after the IEB's grant was made, and his successor, W. N. Keesom, apparently lacked his ability to inspire outstanding work. The complexity of the laboratory technology and the emphasis on exact measurement also encouraged a narrow outlook. In 1926, Trowbridge reported that the laboratory attracted "the plodder, who wishes to obtain routine data with the exceptional facilities offered by the Lab."[14] As with Siegbahn, the data produced by Keesom's group was very much in demand by theoretical physicists and chemists, but the laboratory was not an inspiring place, not a good place for IEB fellows. Also, Trowbridge suspected that the Dutch government's subsidies for foreign visitors attracted applications from mediocre scientists who chose Leiden not because it was the best place but because it was a paying place.[15] Trowbridge preferred the competitive mechanism of the IEB's fellowship program.

Jurisdictional rivalries between disciplines also complicated Trowbridge's work. These are especially apparent in projects in biology and chemistry, disciplines more factionalized than physics. At Copenhagen, for example, the physical chemist Johannes Brønsted had a laboratory in the Technical College, where he enjoyed the unusual privilege of being a research professor with no routine teaching duties. A privileged position at the margin was one solution to the problem of a productive scientist who was too specialized for a chair and institute but overqualified for the faculty of the Technical College. A side effect of this arrangement, however, was that neither the university nor the college regarded Brønsted as its responsibility, and years of effort had failed to improve his wretched quarters—a few small basement rooms crowded with foreign workers.[16]

The Svedberg had a similar situation at Uppsala. Awarded the Nobel Prize in 1926 for his work on colloid chemistry, Svedberg was the university's most distinguished chemist but too specialized for the chair of chemistry. He occupied a personal chair in physical chemistry (the first in Sweden) created for him in 1912. In the early 1920s, Svedberg had

[13]W. N. Keesom to Rose, 22 Apr. 1924; H. Kamerlingh Onnes to Rose, 22 July 1924; both in IEB 1.2 35.498.

[14]Trowbridge to Rose, 29 May 1928, IEB 1.2 36.499.

[15]Ibid. Tisdale diary, 20–21 Oct. 1927; Trowbridge to W. W. Brierley, 28 Oct. 1927; Trowbridge diary, 17 Jan. 1928; all in IEB 1.2 36.499.

[16]Trowbridge diary, 25 Sept. 1925; Trowbridge to Rose, 25 Mar. 1926; Jones diary, 25–28 June 1926; Brønsted to Rose, 22 Jan. 1927; all in IEB 1.2 28.406.

embarked on an ambitious program to develop instruments for studying the size-distribution of colloids. By 1925, however, he had become fascinated by the most novel of these instruments, and the most challenging from an engineering standpoint, the ultrahigh-speed centrifuge. When Trowbridge arrived at Uppsala in 1926, Svedberg was working almost exclusively on making the ultracentrifuge into a tool for exact measurement of molecular weights of proteins.[17] Svedberg's fruitful diversion from mainstream physical chemistry was possible in part because he was insulated from the routines of the chemistry institute. He was also insulated, however, from access to resources. In 1924, for example, he shocked the professor of chemistry by asking for a new laboratory. It took a Nobel Prize to get the Swedish parliament to come up with the funds.[18]

Biology also had many such cases. Bohr's colleague, the physiologist August Krogh, was another Nobel laureate in an uncertain position. His laboratory was in the science faculty, separate from both zoology and the powerful physiological institute in the medical school. (Although Krogh worked on problems that were relevant to medicine—respiration, capillaries, osmotic processes—he approached them as a biologist, not a medical physiologist.) Like Brønsted and Svedberg, he was free to devote himself to research, but he was cut off from the power and resources that came with medical teaching. His laboratory was antiquated, crowded, ill-equipped, and underfunded. Krogh's unusual approach to physiology brought foreign visitors to his laboratory, but disciplinary politics made it difficult to create a proper space for him among existing institutes.[19]

It is easy to see why Trowbridge and Rose had trouble deciding whether or not to support eminent individuals in marginal positions. They had to judge if a specialist was likely to expand his horizons and if his laboratory had the potential to do more than work a highly specialized technique. Rose displayed none of the enthusiasm for Krogh that he did for Bohr. He and Trowbridge agonized over Brønsted, even though they agreed he was one of the five or six best physical chemists in Europe. Trowbridge was cool to Hans von Euler's project for an institute of enzymology at Stockholm's Tekniska Högskola, because it appeared to him that von Euler was more interested in bettering his

[17]S. Claesson and K. O. Pederson, "The Svedberg," *Biog. Mem. Fell. Roy. Soc.* 18 (1972): 595–627. Arne Tiselius and S. Claesson, "The Svedberg and fifty years of physical chemistry in Sweden," *Ann Rev. Phys. Chem.* 18 (1967): 1–8.

[18]Claesson and Pederson, "Svedberg," p. 600.

[19]August Krogh to Vincent, 16 Apr. 1923; Krogh to Pearce, 23 June 1923; Pearce to Rose, 2 Mar. 1924; all in IEB 1.2 28.404.

own circumstances than in building a permanent institution.[20] And Trowbridge was not at all certain about the ultracentrifuge. He was not impressed by Svedberg's work on his first visit to Uppsala, noting that the work on sedimenting proteins was routine, a misjudgment of the first order. He even forgot Svedberg's name.[21]

Krogh's project was one of the first to be approved, in part because Richard Pearce and Flexner wanted to make it a part of their plan to aid the University of Copenhagen's medical school.[22] With Svedberg, Trowbridge learned from experience. A stream of IEB fellows, a succession of ever more powerful, beautifully engineered instruments, and world applause for the research on proteins made Trowbridge an enthusiastic advocate of Svedberg's work. In 1928 he made sure that Svedberg got an application in before the IEB went out of business, even though Svedberg himself, preoccupied with planning his new institute, seemed to have little sense of urgency.[23]

Trowbridge took an active personal role in all these projects. The files are full of wish lists for laboratory equipment, which he had heavily red-penciled. (His experience as an experimentalist served him well.) He grew more confident in his judgments of people in disciplines new to him, and found the mid-sized projects of the Northern provincial universities eminently doable. Rose, meanwhile, was turning from these limited projects to his dream of great cosmopolitan centers for all the sciences.

Planning Science: Paris versus New York

Large projects in cosmopolitan universities (Rose's "primary centers") provided some of the rare occasions when Rose and Trowbridge differed fundamentally over policy. More than most projects, these looked quite different from the perspectives of Paris and New York, from the field and from the president's office.

Rose was usually the one pushing to make small projects into big ones. He had always favored international over national projects. As he put it to his staff in 1925, the IEB had "science in Europe as its major

[20]Trowbridge diary, 26 Apr. 1927; Rose to Trowbridge, 18 Apr. 1927; Trowbridge to Rose, 26 Apr., 13 June, 22 Sept., 27 Oct. 1927; Hans von Euler to Trowbridge, 20 July 1927; Trowbridge to von Euler, 1 Sept. 1927; all in IEB 1.2 42.588.

[21]Trowbridge diary, 2 Oct. 1925, IEB 1.2 41.587.

[22]Rose diary, 16 May, 2 Oct. 1923; Pearce to Krogh, 17 May 1923; Pearce to Vincent, 11 July 1923; Rose to Krogh, 25 Mar. 1924; all in IEB 1.2 28.403–404.

[23]Trowbridge diary, 2 Oct. 1925; Trowbridge to Rose, 22 Sept. 1927; both in IEB 1.2 41.587–42.588. Trowbridge to Svedberg, 12 June 1928; Svedberg to IEB, 17 June 1928; Trowbridge to Arnett, 7 July 1928; all in IEB 1.2 42.590.

interest, and as its minor interest, that Italy, etc. might develop biology, chemistry, etc. for its own needs."[24] In 1927–1928, however, Rose began to lean hard on the Paris office to think big. Politics fueled Rose's new sense of urgency. Millions remained of the large endowment that Rockefeller, Jr., had made to the IEB in 1926, and Rose wanted to spend it fast, in large lumps. Rose was approaching retirement age and was aware of the plans being laid for consolidating the Rockefeller boards. He wanted to make sure that the IEB's programs would live on, in whatever institution emerged from the reorganization. He wanted to leave behind a few grand showpiece projects as demonstrations of the IEB's methods and as precedents for future programs. Trowbridge was less alert to the necessity of spending money fast while the IEB was still in business, perhaps because Rose liked to keep high politics to himself and did not keep Trowbridge informed about the impending reorganization.

Rose was also emboldened by his conviction that he had discovered a solution to the problem of weak departments: namely, to make sure that people from strong disciplines were in charge of deciding what the weaker ones needed. Committees of all the natural sciences would be created to survey departments' strengths and weaknesses and to formulate plans for long-term development of whole groups. Survey and planning committees, Rose believed, would enable experienced and capable scientists to educate the rest about modern scientific practice. By paying for surveys and holding out the reward of a large grant, the IEB could encourage everyone to put the common good ahead of parochial disciplinary interests. It was a quintessentially American vision, and in fact embodied Rose's satisfaction with an early cooperative project at Princeton (see Chap. 8). Rose saw no reason why the method should not work just as well in Europe.[25]

Rose's enthusiasm for large planned projects took Trowbridge somewhat by surprise. He was not so sure that Rose's grand comprehensive schemes were a good idea. Policy was not the issue so much as practicality: Trowbridge always preferred to do what was doable. He tended to be more sympathetic than Rose was with local and national aspirations—a result, no doubt, of being exposed daily to the arguments of local aspirants. Field experience also provided regular reminders of the difficulties of comprehensive planning in universities that had grown chaotically out of the efforts of headstrong individuals. Trowbridge knew about the factionalism and the difficulty of finding someone who

[24]Officers conference, 30 Nov. 1925; Trowbridge to Rose, 12 Dec. 1925; both in IEB 1.1 24.343. Trowbridge to Rose, 29 Oct. 1925; IEB 1.1 43.605.

[25]Trowbridge diary, 23 Feb. 1928, IEB 1.2 34.479. Rose to Trowbridge, 11 Jan., 21 May 1928; Trowbridge to Rose, 1 May 1928; all in IEB 1.2 235.491.

could rise above it. He knew about the problems of weak departments. Unconstrained by the practical difficulties that Trowbridge faced everyday, Rose could dream freely of a system of cosmopolitan centers. Trowbridge could not. Trowbridge was not reassured by Rose's optimistic assertions that planning committees would solve the problem of weak departments. He had firsthand experience of the Princeton project and felt that cooperation had worked there because it had been spontaneous. And he doubted that the American style of planned cooperation would work in the very different context of European academic politics. Field experience had taught him to get on with projects that he knew would work and not risk everything for a larger plan.

Rose did not agree: in his view, small projects were worth doing only if they were a step toward bigger things, and he took every opportunity to try to scale up. Usually it worked like this: Trowbridge or Claude Hutchison would work up a proposal in a discipline X at University Y and send it on to Rose. Rose would reply, expressing eager interest in University Y as a center and directing the Paris office to send him complete information on all the science departments and the prospects of their cooperating in a comprehensive plan. For Rose, a proposal from any important university was an occasion for exploring an overall plan.[26] Sometimes Trowbridge was infected by Rose's enthusiasm; other times he tactfully tried to get Rose to settle for what could be done then and there. By 1927–1928 it had become a New York–Paris ritual.

The issue first arose in connection with Brønsted's proposal for a new institute of physical chemistry. Trowbridge was discouraging at first, on the grounds that Brønsted was not quite the superstar that Bohr was. So he was surprised when Rose took a keen interest in the project, as a possible step toward turning the University of Copenhagen into a center of science in Scandinavia. Rose instructed Trowbridge to assess Brønsted's proposal not just on its own merits but "in terms of its scientific setting." Agreeing that it would be hard to justify an institute just for Brønsted, Rose argued that the strength of the ensemble at Copenhagen—Bohr, Brønsted, Krogh—was the decisive factor, even if chemistry and biology were not up to the exalted standard of physics.[27]

Trowbridge acquiesced in Rose's wish but with some misgivings. He feared that if the IEB started trying to make second-rank into first-rank institutions, it would find it very hard to know where to stop. There were lots of *A*-minus institutes in Europe, and it was all too easy for ambitious academics to magnify their claims by pointing to the achieve-

[26]Rose to Trowbridge, 11 Jan., 21 May 1928, IEB 1.2 35.491.

[27]Trowbridge to Rose, 6 Apr. 1927; Rose to Trowbridge, 7 Apr. 1926, 3, 14 Mar., 18 Apr. 1927; Rose diary, 21 Apr. 1927; all in IEB 1.2 28.406.

ments of neighboring departments. Trowbridge preferred to think of the grant to Bohr as a grant to a brilliant individual and to tell Brønsted that the IEB did not support enterprises that had not already proved themselves by attracting local support. (That strategy left the door open but put the responsibility clearly on the university.)[28] Trowbridge's tactics were shaped by the practical demands on a field manager: they eased the task of making hard choices and bearing bad news.

The small Northern countries were not the best places to build large centers, Rose and Trowbridge agreed. Copenhagen was unlikely to become a center of science for all of Scandinavia, since each country was bound to want its own national center. So, too, with Uppsala and Stockholm.[29] In the Netherlands, ancient provincial rivalries had produced three feuding universities (Leiden, Amsterdam, and Utrecht), each trying to outdo the others and become the national center.[30] Consequently, no one institution had sufficient concentration of scientific talent to be a major scientific center. The larger provincial universities of Germany had more appeal, and activity there did finally pick up. At Munich, for example, Trowbridge had previously discouraged a request from the zoologist Karl von Frisch, despite his outstanding research in animal behavior. In early 1928, Rose asked Trowbridge to reassess von Frisch's plan in the larger context of making Munich a regional center, but time ran out before anything could be done.[31] Göttingen was another possibility for rounding out a group of departments, and Rose also had his eye on Zürich when the IEB went out of business.[32]

But these were sideshows. The big opportunities in Rose's view were in Paris, London, and Cambridge, where movements were already underway to exploit the advantages of size and eminence by systematic planning and rationalization. Such movements seemed to fall in perfectly with Rose's strategy of making small projects into big ones.

Paris and London

The IEB's projects at Paris began early and never quite became a coherent whole. Four substantial grants were made: a building for the

[28]Trowbridge to Rose, 6 Apr. 1927, IEB 1.2 28.406.

[29]Rose memo, 21 Apr. 1927; Trowbridge memo, 7–8 Dec. 1926; both in IEB 1.2.

[30]Trowbridge to Rose, 8 June 1926, pp. 34–35, IEB 1.1 24.345.

[31]Trowbridge diary, 5 July 1926, 4 May 1928; Trowbridge to K. von Frisch, 16 Sept. 1926, 28 June, 16 July 1928; Lindstrom diary, 15 Nov. 1927; Trowbridge to von Frisch, 3 Jan., 1 Feb. 1928; Rose to Trowbridge, 11 Jan., 31 May 1928; all in IEB 1.2 35.491.

[32]Rose to Trowbridge, 21 May 1928, IEB 1.2 35.491.

mathematics institute, endowment of a new chair in mathematical physics, equipment for the physical chemist Jean Perrin (to cover a shortfall in his new institute), and general support for the Jardin des Plantes. Smaller grants were made to the library of the Societé de Biologie, the major library for the university's biologists, and to the two French marine biological stations, which were operated by the university. Extensive negotiations were carried on with the astronomical observatory—the most bizzare of all the IEB's projects, involving a charming but quite untrustworthy Indian gentleman, a Mr. Dina, and his wealthy and sensible American wife. Support for the institute of geophysics was discussed but abandoned when the university authorities showed no interest. Thus, almost all the natural sciences were supported or considered for support. A comprehensive, multimillion dollar project in biology was being discussed when time ran out in 1928.[33]

The complex shape of the IEB's projects at Paris reflect the complex structure of the university itself: a congeries of quasi-independent fiefdoms, jealously defending their prerogatives and competing for resources. The variety of cultural institutions in Paris had always afforded unusual opportunities for improvisation, especially in the natural sciences. Bitter opposition by the humanistic and literary disciplines to the expansion of the science departments in the late nineteenth century made it easier to create new institutions extramurally than to fight vested interests within the university. As a result, scientific departments were scattered in different schools, often in makeshift quarters and with minimal operating budgets.[34] Similar patterns prevailed at other large urban universities, but Paris was extreme even by the standards of Harvard, London, or Cambridge.

It was precisely that lack of central organization that challenged and inspired Trowbridge and Rose. From the start they saw each grant to the University of Paris as part of a comprehensive development of all the natural sciences. As early as 1926 Rose instructed Trowbridge to regard all proposals from Paris as "being parts of an important scientific center." Not just the strong departments: Rose was excited by the challenge of building up the weaker departments and by bringing "scattered forces together."[35]

[33]Trowbridge to S. Gunn, 14 Nov. 1928, for a summary of IEB actions, IEB 1.2 34.471. For the Dina story, see IEB 1.2 33.469.

[34]George Weisz, *The Emergence of Modern Universities in France, 1863–1914* (Princeton: Princeton University Press, 1983). Terry Shinn, "The French science faculty system, 1808–1914: Institutional change and research potential," *Hist. Stud. Phys. Sci.* 10 (1979): 271–332. Trowbridge diary, 21 Sept. 1926, IEB 1.2 34.473.

[35]Rose to Trowbridge, 2 Oct. and 23 June 1926, IEB 1.2 234.473 and 233.470.

They were not alone. Efforts had been going on for years and especially since World War I to consolidate the scattered scientific departments into a cluster of new buildings on the Rue Pierre Curie. On the scientific side, the chief activists were the physicists and chemists of the powerful and well-connected Curie circle. Not surprisingly, they had also been the chief beneficiaries. Chemistry, physical chemistry, and the radium institute had got fine new buildings, and buildings for mathematics and physics were being planned when Trowbridge arrived on the scene in 1925.[36]

Rose was thrilled by Trowbridge's reports of the new science campus—evidence, he felt, that the French "appreciate the importance of scientific cooperation." It seemed an ideal chance to realize his dream of a great international science center. More realistic about French individualism, Trowbridge advised Rose not to try to impose American ideals of cooperation but to let the French work out organizations that suited their cultural traditions.[37] They agreed, however, that it was best to begin with the strong and unproblematic sciences and work up the more difficult weaker ones. The mathematicians were the first and easiest, their prestige and international draw matched only by the Göttingen school. A grant for a new building for mathematics was quickly negotiated. Easy terms regarding matching funds got around the problem of a currency that was not yet stabilized.[38]

Planning and negotiation were greatly facilitated by Emile Borel, whom Trowbridge knew from the war (Borel had also been in charge of artillery range finding). Borel was an experienced and adept politician of science, a kind of hybrid of Robert Millikan and Elihu Root. A core member of the Curie circle, he had served as a deputy in the new left–liberal government in 1924 ("the Republic of Professors"), from which vantage he had helped lead the movement to increase state support of science. He was instrumental in framing a law that set aside for research a small addition to the apprenticeship tax—a crucial turn in the politics of science. He was, as Trowbridge put it, "distinctly of the organizing type." Borel was the perfect counterpart for Trowbridge: capable and smooth, experienced in institution building, well connected politically, and thus in the know about the government's plans to stabilize the cur-

[36]Weart, *Scientists in Power,* ch. 2. Trowbridge diary, 21 Sept. 1926; Trowbridge to Rose, 19 Oct. 1926; both in IEB 1.2 34.473. Trowbridge diary, 28 Dec. 1927, JEB 1.2 33.470. Trowbridge to Rose, 31 Jan. 1928, IEB 1.2 34.471.

[37]Rose to Trowbridge, 2 Oct. 1926, IEB 1.2 34.473. Trowbridge to Rose, 8 June 1926, IEB 1.1 24.345.

[38]Trowbridge memos, 29 May 1926, 28 Dec. 1928; Trowbridge diary 12 Oct. 1926; Brierley to Trowbridge 23 Dec. 1926; Trowbridge to Rose, 7 Jan. 1927; all in IEB 1.2 22.470.

rency and adjust university salaries. Without that inside knowledge, Trowbridge would have been hard put even to ascertain the value of matching funds.[39]

The 1926 grant to the mathematics institute was the first step of a plan to bring mathematics and physics closer together. The Parisian style of mathematics was rigorously abstract (what Borel thought of as applied mathematics—rational mechanics and statistics—would not have been considered "applied" anywhere but in France). Paris had no chair of mathematical physics and no one qualified for such a chair. The physics department was weak and in need of rejuvenation, and the mathematicians were not interested in cooperation. It is not surprising that endowment of a chair in mathematical physics, the second step in Borel and Trowbridge's plan, was not consummated until 1928, despite Trowbridge's keen personal interest.[40] The first incumbent of the chair was Emile Borel.

When it came to the weaker departments, Trowbridge was much more cautious. In late 1927, for example, he told the authorities that the IEB would only aid outstanding departments. Rose, however, gently but firmly reminded him that he was supposed to be developing all the sciences at Paris and that the strong group of mathematicians was an "admirable environment for the development of a real Department of Physics." Rose let it be known that he would welcome a proposal from physics and from biology, though the authorities had not raised that possibility.[41]

Biology presented the greatest difficulties for Rose's plan. the main problem was an ancient rivalry between the university and the Jardin des Plantes, dating from the founding of the science faculty in the 1880s. (Trowbridge could sense the lingering resentments.) Although they were neighbors on the Left Bank, the two institutions represented quite different scientific cultures. The jardin was a research organization, established to develop science at a time when the university refused to do so. Since the creation of the Faculty of Sciences, however, the jardin had become a sinecure for aging natural historians, who gave occasional public lectures and devoted themselves to preserving the vast accumulation of specimens in the herbarium and natural history museum. Combining the jardin's facilities and the university's teaching departments could make Paris a world center in biology. But despite

[39]Trowbridge, "Memorandum of conference," 29 May 1926, IEB 1.2 33.470. Weart, *Scientists in Power,* pp. 15, 24–26. "Émile Borel," *Dict. Sci. Biog.*, vol. 2.

[40]Trowbridge, "Memorandum of conference." Trowbridge to Rose, 31 Jan., 8 June 1928; Rose to Trowbridge, 31 Jan. 1928; all in IEB 1.2 34.471.

[41]Trowbridge diary, 28 Dec. 1927; Rose to Trowbridge, 12 Jan. 1928; both in IEB 1.2 33.470–34.471.

high-level political support for consolidation, rivalry between the two groups made cooperation virtually impossible. Informal negotiations began in mid-1927, but after a frustrating year no one expected anything good to come of them.[42]

It was a situation, Trowbridge realized, in which an outside mediator (especially one with money to spend) might break the deadlock. His task was complicated, however, by the director of the jardin, Louis Mangin, a political naïf who was as different as could be from the savvy and experienced Borel. Mangin wanted the IEB to rebuild the run-down and impoverished herbarium, and he could think of nothing else. Trowbridge was not interested in museums or systematic botany. He tried to open Mangin's eyes to the benefits of cooperation with the experimental biologists, but Mangin seemed incapable of taking Trowbridge's cues. He seemed never to know how to take the next step. Eventually, the impasse was resolved at the top, between the rector of the university and the minister of education. (The IEB agreed to contribute half the funding for a new herbarium, on the understanding that it was a first step toward a systematic reorganization of biology in the two institutions.)[43]

Rose's ambitions for Paris did not stop there, and he wrote Trowbridge to think big, perhaps as big as $5 million. With some misgivings, Trowbridge approached the Dean of Sciences Charles Maurain with the idea of a comprehensive survey of the sciences in Paris and a master plan for development in the American style. To Trowbridge's surprise, Maurin did not resent the intrusion of American ideas and promised to take the matter up with the Minister, an old chum from the Ecole Normale.[44] Trowbridge felt that nothing stood in the way of the IEB (or RF, the IEB would expire in a few months) assisting in the buildings of a central biology laboratory.[45] But it did not come to pass. With Trowbridge no longer there to guide the negotiations, the IEB's diverse grants in Paris remained fragments of what Rose had hoped would become an integral whole.

The University of London had all the same problems as Paris and more. An umbrella for dozens of diverse institutions (King's College, Imperial College, and University College were the largest), it was fragmented by obscure and ancient feuds that a weak central administra-

[42]Trowbridge diary, 27 Oct. 1925, and memo, 29 Oct. 1925; Trowbridge to Rose, 28 Mar. 1928; all in IEB 1.2 34.479.

[43]Trowbridge diary, 26 Apr., 30 Sept., 7 Nov. 1927; Trowbridge to Rose, 16 Oct. 1926, 4 Apr. 1927, 28 Mar. 1928, p. 8; all in IEB 1.2 34.479.

[44]Rose to Trowbridge, 9 Feb. 1928, IEB 1.2 34.471. Trowbridge diary, 23 Feb. 1928, IEB 1.2 234.479.

[45]Trowbridge to Gunn, 14 Nov. 1928, IEB 1.2 34.471.

tion could do nothing to suppress.[46] The biological sciences were especially divided and contentious. And there was in London no dominant elite like the Curie circle, who might at least have imposed their vision on the rest. The biomedical scientists of University College had experience with the RF but seemed unable to use that experience effectively with the IEB.

One of the periodic efforts to unite the stronger colleges into one university was in progress when Trowbridge arrived on the scene. The plan was to acquire a large tract of land in Bloomsbury and to build on it a great university, into which the separate colleges would be merged. The Rockefeller Foundation had pledged $410 million toward purchase of the land, and Rose, too, was very interested in the project, though Trowbridge was "somewhat appalled at [the] intricacy of [the] situation." With good reason: although the plan resembled the one then taking shape on the Rue Pierre Curie, the problems of merging distinct institutions probably doomed the London project from the start. When King's College refused to be merged, the plan was abandoned.[47] When it became clear that real consolidation was impossible, the IEB focused its attention on University College. But even on that more local scale, formulation of a comprehensive project was hampered by too many players and too many plans, and there was no Emile Borel to cut the Gordian knot.

The central figure in the negotiation was the professor of zoology, D. M. S. Watson; the initiative, however, came from the group of powerful professors in the medical sciences led by the anatomist Grafton Elliot-Smith. These well-connected and seasoned entrepreneurs had led the earlier reorganization of the medical faculty, toward which the RF had made large contributions.[48] They appealed to the IEB to complete that task by bringing comparative anatomy and zoology into the newly built central laboratory. Elliot-Smith laid Siege to Trowbridge with an aggressive assurance that Trowbridge found decidedly irritating. Lured to a conference without being forewarned, he was lectured (and misinformed) about the RF's generosity, then ignored when he repeatedly advised against asking the IEB to bear the full cost of a new building.[49] Previous investment by the RF did make a project more attractive to

[46]Rose and Trowbridge memo, 13 June 1927, IEB 1.2 29.419.

[47]Trowbridge diary, 28 Dec. 1926, 2–3 Jan. 1927; Rose to Fosdick, 6 Jan. 1927; Vincent to Rose, 6 May 1927; Embree to Fosdick, 7 Dec. 1926; all in IEB 1.2 29.419–420. On the Bloomsbury project, see RF 401A 32.408–412.

[48]Donald Fisher, "The impact of American foundations on the development of British university education" (Ph.D. diss., University of California, 1972). Fisher, "The Rockefeller Foundation and the development of scientific medicine in Great Britain," *Minerva* 16 (1978): 20–41.

[49]Trowbridge diary, 15 and 26 Apr. 1926, IEB 1.2 30.432.

Rose. On the other hand, the case for zoology was not strong, and for Trowbridge that was crucial.

In 1926 Trowbridge commissioned Frank Lillie to investigate the prospects for a joint project in biology at the three London colleges. What Lillie found was not encouraging. Each department tried to cover the whole of zoology with a small staff, all of which were badly housed and equipped. None of the senior professors, including Watson, were outstanding, in Lillie's opinion. There was no chance of consolidation and few signs of cooperation. Lillie recommended that the IEB start by trying to develop experimental zoology at University College, where anatomy and physiology at least were already strong. Rose agreed. Trowbridge did not: he thought the Edinburgh zoologists were a better group, and he did not relish the prospect of making the headstrong medical scientists understand that the IEB had its own ideas about making projects.[50]

Watson's lack of entrepreneurial experience was also a problem. Though sympathetic to modern experimental biology, he himself was a paleontologist. (His medical patrons protested that Watson was a "real zoologist," but Lillie thought him "more paleontologist than zoologist.") Lacking an insider's vision of experimental biology and inexperienced in devising programs, Watson tended to base his plans on his personal interests rather than on a larger vision of biology as a whole. Trowbridge tried to give Watson some basic instruction in grant-getting, but Watson continued to insist on his need for a building (no doubt his medical advisors were urging him on). With help from Gary Calkins, however, a scheme for zoology and comparative anatomy was finally worked out.[51]

The real problems, however, were not programmatic but financial. Provost Gregory Foster was reluctant to undertake more than had been originally planned for the Department of Zoology in their fund drive. To adopt the IEB's comprehensive plan, Foster would have had to get approval from the university senate, a body made up of jealous rival factions who were bound to make trouble if given half a chance. Also, Foster had been unable to raise any money for zoology in the past, and he did not expect to do better in the future. He had not even managed to make good the university's promise of matching funds for the RF's

[50]Lillie diary, 8–23 Mar. 1926; Rose to Trowbridge, 5 Apr. 1926; Trowbridge to Rose, 14 Sept. 1926; all in IEB 1.2 30.432. Trowbridge diary, 2–3 Jan. 1927; Rose to Trowbridge, 13 Jan. 1927; both in IEB 1.2 29.419.

[51]D. M. S. Watson to IEB, 30 Aug. 1926; F. R. Lillie memos, 21 Mar., 21 Oct. 1926; Lillie to Rose, 1 Nov. 1926; Trowbridge diary, 26 Apr. 1926; all in IEB 1.2 30.432. D. M. S. Watson "The policy of the Zoological Department," 7 Feb. 1927; IEB 1.2 30.433. Trowbridge diary, 2–3 Jan. 1927; IEB 1.2 29.419.

grant to anatomy and physiology. Tempted by the prospect of IEB funds, Foster reluctantly agreed to undertake the IEB's plan.[52]

Trowbridge also worried that the IEB might tempt University College to take on more than it could handle, and did his best to restrain others from still more ambitious schemes. (For example, he headed off a plan to unite zoology with anthropology and psychology, the brain child of Elliot-Smith and Edwin Embree, who hoped to make it a center of a program in human biology.)[53] Trowbridge also stood fast when, late in the game, Rose asked for information on the physical science departments at University College, with an eye to a comprehensive plan for all the sciences. Chemistry was excellent and well funded, Trowbridge protested; physics was weak and had no definite ideas about how to improve. To try to make University College a comprehensive center for the sciences, in his opinion, would only jeopardize the zoology project. Rose did not press the matter.[54]

Even then, the IEB's plan for zoology was too ambitious for University College's limited financial base. Foster found no benefactors for zoology, and Rose's vision of a great center of science in London was not realized even in part when Trowbridge retired.

Cambridge

With the Cambridge project it was a different story: it was Rose's masterpiece. It evolved as usual from small beginnings: a minor project in agricultural entomology, in which Rose perceived the opening gambit of a grand scheme for the natural and agricultural sciences. At once he instructed Albert Mann to put the entomology project on hold until Trowbridge could survey all the science departments at Cambridge.[55] Trowbridge and Mann were less than pleased. They had done all the work and even got the minister of agriculture to come up with matching funds. The bird was in the hand, even if it was a smallish one.

Trowbridge doubted that the American style of cooperative plan-

[52]Trowbridge to Foster, 9, 16 Mar. 1927; Foster to Trowbridge, 24 Mar., 1927; Trowbridge and Gunn to Vincent and Rose, 15 Apr. 1927; Watson to Trowbridge, 5 Apr. 1927; all in IEB 1.2 30.433–434.

[53]G. Elliot-Smith et al., "Proposal to make fuller provision for research in experimental psychology and anthropology," Jan. 1927; Foster to Trowbridge, 12 Mar. 1927; Trowbridge diary, 2–3 Jan., 15 Apr. 1927, p. 9; all in IEB 1.2 30.433–434. On Embree's project, see RF 401A 33.418.

[54]Rose to Trowbridge, 7 Mar. 1927; Trowbridge diary, 5 Apr. 1927; Watson to Trowbridge, 5 Apr. 1927; all in IEB 1.2 30.433–434.

[55]Rose to Mann, 2 Nov. 1925, 8 June 1926; Trowbridge diary, 19 Apr. 1926; Trowbridge to Rose, 3 May 1926; all in IEB 1.2 30.436.

ning would work at Cambridge. Ernest Rutherford's galaxy of physicists and mathematicians were already well funded and had never shown the slightest interest in the IEB (much to Trowbridge's regret). The Cambridge biologists were not outstanding (Lillie rated them below those at Edinburgh and London) and were pursuing with some success their own quite different strategy of building connections to agriculture rather than to physics and chemistry. The chemists were the natural intermediaries between biology and physics, Trowbridge thought, but he saw no sign that they wished to play such a role.[56] Like chemistry departments everywhere, they seemed amply funded by industry and indifferent to American foundations. In the absence of any spontaneous movement for cooperation, Trowbridge was loath to intercede with an idea that was alien to local custom.[57] Trowbridge agreed to investigate further, but reluctantly.

What he and his advisors found was even worse than he had expected. Everyone he talked to remarked on the extreme individualism of the Cambridge scientists and their antipathy to collective action. The Balkanization of science was especially advanced in biology and agriculture. Surveying entomology, Royal Chapman reported an intense rivalry between the zoology department and the School of Agriculture. Even worse was the conflict over protozoology between zoology and the independent Molteno Institute.[58] Zoologists with whom Gary Calkins talked blamed Stanley Gardiner, the professor of zoology, while Gardiner blamed the anatomists, botanists, geneticists, and physicists for building up walls around themselves. Calkins saw parochialism on all sides and concluded that nothing could change until Gardiner retired.[59] He had similar stories to tell of other departments of the School of Agriculture, where senior professors like the geneticist Reginald Punnet ran their laboratories like little fiefdoms and put more energy into defending their perimeters than in cultivating the ground they had.[60]

Most of the people Trowbridge talked to thought that the biologists' territoriality was a matter of personality. No doubt, but such pervasive patterns of behavior must also have been structured by the manner in which that generation of experimental biologists acquired their privileged enclaves. For departments of zoology and botany, the new

[56]Trowbridge to Rose, 12 July 1926; Mann to Rose, 6 May, 29 June 1926; all in IEB 1.2 30.436.

[57]Trowbridge to Rose, 12 July 1926.

[58]R. Chapman to Hutchison, 15 Dec. 1926, IEB 1.2 31.436.

[59]Calkins report, 10–15 Jan. 1927, p. 6, IEB 1.2 31.436.

[60]Hutchison to Rose, 21 Mar., 18 Oct. 1927, IEB 1.2 31.437. Calkins report, 10–15 Jan. 1927, IEB 1.2 31.436.

experimental specialties of biochemistry, genetics, and plant physiology were too important to be allowed to go their separate ways, but not important enough to make space for them in the mainstream of plant and animal morphology. Cut off from departmental resources, the newer specialties tended to be developed in enclaves outside departments on soft money from private benefactors or the government's research councils. There, separation nourished individualism and faction. Zoologists were angry that important fields had slipped from their control; those who had escaped, cut off from access to the resources of teaching departments, nursed bitter memories of indifference and neglect.

This experience was especially intense for the generation who came of age in the 1900s and 1910s, and as a result selfishness and territoriality were especially marked among scientists at the top of the pecking order circa 1925. Hutchison, Trowbridge, and their advisors all remarked how different the younger generation was in personality and behavior: more open to cooperation and less given to defensive individualism. In every field there were bright, productive researchers, eager to cooperate: M. S. Pease (Punnett's chief assistant); Frederick F. Blackman's assistant, G. E. Briggs; and the comparative physiologist James Gray, who impressed everyone as the obvious person to succeed Stanley Gardiner as professor of zoology. Here were people, Trowbridge thought, who could make a large cooperative project work; unfortunately, they were more or less oppressed by the senior professors and were powerless to plan or negotiate with the IEB.[61]

The only place at Cambridge where individualism did not rule was the Dunn Institute of Biochemistry. Under the direction of Frederick G. Hopkins, it was the center of an extended interdisciplinary family that included protozoologist David Keilin from the Molteno Institute, nutritional biochemist Thomas B. Wood and his group in the Field Nutritional Laboratory, Blackmann's group of plant physiologists in the School of Agriculture, and others. As for the biologists, the biochemist J. B. S. Haldane told Calkins there was "not even a dream of a closer cooperation in the future."[62]

Rose seemed blithely untroubled by these tales of fragmentation and dissension. But then he knew some things that Trowbridge did not, for example, that Whitney Shepardson's long-awaited report on agricultural education was a brief for concentrating on the basic sciences

[61]Hutchison to Rose, 18 Oct., 21 Nov. 1927; Hutchison diary, 28 Oct. 1927, p. 73; all in IEB 1.2 31.438.

[62]Ibid. R. E. Kohler, "Walter Fletcher, F. G. Hopkins, and the Dunn Institute of Biochemistry: A case study in the patronage of science," *Isis* 69 (1978): 331–355. Kohler, *From Medical Chemistry to Biochemistry* (Cambridge: Cambridge University Press, 1982), ch. 4.

underlying agriculture. This program would shortly be put before the IEB and GEB boards, and Rose was confident that it would be approved. He was also aware, as the Paris office was not, that Cambridge University had asked Rockefeller, Jr., for $500,000 for a new central library, and that Rockefeller, Rose, and Fosdick were inclined to make the library a part of a master grant from the IEB for all the sciences at Cambridge.[63] Rose never shared Trowbridge's fear that the IEB might not know what was best for Europeans. So while Trowbridge awaited signs of a spontaneous greening of the cooperative spirit, Rose prepared to force the issue with American ideas of organization and a very large infusion of American dollars. He sent Whitney Shepardson to Paris as his personal emissary and set aside $15,000 to pay for a full-scale survey of Cambridge science.[64]

The question was, Who at Cambridge had the wisdom and clout to make a plan and get all parties to agree to it? The right person had to be found fast: word was getting around that something big was in the air. With half-a-dozen IEB people asking pointed questions, it could hardly be kept secret, and rival schemes were beginning to appear. T. B. Wood and Stanley Gardiner, with Hutchison's help, had already drawn up a proposal for the School of Agriculture, emphasizing nutrition, agronomy, and farm management, local demand for which was growing rapidly at the time. Rose told Hutchison to defer Wood's proposal until the machinery was ready for the big project.[65]

The Hopkins circle also got wind of what was happening, learning of it from Wood and William B. Hardy, Hopkins's friend and colleague. Hardy was a natural communicator, well connected in British scientific and political circles, and he knew far better than Wood and Gardiner how to approach American philanthropists.[66] In October 1927, Hardy and Wood gave Hutchison an elaborate account of the needs of all the biological departments. Not surprisingly, Hardy put biochemistry and biophysics at the center of his plan, connected on the medical side to neurology, radiology, and immunology, and on the other side to the agricultural sciences. The most urgent need, Hardy thought, was a

[63]Hutchison to Rose, 18 Oct. 1927; Rose to Hutchison, 27 Oct., 6 Dec. 1927; Rose to Trowbridge, 11 Jan. 1928; all in IEB 1.2 31.438. Whitney H. Shepardson, *Agricultural Education in the United States* (New York: Macmillan, 1929).

[64]Hutchison to Rose, 21 Mar. 1927, IEB 1.2 31.437. Rose to Hutchison, 6 Dec. 1927, 24 Jan. 1928; Rose to Trowbridge, 11 Jan. 1928; Docket, 9 Feb. 1928; Shepardson to Rose, 27 Feb. 1928; Trowbridge diary, 25 Feb. 1928; all in IEB 1.2 31.438.

[65]Hutchison to Gardiner, 17 Mar. 1927; Wood and Gardiner to Hutchison, 22, 31 Mar. 1927; all in IEB 1.2 31.437. Hutchison to Rose, 1 Apr., 18 Oct. 1927; Wood to Hutchison, 26 Mar. 1927; Rose to Hutchison, 27 Oct. 1927; all in IEB 1.2 31.438.

[66]F. G. Hopkins and F. E. Smith, "William Bate Hardy 1864–1934," *Obit. Notices Fell. Roy. Soc.* 1 (1934): 327–333. Kohler, "Walter Fletcher."

second professorship within biochemistry, to relieve Hopkins and create space for biophysics (a field close to Hardy's own research on the physical chemistry of proteins).[67] It was quintessential Hardy: a grand rational vision, but perhaps more rational than practicable. It did, however, bring the IEB group closer to the active center of Cambridge science.

The key person in the Cambridge project was another member of the Hopkins circle, the master of Gonville and Caius, Sir Hugh K. Anderson. An eminent physiologist turned administrator, Anderson was an academic politician second to none. As an influential member of the university council and planning committees, he had taken part in the various efforts to develop the sciences at Cambridge; he knew the intimate details of each department, their resources, their personalities, their Byzantine quarrels. No one knew better how to skirt the political mine fields and dispel academic smoke screens. Most important, Anderson's seat on the university's finance board assured his influence in all funding decisions. He was party to the negotiations with Rockefeller, Jr., and he knew the whole picture of university funding and priorities, just as Emile Borel did in the Curie circle and Millikan in the NRC.

Trowbridge and Shepardson were pointed in Anderson's direction by James Gray and the physicist William Bragg. (Hardy, Bragg noted, was "a bit too much of an enthusiast" for such a delicate job.)[68] Trowbridge was still uneasy about getting American-style planning to take root on the banks of the River Cam—in fact, he had no idea where even to begin with such a complex project. But he agreed to invite Anderson to a discrete tête-à-tête in London. If Anderson showed a spontaneous interest in the IEB's ideas and was willing to take charge, then and only then would Trowbridge proceed. Anderson more than lived up to his reputation for political sagacity. The time was ripe for comprehensive planning, he told Trowbridge and Shepardson, but he warned them to give up any idea of a committee of department heads. That, he was sure, would only make trouble "because of the personalities of certain men who would have to be included." He also urged the IEB trio to stay away from Cambridge for awhile. Anderson would discretely get the support of a few powerful and broadminded people (including younger faculty like James Gray and the biophysical chemist Erik K. Rideal), then bring together a small planning group within the university council. Anderson hoped, thus, to prevent department

[67]Hutchison diary, 28 Oct. 1927; Hutchison to Rose, 21 Nov. 1927; Wood to Hutchison, 31 Jan. 1928; all in IEB 1.2 31.438.

[68]Trowbridge diary, 25, 26 Feb. 1928; Shepardson to Rose, 27, 28 Feb. 1928; all in IEB 1.2 31.438–439. C. S. S[herrington], "Sir Hugh Kerr Anderson 1865–1928," *Proc. Roy. Soc.* B104 (1929): xx–xxv. Kohler, "Walter Fletcher."

heads, especially those in the weaker disciplines, from turning the planning process into a political logrolling.[69]

In fact, Anderson managed the whole thing himself. He got statements of "needs" from over a dozen departments and trimmed them ruthlessly; he got the financial board to make the raising of matching funds its top priority. He ironed out territorial disputes and arranged for cooperation among departments. He drafted the proposal to the IEB. It does not appear that any formal planning committees were appointed or any formal surveys made; it was all done by one expert hand, quickly and decisively, before interests could mobilize and conflicts ripen.[70]

Anderson's plan, which was accepted virtually as it stood, was a masterpiece of academic diplomacy. Anderson gave priority to rounding out fields of biology that were already strong, and to developing new fields like biochemistry and biophysics. Thus he satisfied the IEB's desire for planning and coordination and put resources where there was reason to believe they would be used effectively.[71] Anderson excluded all the physical sciences as not in need or not relevant to biology, with the exception of mathematical physics—a consolation prize, it seems, for both sides. (Trowbridge and Anderson had pressed Rutherford to cosponsor the development of biophysics, but the physicists had declined to have biologists within the precincts of the Cavendish.)[72]

In biology, morphological fields were excluded and experimental fields encouraged. (Anderson cut botany, paleobotany, and anatomy; the IEB eliminated the zoological museum.) Zoology acquired protozoology from the Molteno Institute and entomology from the School of Agriculture, thus restoring part of what previously had been lost through raids on its frontiers.[73] In agriculture practical items like animal pathology and nutrition were ruled out as local responsibilities, and major investments were made in basic plant physiology and genetics.[74] The biochemists got most of their modest requests, but of the physiologists' immodestly long list of "needs" Anderson selected only general physiology and neurology, which happened to be in the hands

[69]Trowbridge diary, 25 Feb. 1928; Shepardson to Rose, 27, 28 Feb. 1928; Hutchison to Rose, 28 Feb. 1928; all in IEB 1.2 31.438–439.

[70]Hutchison to Rose, 23 Mar., 24 Apr. 1928; Memorandums A–N and correspondence Mar., Apr. 1928; "Cambridge proposal," in Hutchison to Rose, 24 Apr. 1928; T. Arnett, memos, 29 June, 5 July 1928; all in IEB 1.2 31.439–440.

[71]"Cambridge proposal," p. 2.

[72]Ibid., pp. 13, 23–24.

[73]Ibid., pp. 6, 10–12, 15–19, and memorandum N. James Gray, "Scope of experimental zoology," n.d.; Hutchison to Rose, 30 May 1928; all in IEB 1.2 31.439–440.

[74]"Cambridge proposal," pp. 3–5, 6. F. F. Blackman, "Botony and plant physiology," n.d.; A. C. Steward, "Botany," 18 Mar. 1928; both in IEB 1.2 31.439.

of able and underprivileged young faculty.[75] Anderson wanted to create a separate department of biophysics, but no one liked that idea, so provision was made for it within biochemistry and physical chemistry, where in fact varieties of biophysics were already being practiced.[76]

Most important, perhaps, Anderson gave first priority not to the chief professors but to younger faculty like James Gray and Eric Rideal. Endowments were earmarked for senior lectureships, which gave fresh talent independence and control of resources without the trouble of unseating the old guard. Everyone was saying that change would occur when a new generation succeeded to major chairs. Anderson and the IEB in effect speeded up the succession by opening up new positions just below the level of professor. Consciously or not, Anderson and the IEB introduced into a European university the American system of multiplying professors and specialties inside expandable departments.

The Paris, Cambridge, and London projects reveal how similar and yet how different they were from smaller projects in provincial centers. The basic issues—finance, organization, program—were much the same. But scaling-up made Trowbridge more dependent on a few scientific statesmen who were movers and shakers at the highest level of university politics. Risks went up with the stakes, and Hugh Andersons and Emile Borels were rare. Large cosmopolitan universities found it difficult to realize the benefits of large-scale organization. Weak departments and disciplines remained a problem. The IEB's process of making projects was best suited to helping excellent secondary centers join the elite of international science.

Madrid: On the Fringe of Scientific Europe

Despite Rose's genuine interest in the undeveloped regions of Europe, only two projects were ever carried out there—the Institute of Physics and Chemistry in Madrid, and the Agricultural College in Sofia, Bulgaria. It did not take Trowbridge long to discover how hard it was to do things in countries where higher education was narrowly based and vulnerable to volatile, highly partisan politics. The usual methods of handling fellowships, for example, did not work in places like Poland,

[75] "Cambridge proposal," pp. 14–15. "Physiology," n.d., IEB 1.2 31.439.

[76] "Cambridge proposal," pp. 6–9. Memorandum G, 23 Mar. 1928; Hardy, "Biophysics" (memorandum F); Anderson, "Proposed development of colloidal physics," 5 Apr. 1928 (memorandum I); Hopkins, "Memo on present needs and future development of the Department of Biochemistry, 29 Feb. 1928; all in IEB 1.2 31.439.

Spain, and the Balkans. Candidates from these countries were less well trained and had to be judged by local, not international, standards. Academic jobs were very scarce, and Trowbridge had to insist that professors who sponsored fellows guarantee that they would have a job when they returned from study abroad. With those who were already studying in the west and who might be prevented from returning home by their politics or ethnic background, Trowbridge simply had to take calculated risks.[77]

The prospects for institutional grants in such countries were even less promising. "I judge," he wrote, "that racial, religious and political differences will complicate the handling of most of the projects which are likely to arise."[78] In fact, Trowbridge received few proposals from Southern and Eastern Europe. Few scientists there even knew that American foundations existed, much less how to approach them. There were no politically connected scientific elites comparable to the Curie, Hopkins, and Hale circles. Trowbridge would have liked to help with small emergency grants, but Rose held firm for concentrating on big opportunities, and of those there were precious few.[79]

It was not material resources or political tolerance that made Spain the exception but the potential for leadership by the Junta para Ampliación de Estudios. This unusual institution was created in 1907 by the educational reformer Francisco Giner and his disciple José Castillejo. The junta was a board or commission of cultural leaders, representing all shades of the political and religious spectrum. Sponsored and funded by the ministry of education but separate from the educational bureaucracies of both church and state, the junta was intended to be a nonsectarian force for reform of education, from primary schools to universities. Its aim was to bring Spanish education and culture into the mainstream of Western Europe. Bureaucratic inertia and the opposition of the Catholic church made it difficult to change universities from the top down, and the junta rested its hopes on grass-roots improvement of university teachers. It had two key programs. Fellowships enabled young scholars to study in the universities of Northern Europe. Then small grants were made to returning fellows to help them set up laboratories and continue research begun during their apprenticeships abroad. There were some twenty such minilaboratories in the

[77]Trowbridge to Rose, 29 Oct. 1925, IEB 1.1, 43.605.

[78]Trowbridge to Rose, 8 June 1926, p. 22, IEB 1.1 24.345.

[79]On Bulgaria, IEB 1.2 26.369 ff. Trowbridge to Rose, 13 June 1925, 1 Mar., 30 Dec. 1927; Hale to Trowbridge, 9 Dec. 1927; Rose to Trowbridge, 22 Jan. 1927; Rose to Hale, 9 Feb. 1927; all in IEB 1.2 36.515. Trowbridge to Rose, 27 Mar. 1928, IEB 1.2 26.367. Hutchison to Rose, 11 Jan. 1928; Rose to Hutchison, 26 Jan. 1928; both in IEB 1.2 38.537.

mid-1920s and a community of several hundred former fellows, including sixty-six in chemistry and physics. By providing cosmopolitan experience and improving somewhat the material poverty of academic careers, the junta hoped to create a new generation of scholars who would participate as equals with French, Germans, and British. Experimental science was especially important to Castillejo, who saw it as the best antidote to bookish and dogmatic religious education.[80]

It is not surprising that the junta appealed so strongly to Rose, since its ideology and methods were strikingly like his own. Castillejo and Rose shared a common faith in science as a moral force for change. The junta's programs of international fellowships and grants of equipment for budding centers of research were ready-made for IEB projects. Rose was less enthusiastic about the specific scheme for which Castillejo sought the IEB's aid. Castillejo wanted to build a central research laboratory in Madrid, where the junta's scattered laboratories could be brought together in a single national research institute. (The staff would be full-time researchers but also give advanced training to science teachers.) Rose preferred to invest in people rather than bricks and mortar. Also, he was troubled by the fact that the proposed facility would be separate from universities, where the new generation of scientists would be trained and employed. Rose wanted to buy time with small grants-in-aid for equipment in the hope that a spontaneous reform movement would develop within the University of Madrid.[81]

Trowbridge thought such a movement unlikely and was inclined to go with Castillejo's plan. Universities, he thought, were the most conservative institutions in the world (except for the church), and Spanish universities were ultraconservative.[82] Trowbridge could hardly imagine a less likely base for scientific development than the University of Madrid. Its laboratories were "not worth considering." Chemistry had no equipment to speak of—Trowbridge had "never seen anywhere worse conditions"—and physics was not much better. The university's

[80]José Castillejo, *Education and Revolution in Spain* (Oxford: Oxford University Press, 1927), pp. 16–20. Thomas F. Glick, "La fundación Rockefeller en España: Augustus Trowbridge y las negociaciones para el Instituto Nacional de Física y Química, 1923–1927," in José Manuel Sánchez Ron, ed., *La Junta para Ampliación dé Estudios e Investigaciones Científicas 80 Años Después* (Madrid: Consejo Superior de Investigaciones Científicas, 1988), vol. 2, pp. 281–300. Castillejo to Embree, 12 Aug. 1919; Castillejo, "Character and functions of the Board of Extension of Studies," n.d.; all in Embree diary, vol. 5, RF. Glick, *Einstein in Spain* (Princeton: Princeton University Press, 1988), pp. 8–16. Castillejo to Rose, 20 June 1923, 21 July 1924, IEB 1.2 41.577. I am grateful to Thomas Glick for an English translation of his article.

[81]Castillejo to IEB, 21 July 1924; Rose to Castillejo, 27 Apr. 1924; Castillejo to Trowbridge, 3 Sept. 1925; all in IEB 1.2 41.577–578.

[82]Trowbridge diary, 25–26 Jan. 1927; p. 29, IEB 1.2 41.580.

School of Pharmacy had a better chemistry laboratory (thanks to junta grants for equipment and upkeep), but its small size forced most applicants back into the "intolerable conditions" of the university's faculty of science. In physics and physical chemistry, the junta had been obliged to rent facilities and equip small laboratories outside the university. The only hopeful sign was that academic scientists seemed less hostile to the junta than Trowbridge had expected.[83] Conditions in the University of Barcelona were somewhat better but still impossible, and the chemists and physicists there were thoroughly discontented and demoralized. Physicist Estaban Terradas was of the "organizing type," and the organic chemist Antonio Garcia Barús was an active researcher, but they were discouraged by lack of support. These former junta fellows felt that academic careers were so poor that foreign study only made ambitious Spaniards more discontented with their lot at home.[84]

Always practical, Trowbridge thought a research institute was the best bet for the IEB in Spain, because it seemed the most likely to succeed in the short run. That was vital, since Castillejo was nearing retirement and had groomed no successor. Trowbridge thought it wise to move ahead quickly while the junta was still an effective force. Also, the foreign-trained alumni of the junta's fellowship program were available to staff an institute. Trowbridge urged Rose to seize the opportunity, with all its risks, and not wait for a movement within the university that might never happen.[85]

Another reason for pushing ahead with an institute was to force the Madrid scientists to be more active in their own behalf. Trowbridge was struck by the inability of the physicists and chemists to take any definite action. They conceived of the institute as a vague dream, an ideal that might be realized in the indefinite future but not now. They seemed incapable of taking the first steps toward making their dream a concrete reality. Trowbridge hoped that having to make concrete decisions about architectural plans might force the scientists to realize they could help themselves. Without such stimulus from the IEB, Trowbridge feared, they might never shake off their pessimism and inertia.[86] Trowbridge saw the institute project as a lesson in entrepreneurship; with his extensive experience in building projects, he was an ideal teacher.

The Madrid scientists' passive and despairing attitude was clearly a

[83]Trowbridge to Rose, 4 May 1925, pp. 3–6, 16 May 1928, IEB 1.2 41.577. Trowbridge diary, 21 May 1928, IEB 1.2 41.580.

[84]Trowbridge diary, 25–26 Jan. 1927, IEB 1.2 41.580. Castillejo, *Education and Revolution,* p. 20. Glick, *Einstein in Spain,* p. 27.

[85]Trowbridge diary, 27 Apr., 4 May 1925, IEB 1.2 41.577.

[86]Ibid.

habit of people who were not accustomed to being able to make things happen. One of the most interesting things about the Madrid project is the light it sheds on the behavior of scientists in societies that gave them little support. Trowbridge became a shrewd and sympathetic analyst of underdevelopment, as did also his advisor in physics, Charles Mendenhall.[87]

The most striking thing to Trowbridge was how totally inexperienced Spanish scientists were in planning and designing laboratories—not surprisingly, given how seldom they had the opportunity. Mendenhall was amazed that the physicists did not know to mount machinery on concrete piers to avoid vibrations, or to build removable interior walls so space could be altered to fit changing research needs, or to locate the machine shop and electrical switchboard close to the laboratories and not in a separate building. Trowbridge found that the scientists on the building committee shrank from making decisions, always trying to defer to Trowbridge or other foreigners. Trowbridge realized they would "need hypodermics from time to time," but he did not relish the job of administering them, fearing that the IEB would be accused of "exerting pressure and forcing [the] Junta into an infra dig position." Making decisions on purely technical matters, he hoped, would give the scientists confidence in more difficult matters of policy; but it was almost impossible to get them to be decisive about anything, however small.[88]

Chronic lack of facilities and support also affected the way the Spanish scientists selected research problems and did scientific work. Mendenhall observed a tendency to pursue narrow lines of research, a taste for elegant experimental apparatus, and an unwillingness to adapt to changing fashions in research. Blas Cabrera, Spain's most eminent physicist, made a life's work out of a single problem in the theory of magnetism. Mendenhall reported:

> He impressed me as experimentally ingenious, but perhaps somewhat too much interested in working out nice arrangements of apparatus which could be used with the maximum of convenience by the experimenter during a long series of observations. He showed me a number of very beautifully constructed instruments which had been made in the shop of the laboratory, but I

[87]Glick, "La Fundación Rockefeller i Espanya: la crisi dels laboratoris," in Luis Navarro Veguillas, ed., *Trobades científiques de la Mediterránia: História de la Fisica* (Barcelona: Generalitat de Catalunya, 1988), pp. 367–372.

[88]Trowbridge diary, 14 Jan. 1927, pp. 8–9; Rose to Trowbridge, 31 May 1928; both in IEB 1.2 41.580.

> saw little or nothing which indicated much interest in improving apparatus or in trying out new ideas.[89]

It was a lifetime investment in one piece of equipment and one special skill.[90]

Miguel Angel Catalán also kept to the narrow field in which he was trained; a capable experimentalist, he was unwilling to risk any departure from familiar routine. He told Mendenhall of his dream of a larger spectroscope, which he dared not begin to build because it would take two years to perfect. If he wrote no papers for two years, he feared, younger men would be promoted over his head. Mendenhall suggested that rather than build the perfect instrument, Catalán could improvise a smaller one and gain experience in its use, then scale-up. But such risk taking did not come naturally to the Spaniards. Mendenhall remarked on the inability of the Spanish physicists to skim the cream from hot new research problems, trying out lots of ideas with improvised equipment, doing the ones that worked and then moving on. There was no creative symbiosis between experimentalists and nimble-minded theoreticians. Both physicists and chemists, Mendenhall noted, "were too prone to get into a stereotyped type of research . . . [and] needed for a time the stimulus of someone prolific of suggestions and with the technical ingenuity to quickly improvise a test of these suggestions, someone who can throw apparatus together in a hurry and carry out a qualitative experiment."[91] That was how it was done at the Cavendish, Harvard, or the University of Wisconsin. But chronic scarcity and customs conditioned by scarcity made the productive practices of scientific centers ill-suited to underdeveloped regions like Spain. As Thomas Glick has pointed out, Cabrera's life investment in one instrument was a rational and adaptive strategy in a system where funding for new ventures could never be counted on.[92]

Mendenhall's observations are as revealing of the peculiar new scientific culture of European and American physics as they are of the scientific culture of Spain. Intensive cultivation of a single specialized instrument and a single problem was hardly peculiar to Spain. Until quite recently it probably had been the usual strategy of making careers in most countries. Certainly, it had been common enough among American scientists in the late nineteenth century. As we have seen, similar patterns of behavior were common in the specialized laborato-

[89]Mendenhall report, 24 Mar. 1926, pp. 1–2, IEB 1.2 41.579.

[90]Glick, "crisi dels laboratoris." Glick, *Einstein in Spain,* pp. 26–32.

[91]Mendenhall report, 24 Mar. 1926, p. 6.

[92]Glick, "Augustus Trowbridge," pp. 292–294.

ries of Scandinavia and the Netherlands, though scientists there were more interested in improving the design of laboratory instruments. The fast-paced, opportunistic style of atomic physics was a special adaptation to the culture and resources of scientific centers, with their access to resources and to systems of rapid, informal communication. Without such infrastructure and customs it was impossible to participate. Older modes of practice lingered on in Spain because, in the conditions there, monopolizing the production of exact data in specialized, not too fashionable fields offered at least the possibility of acquiring a modest international reputation.

No group of academic scientists was more demoralized than the Spaniards, and Trowbridge saw the same self-defeating lack of confidence everywhere in Spanish society. Madrid, he observed, was full of grandiose projects half completed: great boulevards that led nowhere and houses half-built. In Barcelona, too, Trowbridge observed the same lack of confidence, the same sense that foreign scientists were better; he heard the same laments of isolation from the European centers of scientific culture. Trowbridge felt, however, that the Spanish scientists were unduly pessimistic about their abilities. Their sense of inferiority and isolation belied, it seemed to him, the record of their accomplishments. Blas Cabrera, for example, was part of the international network emanating from Arnold Sommerfeld's institute at Munich. Mendenhall was impressed by the ability of the Spanish physicists and chemists to produce good work under conditions that Europeans or Americans would have found intolerable. He considered Spain a most promising opportunity for the IEB.[93] If the Spanish scientists would only stop feeling sorry for themselves, Trowbridge thought, they might get more support and recognition.[94]

Although they were undoubtedly right about the psychological dimension of backwardness, Trowbridge and Mendenhall underrated the difficulties of getting things done in a hierarchical social order, in which every action was ideologically charged. These practical difficulties were soon brought home to Trowbridge. Delays were endless, and the bulky files of correspondence reveal how hard Trowbridge had to work at every step. Even so simple a matter as selecting an architect was not a purely "technical" decision, as it had always been in Trowbridge's experience. In Spain, he discovered, it was necessary to hold a public, government-sponsored competition and award two (or better three) prizes to the best designs. Castellejo explained that if he selected the architect himself it would be taken as a personal choice and would make

[93]Mendenhall report, 24 Mar. 1926.

[94]Trowbridge diary, 26 Jan. 1927, pp. 29–30, IEB 1.2 41.580.

enemies of all the architects and engineers not chosen. Since these guild-like communities could make trouble for the project later on, it was better to let the government take the blame. Trowbridge was unhappy about the prospect of long delays and complained of inaccurate cost estimates by inexperienced young architects. But there was no alternative.[95]

National politics also intruded, in a far more serious way. In 1923, after a period of extraordinary political instability (ten governments in five years), General Miguel Primo de Rivera became dictator. His policies were a peculiar mixture of the progressive and reactionary. He was strongly supportive of public education, but when opposed by student strikes declared that Spain could do without universities. He hoped to bring Spain more into the mainstream of European culture, but sided with reactionary church and nationalist movements. He wanted to modernize Spain's economy but suppressed the most advanced industrial region, Catalonia.[96]

Primo's actions toward the junta betray the same dual character. In June 1926 a royal decree gave the government the right to appoint half their number. Trowbridge was naturally alarmed by the junta's loss of independence, especially since he had learned of the coup by chance from a Paris newspaper. Castillejo tried to reassure Trowbridge that this was nothing new for the junta, that opposition in Spain always took the form of royal decrees, which were then quietly forgotten. In fact, he had been told by Primo that the decree had been forced on him by the church party and that his real intention was to give the junta greater administrative freedom, not less. Castillejo did not mention that Primo had also suspended payments to the junta until the reorganization was effected.[97] Trowbridge was not reassured and sought a formal decision from New York to withdraw or proceed. (It was just one month after the IEB had voted $420,000 for the institute.) Rockefeller counsel Thomas Debevoise considered the IEB had a moral obligation (though not a legal one), and the board decided to take a calculated risk and proceed.[98]

Although Trowbridge usually stayed clear of national politics, this time he could not. His most useful contact was one of the more remarkable members of the junta, the Duke of Alba. A descendant of the

95Mendenhall report, 24 Mar. 1926. Trowbridge diary, 13 Jan. 1927, pp. 4–8, IEB 1.2 41.580.

96Raymond Carr, *Modern Spain* (New York: Oxford University Press, 1980) chs. 6–7.

97Trowbridge to Rose, 14 Oct. 1926; Trowbridge to Castillejo, 1 July 1926; Castillejo to Trowbridge, 25 Sept. 1926; Jones diary, 12 Oct. 1929; all in IEB 1.2 41.580.

98T. M. Debevoise to Rose, 1 Dec. 1926; Brierley to Cajal, 23 Dec. 1926; both in IEB 1.2 41.579. Rose to Trowbridge, 7 Mar. 1927, IEB 1.2 41.580.

English Stuarts, the duke had been educated in England and had the speech and manners of a cosmopolitan Oxbridge gentleman. In the ducal palace, Trowbridge dined off solid gold plates, surrounded by Old Master portraits of Alba's ancestors, and had to keep reminding himself that he was dealing with a Spanish grandee. Alba was knowledgable and sympathetic with English ideals of organization, progressive socially, and liberal in politics. He was an intimate of King Alfonso and a boyhood chum of General Primo. Fearful that the IEB might back out of the project, the duke offered to arrange an interview for Trowbridge with the dictator. This he did. With the duke observing silently, Primo assured Trowbridge that the reorganization of the junta was not politically motivated but was only meant to bring in new men with broad administrative experience, thus preparing it for greater autonomy. Trowbridge judged the dictator to be sincere and was assured by others that there would be no trouble from the political right so long as Primo remained in power.[99]

From Primo and others, Trowbridge gradually became aware that there was more to the matter than he first thought. Primo's profession of support for the junta probably was sincere. At the same time, it is disingenuous to think that his actions (or any in Spain at the time) were not political. The church party had never liked the junta, which threatened the Jesuits' control of education, and liked even less Castillejo himself, who made no secret of his anticlerical convictions. Conservatives saw the dictatorship as a chance to settle some old scores, and the junta was one of them.[100] Apparently, Primo was trying to keep his political fences mended with his conservative backers, but without destroying an institution devoted to modernizing Spanish science and education. In fact, government payments to the junta did increase after the reorganization, as Primo had promised they would.

Trowbridge also learned of opposition to Castillejo within the junta. Several members told him that the junta had over the years become virtually a rubber stamp for the permanent secretary. Half the members never came to meetings, and the other half voted "yes" to whatever Castillejo proposed. Castillejo ran the junta like a dictator, attending to every detail himself and not delegating authority. He did not inform members of what he was doing: the building committee, for example, complained that information had been withheld from them. Castillejo was adept at keeping his political fences mended with the educational bureaucracy but was less effective as an administrator of the junta's projects. Loving the junta too well, Castillejo made himself vulnerable, not

[99] Trowbridge diary, 13, 23, 24 Jan. 1927, IEB 1.2 41.580.

[100] Castillejo, *Education and Revolution,* p. 20. Glick, "Augustus Trowbridge," pp. 295–296.

just to the conservatives but also to moderates who wished the junta well and thought new members would improve it. Even Trowbridge came to believe that revolving membership was a good thing. By getting rid of Castillejo, Primo placated the conservatives and got new talent into the junta.[101]

Most of the new members of the junta were young men from manufacturing families, experienced in managing large industrial organizations. Orthodox in religion, they tended to be culturally cosmopolitan, educated abroad, and sympathetic with the junta and its mission. They were selected for qualities that Castillejo lacked—administrative know-how, political tact, and social connections with industry rather than with the educational bureaucracy. The new members were also alert to the need to win back the confidence of their American patron. Juan de la Cierva called on Trowbridge in Paris (at the request of Primo and the junta president Santiago Ramón y Cajal) to reassure him of the government's goodwill and to get a promise that the IEB would not back out. So too did José María Torroja, who especially impressed Trowbridge as able and forceful, quite unlike the timid and indecisive academics. One of the first to receive a fellowship from the junta, Torroja had studied in Germany but on his return had gone into industry (he was an engineer) rather than teaching. An able manager and entrepreneur, he was clearly being groomed to succeed Castillejo—and did a year later.[102]

The building project proceeded in the way all projects did in Spain, every step having to be taken with the consultation of every official with the slightest interest in it. For Trowbridge it seemed endless, with long periods of no visible progress and endless correspondence and consultations. An architect was eventually selected, and construction began in 1928. In 1932 Lauder Jones admired the impressive building, its iron framework clad in burnished copper, its interior mahogany trim, a luxuriously appointed machine shop, and active laboratories safely insulated from the routine of the university departments.[103] The junta continued its activities until the revolution of 1936–1939 when, with its policy of toleration, it was ground out of existence by zealots of both the right and the left.[104] In 1936, Harry Miller was assured that research was continuing in the "Rockefeller" Institute (as it was famil-

[101]Trowbridge diary, 13–14, 23–24 Jan. 1927, pp. 1–3. 23–24; Trowbridge memos, 6 July, 21 May 1928; all in IEB 1.2 41.580. Trowbridge's informant, it should be noted, was José Torroja.

[102]Trowbridge to Rose, 14 Oct. 1926, pp. 4–5, 7–8, IEB 1.2 41.579. Trowbridge diary, 13 Jan. 1927, pp. 3–4; Trowbridge memos, 6 July and 21 May 1928; all in IEB 1.2 41.580.

[103]Jones diary, 28 Mar. 1931, 10 Mar. 1932, 7–9 Apr. 1932, IEB 1.2 41.580.

[104]Castillejo, *Education and Revolution,* pp. 23–25.

iarly known in Spain), despite the civil war.[105] Perhaps so—Spanish scientists were used to adversity. But Miller's informant, a physicist formerly in the institute, was then a political refugee in France. The institute became an early casualty of the political storm that would soon engulf everything that American philanthropists had helped to build in Europe since 1918.

The IEB and Science in Europe: Conclusions

"Make the peaks higher"—help strategically placed institutions to attain the acme of scientific organization and practice and let competition and emulation do the rest. It was an optimistic and very American style of reform. How effective was it in the fragmented, troubled world of postwar Europe? How much of the vision that Rose laid before the IEB board in 1923 is recognizable in what he and Trowbridge actually did? Clearly, the utopian language of science as the engine of economic and social improvement and national survival does not describe the IEB's real but more modest accomplishments. It was scientific communities that benefited, not agriculture and industry, and they were not transformed so much as assisted to do better what they had always done. Trowbridge and his staff were not utopians but artists of the possible, and IEB projects map the points where IEB intentions coincided with practical opportunities within the diverse world of European science. Opportunities were made by many things: national economies and cultures, international connections, research fashions, but most saliently by location in the European university system. That, above all, is what shaped strategies of competition and uplift, and those strategies were what made or unmade negotiations with the IEB.

Everywhere in Europe universities were adapting to common problems: a chaotic accumulation of institutes; research fronts that changed faster than academic life cycles; new fields in no-man's-lands between disciplines; a flood of students, which made it difficult to combine research and teaching; competitive pressure from below; and rising costs of staying competitive, both in material resources and managerial know-how. Having a patron like the IEB was not yet essential to staying competitive, but it would not be long before it was. The favored strategy in Europe (less so in the United States) was to create specialized research enclaves outside the official university institutes. The most extreme manifestation of this strategy was the system of Kaiser Wilhelm Institutes, but the specialized research chairs and laboratories of the North Sea countries, and the junta's shoestring laboratories were vari-

[105] Miller diary, 3 May 1936, IEB 1.2 41.581.

ants on the same theme. Large cosmopolitan centers might aspire to comprehensive excellence, but for most universities a strategy of concentration was the only practical one, and how it worked depended on location. Regional centers like those of Scandinavia, the Netherlands, and the middling German states could seek a competitive advantage in new and productive fields (not necessarily ones in high fashion). Smaller universities and those on the margins of scientific high culture could realistically hope to assist a few individuals to participate in international science. Such strategies cut across national, economic, and disciplinary lines.

Trowbridge and his staff learned to adapt and improvise within this complex system, and their labors were most likely to bear fruit where local strategies of competition were most like their own. Rose deceived himself in thinking that he had a surefire method for organizing whole science faculties; the reality of science patronage was improvisation and opportunism, as Trowbridge and his colleagues practiced it. Despite Rose's enthusiasm for grand cosmopolitan projects, the record shows that Trowbridge did best with small groups of scientists and with middling, up-and-coming universities in countries that were already supporting scientific high culture. Such projects were clearly defined and doable. Scientists were likely to have had prior experience in building institutions or networks, and that experience served them well in dealing with the IEB. It was not just power and influence that counted but a breadth of experience and vision that transcended individual and disciplinary interests. It was the same kind of vision that Trowbridge had to have to do his job. The crucial skill, as in any social relationship, was the ability to see things from the perspective of the other side. It had to be learned quickly, as the IEB's window of opportunity was open for only a few short years. The prizes went to those who understood the process and knew how to make it work.

The same was true for scientific entrepreneurs and their universities on the other side of the Atlantic.

Chapter Eight

Developing American Science

Although the United States is not divided, as Europe is, into separate sovereignties, American universities were no less diverse in the 1920s than the universities of Europe. There were differences between levels of the institutional pyramid—rank-and-file, upstarts, old elite; between regional subcultures—Northeast, Midwest, South, and West; between rural and urban; private and public. Serving different markets for higher education, university leaders had to employ diverse strategies for keeping up with research fashions, for moving up the pecking order or staying on top. These strategies, which were adapted to local cultural and financial circumstances, in turn determined the likelihood of getting funds from the GEB. In the United States, no less than in Europe, location within institutional systems was the single most important thing in determining who got what.

In fact, American and European universities can be categorized and mapped in much the same way. The up-and-coming group, exemplified by Caltech, are analogous to the secondary centers of the small countries around the North Sea. They were relatively small and could accommodate free-wheeling entrepreneurs; they had dedicated local constituencies. As Trowbridge found his best opportunities in specialized institutes on the fringes of the German scientific world, so too were small private universities the most hospitable places for his colleagues in the GEB. In old elite universities like Johns Hopkins or the University of Chicago, as in the cosmopolitan centers of Paris or London, Byzantine disciplinary politics and the temptations of comprehensive planning made it difficult to capitalize on past eminence. Finally, the development of universities in the American South and West was impeded by regional isolation and reactionary cultural politics, just as were the universities of Eastern and Southern Europe.

Caltech and Copenhagen, Paris and Chicago, Virginia and Madrid were more alike than we might have expected.

Look at the distribution of the GEB's projects (table 8.1). All but two were in private universities, and the best represented group (Caltech, Princeton, Stanford, and Cornell) were (mostly) strong undergraduate colleges pushing hard to become centers of graduate teaching and research. Among old elite universities the record is mixed: Harvard and Chicago did well, Yale less so, and Johns Hopkins poorly. The large Midwestern state universities are conspicuously absent, and the few grants to Southern universities (North Carolina and Virginia) were tentative experiments in aiding science in regions where academic high culture had not flourished for many decades. The distribution of GEB projects, in short, is quite similar to the IEB's projects, and it can be understood in the same way as adaptations of local institutional strategies to the GEB's process of grant-giving. With some the fit was good, with others not.

As we saw earlier in connection with the NRC fellowships, many universities were trying to become centers of research and graduate training in postwar America. For most, the strategy of choice was to improvise piecemeal, department by department, using eminent researchers as bait to attract patrons, or using the promise of foundation support to lure large-caliber talents around whom programs of training and research would grow. This bootstrap strategy required skill and timing, and some eminent researchers never quite got the knack of it. Johns Hopkins physicist Robert W. Wood, for example, complained to Rose that he was prevented by the lack of good facilities from attracting big names, and by the lack of big names from raising funds for new facilities. (Rose tried to tell him how to do both at once, but to no avail.)[1] Yet everyone seemed to be trying it in the mid-1920s, at every level of the university system, in large and small ways.

The Caltech troika were, of course, the masters of this strategy, applying it to create entire departments and groups of disciplines in a systematic way. (They had an advantage starting from scratch.) More typically, the strategy was used to develop one department, which for one reason or another happened to have a particular advantage, and which, it was hoped, would cause other departments to emulate it. Further down the pecking order, where there could be no hope of making entire departments over, one or two token individuals might be imported and given special privileges (funds, reduced teaching) to keep a line of research alive. It would be a visible sign of greening, even if in an academic hothouse.

[1]Rose diary, 9 Jan. 1928, GEB 1.4 588.6258.

TABLE 8-1

General Education Board Appropriations in Natural Science (in $1,000s)

Institution	*Field*	*1923–24*	*1924–25*	*1925–26*	*1926–27*	*1927–28*	*1928–29*	*1929–30*	*1930–31*	*Totals*
Rochester	Nat. sci.	1,000		750						1,750
Caltech	Physics, biol.		529		1,050	1,000			500	3,079
Vanderbilt	Nat. sci.		343	350						693
NRC	Publica.		30			25				55
NRC	Forestry		50							50
Princeton	Nat. sci.			1,000			1,000			2,000
N. Carolina	Nat. sci.			15						15
Harvard	Phys., chem., astron.			500	400	200		75		1,175
Woods Hole	Biology			300			200			500
Columbia	Biophys.			10						10
Chicago	Nat. sci.				1,500		298			1,798
Cornell	Nat. sci.					1,500[a]				1,500
Stanford	Nat. sci.					810[a]		30	30	870
Texas	Genetics					65				65
NRC	Oceanogr.					75				75
NRC	Biophys.						63			63
Yale	Behav. sci.						500			500
Totals		1,000	952	2,925	2,950	3,675	2,061	105	530	14,198

[a]Pledges not redeemed.

These middling American universities offered the GEB a range of opportunities not unlike those that Trowbridge encountered in Germany or Scandinavia. The more systematically this strategy of reform was applied, and the better funded it was locally, the more likely it was to appeal to the GEB. Where there were only token individuals, as was usually the case in the less developed regions of the university system, investment was far more problematic. The question there was the one that Trowbridge asked about specialized institutes: might they develop into regional centers of research and training, or would they remain one-man shows strictly of local interest?

In the older elite of research universities, the 1920s saw a rather different strategy being employed. Rather than piecemeal reform, university leaders sought to stay on top of the heap by boldly reviving their founders' vision of a purely graduate university, excellent in every department and serving as a magnet and feeder of research talent for the whole country. The University of Chicago, for example, envisioned a new graduate university separate from the undergraduate college, while Johns Hopkins planned to get rid of its college altogether. Like the sprawling cosmopolitan universities of Europe, Harvard hoped to take advantage of its size and diversity by bringing together scattered research operations (five or six in biology alone) into more efficient and imposing multidepartmental divisions. As in Europe, however, political and financial problems made this strategy less appealing in practice than in theory. Grand schemes of "real" universities, as it turned out, only delayed the consummation of more doable projects in particular departments: hence, the uneven record of the old elite universities on both sides of the Atlantic.

There were differences too, of course, between the United States and Europe. Financial and legal impediments made it very difficult for public universities in the United States to accept endowment from private foundations, especially when matching funds were required. The University of Wisconsin was actually prohibited from taking foundation money. Nothing like that obtained in the state-supported universities of Europe, where Trowbridge had only to cope with the usual ministerial logrolling. Rose wanted very much to aid state universities but was unable to get very far. The dependence of private universities on collegiate alumni also created difficulties in raising funds to match GEB grants, since collegiate loyalties seldom extended to research and graduate training—another problem that Trowbridge never had to face.

Differences between American departments and the European system of chair and institute also gave GEB and IEB projects a different shape. There were almost no projects in the United States like the IEB's modest grants to specialized, extradisciplinary research institutes, for example, for the simple reason that such institutes were rare in the United States. In the American system, new research specialties were accommodated within departments, which were far more expandable than European institutes. Thus, large GEB grants to strong research groups were more likely to affect entire departments; by the same token, however, small grants to individuals were more likely to be co-opted without affect into departmental routine. That was a common fate of GEB experiments with faculty grants-in-aid funds, as we shall see.

All of these things help explain the characteristic pattern of GEB projects: why grants in the United States were fewer in number but larger than in Europe, more evenly distributed among disciplines (including the humanities) and more concentrated in a few institutions (see table 8.1). This diversity of experience in different kinds of institutions also goes to show the fundamental importance of location in determining who got what. The experience of making projects was diverse because the ecology of university systems afforded diverse niches for project makers.

Policies and Projects

Although the IEB was created to give the GEB an arm in Europe, in practice the GEB was an appendage of Rose's and Trowbridge's European operations. Rose was more interested in Europe, and the New York office had no staff like Trowbridge and his group in Paris. Trowbridge's New York counterpart, Halston J. Thorkelson, had been a professor of engineering and business manager (a university "handyman"). The GEB's fellowship program was farmed out to NRC committees, not integrated with institutional grants. Surveys were less grand and systematic on the American side. The GEB's distinguished past was ballast that made it hard to change direction. Thorkelson's program in science had to compete with prestigious ongoing operations like Abraham Flexner's program in medical education. In Europe Rose made a fresh start; in the United States he had to tread more carefully. The IEB tail definitely wagged the GEB dog.

It was not that the GEB trustees disliked what Rose was doing, quite the contrary. It was obvious to everyone that the GEB's program of general endowment of colleges had outlived its purpose; the system of mass higher education had simply outgrown it. Concentration was essential, and training an elite of teacher-researchers seemed an eminently good thing for the GEB to do.[2] Trustees welcomed Rose's new agendas. Re-

[2]Fosdick to Rockefeller, Jr., 15 Feb. 1927, RF 900 17.123. GEB board minutes, 24 May 1923, 20 Nov. 1924, vol. 1, pp. 40–41, 103–108, GEB. Fosdick, *Adventure in Giving the Story of the General Education Board* (New York: Harper and Row, 1962), ch. 10.

sistance was not principled but inertial: old responsibilities could not be disavowed, familiar routines were hard to alter. Programs in general education and Southern education continued. The new program in science was organized within the old Division of College and University Education, gradually dominating as collegiate projects were phased out. The transition was gradual. A 1924 pledge of $2 million to the University of Chicago was framed as a capitalization of an earlier grant for salary increases, but it was tacitly understood that it would be used for graduate training and research.[3] Several pledges in the new science program were formulated in terms of general endowment, causing confusion and embarrassment to local fund raisers.[4]

There was inertia in the GEB's machinery of matching endowment. In 1919 the GEB had $9.4 million in pledges still outstanding, one-third of its total appropriations since 1902.[5] This heavy mortgage on the GEB's future limited Rose's ability to plan new initiatives. There was inertia, too, in the accumulated experience of the board, who had been appointed for their expertise in institutional administration and finance, not their knowledge of science. Naturally, they were unreceptive to projects not framed in familiar terms. Thus, the officers always had to reshape applications into requests for matching endowment, which the trustees could understand.[6] This was necessary even though some clients would have been better served by other kinds of support. (Several worthy projects failed when universities were unable to raise matching funds.)

As Augustus Trowbridge shaped Rose's European projects, those in the United States reflected the character of Halston J. Thorkelson. Born in Racine, Wisconsin, in 1875, Thorkelson was a first-generation American (both parents had immigrated from Norway) and upwardly mobile in the Midwestern manner, through education. He received a bachelor of engineering degree from the University of Wisconsin and a year later became a professor there in mechanical engineering. Forced by a growing family to give up teaching, in 1914 he became the university's business manager. Thus he entered the growing class of "university handymen" from which so many foundation officers were recruited. His business ability and attractive personality impressed many people, among them industrialist and university regent, Walter J.

[3]Brierley to E. D. Burton, 19 Sept. 1924; "Program of development for early realization," Oct. 1924; Thorkelson memo, 11–12 June 1924; all in GEB 1.4 657.6838–6839.

[4]Thorkelson memo, 8 May 1924; Rose to W. Lawrence, 26 May 1924; Brierley to W. B. Donham, 4 Dec. 1924; Brierley to A. L. Lowell, 9 June 1925; all in GEB 1.4 613.6483.

[5]GEB *Ann. Rep.* 1918–1919: pp. 68–69.

[6]Thorkelson memo, 17 June 1926, GEB 1.4 649.6777. Thorkelson to Christensen, 3 Aug. 1926, GEB 1.4 641.6712.

Kohler (who offered him a job in 1921), and Wallace Buttrick, who needed someone expert in university finance to help with the growing college and university program. Thorkelson joined the GEB (half-time at first) in 1921. He was an ardent believer in public education and equal opportunity, values deeply rooted in Norwegian-American culture, and took particular pleasure in the GEB's projects in Southern and Negro colleges. In 1923 he was also placed in charge of the new program in science, no doubt because of his training in mathematics and engineering.[7]

Thorkelson's role in the science program was something of a makeshift, like the program itself. As director of the GEB's division of colleges and universities, Thorkelson also ran projects in honors teaching, humanities, and general college endowments. His main job was auditing college endowments, helping college administrators be more businesslike, and assessing the value of matching gifts. Like Trowbridge, he became a connoisseur of institutions. But Thorkelson was regarded in the New York office as more of a field agent than an executive. He was not expected to inaugurate and develop programs, as were Abraham Flexner or Richard Pearce, and his talents seem to have been underappreciated.[8] Scientists of heavy caliber, like T. W. Richards, Vernon Kellogg, and Hale, preferred to deal directly with Rose. In officers' conferences, he was most vocal on business matters and was a constant advocate of research programs in small colleges.[9] He saw things from the perspective of the grass roots, not the executive office.

Like Trowbridge, Thorkelson spent much of his time on the road. It was a rule in the GEB that officers would present the board with proposals only from colleges with which they were familiar. As the university side expanded, the responsibility for gathering basic information fell mainly on Thorkelson.[10] His first contacts were usually not with scientists but with financial officers, graduate deans, or presidents. From these individuals he gathered information on finances, enrollments, and plans for fund raising and development. He would then be taken around and introduced to the leading scientific faculty. He inspected laboratory facilities (and had a keen eye for good design) and pumped department heads about productive lines of research, rising stars, and material needs. He discretely observed the morale of the institution and the quality of leadership. Though he was an active

[7]*Who's Who in America,* 1928. Martha Thorkelson Riddell Kohler, personal communication, 1987.

[8]Fosdick to Rockefeller, Jr., 6 Oct. 1927, RF 900 17.123.

[9]Officers' conferences, 20 June 1924, 22 Nov. 1925, 28 Dec. 1926, 31 May 1927. vol. 1, pp. 3–5, 11–15, 55, 72, GEB.

[10]Officers conference, 22 Nov. 1924, vol. 1, pp. 11–13, GEB.

participant in shaping projects, his own modesty and the constraints on his role in the New York office made him a less independent and initiating manager of science than was Trowbridge. The GEB's science program was carried on in the style of the old program of institutional endowment.

Making projects was a three-cornered negotiation between scientists, presidents, and Thorkelson. For the process to work, academics had to agree on what they wanted and had to learn from Thorkelson what was doable from the GEB's point of view. Department chairmen looking for grants did not get far before Thorkelson let them know that the GEB only considered institutional, not personal, grants. Presidents looking for general endowment soon learned that the GEB was interested only in science departments.[11] The standard GEB grant was a lump-sum endowment for two or three natural science departments, to get new faculty, reequip laboratories, and support individuals' research. Applications pretty much had to take that form, and Thorkelson spent much effort coaching people how to tailor a winning proposal. Often, he had to act as broker when academic interests could not see eye to eye. Some presidents, like Cornell's Livingston Farrand, learned quickly what the GEB wanted but had difficulty getting individualistic scientists to cooperate. Sometimes scientists got the formula right but were sabotaged by presidents who stubbornly pushed their own agendas, as happened at the University of Michigan with the physicist Harrison Randall and President Clarence C. Little (more on that later).

The Right Stuff: Princeton and Caltech

The arrival of Caltech or Princeton into the elite of research universities was as characteristic of American science between the wars as was the predominance of Johns Hopkins in the 1880s or the University of Chicago around 1900. As newcomers, they were able to take advantage of new modes of scientific production—more organized and programmatic, more resource intensive, more integrated into international networks of postdoctoral exchange and communication. They were able to tap into the new system of patronage—indeed, foundations were essential to their style of practice and helped to demonstrate to the rest of the academic world how well it could work. Caltech, Princeton, and (in the 1930s) Stanford and MIT were not the only places to adopt this new mode of production, but they adopted it in the most deliberate

[11] Thorkelson memo, 10 Dec. 1924, GEB 1.4 625.6593. Thorkelson memos, 11 Jan., 4 Apr. 1928, GEB 1.1 83.734.

and dramatic way. Caltech was, of course, the great success story, metamorphosing in a few short years from a local technical college into a top graduate and research university.[12] For others it was more a matter of retaking lost heights. Princeton and Cornell had been among the first colleges to offer Ph.D. degrees but quickly were outstripped by universities with large graduate and medical schools.[13] So, too, at Stanford, graduate training and research had become secondary to college teaching and became priorities again only when Ray Lyman Wilbur succeeded David Starr Jordan as president in 1916—just when Hale began to transform Throop Polytechnic into Caltech.[14]

Thorkelson's vantage point is a good one from which to get a wider view of the diversity of experience among the up-and-coming. Similar in ambition and upward mobility, Caltech, Princeton, Stanford, and Cornell differed in leadership and resources. All four projects began with the idea of developing a new science of general physiology, then gradually evolved into the GEB's standard grant to a group of departments. How smoothly this process of negotiation progressed depended on the experience of the scientists doing the negotiating, how alert they were to Thorkelson's coaching, and how well they managed their internal politics. The projects at Caltech and Princeton were the result of spontaneous planning by small, multidisciplinary groups of scientists; there was understanding between departments and presidents who mobilized support from alumni or civic constituencies.[15] Grants were consummated with remarkable ease. The failure of the quite similar projects at Stanford and Cornell reveals how essential each of these elements was.

No one was more experienced in grant-getting than the Caltech troika. There was hardly a trace of friction among them, and they showed a remarkable ability to tailor their plans to fit the GEB's machinery. They were always willing to take what could be gotten, even if it was not everything they wanted. Their ability to adapt quickly in negotiations was helped by the peculiar administrative structure of Caltech, in which Millikan, Noyes, and Hale were simultaneously department heads and administrators of the institute.[16] There were none of the

[12]Kargon, "Temple to science," 3–31. John W. Servos, "The knowledge corporation: A. A. Noyes and chemistry at Cal Tech, 1915–1930," *Ambix* 23 (1976): 175–186. Geiger, *To Advance Knowledge,* chs. 4–6.

[13]Willard Thorp et al., *The Princeton Graduate School: A History* (Princeton: Princeton University Press, 1978). Ben F. Wilkins, Jr., "The development of post-graduate studies at Cornell: The first forty years, 1868–1908" (Ph.D. diss., Cornell University, 1964).

[14]Thorkelson memos, 21 Feb., 29 Mar. 1927, GEB 1.4 653.6809.

[15]Thorkelson to Christensen, 8 July 1925, 3 Aug. 1926, GEB 1.4 641.6712. Thorkelson to Farrand, 30 Apr. 1926; Thorkelson memo, 17 June 1926; both in 649.6777.

[16]Kargon, "Temple to science."

conflicts of interest and failures of communication between departments and president that immobilized more traditional universities.

The troika's designs on the GEB in 1923–1928 focused on their plans for a new division of biology and for enlarging Noyes's division to include organic and biochemistry (the basis, Hale hoped, for an elite school of medicine and medical sciences).[17] The proposal that Millikan placed on Rose's desk in early 1925 was a masterpiece of grantsmanship. It included half-a-dozen items, ensuring that there would be something that would fit the GEB's program. Organic chemistry, biochemistry, and biophysics were proposed as the first steps toward a $5 million plan for a complete division of biology, as soon as Hale and Millikan could attract a person of sufficiently heavy caliber (like T. H. Morgan).[18] But just in case these specific items did not appeal, Millikan put at the top of his list additional endowment for the departments of physics, mathematics, and chemistry. It was an insurance policy against the risk of moving too far beyond the GEB's older mode of patronage—a crafty move, as it turned out. The entire project came to $5,150,000, of which the GEB was asked for $1,375,000.[19] Ten days later, Thorkelson asked Millikan if he would be interested in a pledge of $450,000 toward $1,350,000 for physics, mathematics, and chemistry, plus a new chair in organic chemistry. The deal was closed without further ado.[20]

The Princeton project took a more circuitous route to a similar conclusion. It was the brainchild of a small group of scientists, Edwin G. Conklin (biology), Karl T. Compton (physics), Hugh S. Taylor (physical chemistry), and Henry N. Russell (astronomy). All were seasoned scientific entrepreneurs. Conklin, for example, was involved with Frank R. Lillie in getting foundation funds for Woods Hole, and was helping Edwin Embree devise his program in human biology.[21] All four were active in the NRC: Conklin was a leader in the biology division and fellowship board; Taylor, Compton, Russell, and the biologist E. Newton Harvey together chaired seven important NRC committees and were active in planning cooperative, sponsored research programs. Their ambitious ($3.5 million) proposal resonates with the familiar NRC rhetoric of cooperative research and coordination of disciplines. They had been careful, however, to temper collectivist ideals to the realities of academic territoriality. They had thought at first of creating new depart-

[17]Rose memos, 1, 2, 4 Oct. 1923, GEB 1.4 611.6468.

[18]"Application to the General Education Board," 19 Jan. 1925, GEB 1.4 612.6476.

[19]Ibid.

[20]Thorkelson to Millikan, 29 Jan. 1925; Millikan to Thorkelson, 5 Feb. 1925; Brierley to Millikan, 10 Mar. 1925; all in GEB 1.4 612.6476.

[21]Ron Doel, "Department of Biology, Princeton University 1918–1932," (unpublished paper, 1985).

ments of biochemistry and biophysics, in which programmatic research could be done on the interaction of radiation with biological and physical materials. That idea, however, was abandoned before it got to the GEB. The political problems of redrawing departmental boundaries would have been formidable, as the Compton group no doubt realized.

The plan presented to the GEB in 1925 focused instead on strengthening existing departments and lines of research by individuals who were already working on radiation problems. Thus, the physicists would extend their work on spectroscopy and atomic structure, while the chemists would build on work begun by Hugh Taylor on the role of light in chemical reactions and photosynthesis. In biology, Newton Harvey was already the premier specialist in bioluminescence, and Conklin proposed (somewhat vaguely) to see if theoretical chemistry and physics could be applied to cell division and differentiation. By pursuing related lines of research on radiation, each discipline could make use of the expertise of all the others without having to disrupt the established order of disciplines.[22] It was a quintessentially American and postwar style of collective action. It also seems to be one case in which talk of cooperation was not just rhetoric. Visiting Princeton in October 1925, Thorkelson was impressed by the active cooperation between Harvey, Taylor, and Compton. He was impressed, too, by the communal spirit among the graduate students, who were housed in a residential dormitory to counteract the tendency to narrow specialization.[23]

Conklin, Compton, Taylor, and Russell also carried sufficient weight to exclude disciplines that did not fit the programmatic theme, like geology and paleontology (both excellent departments) and botany. This caused some resentment. George H. Shull, an eminent geneticist who was in the midst of a vast research project on primroses, let it be known that he did not like being left out and later tried his own gambit with the GEB.[24] But it was a surprisingly harmonious process, perhaps because it was limited. In other projects, disciplinary self-interest and rivalries made concerted action difficult. It is no wonder that Rose pointed to Princeton as the paradigm of spontaneous and cooperative planning.

Cooperation between the Compton group and Princeton's president, John G. Hibben, was less harmonious. The GEB project was a

[22]Compton, Conklin, H. Fine, "Memorandum for Dr. Wickliffe Rose," 22 May 1925; H. A. Smith to Rose, 13 June 1925, GEB 1.4 675.6990.

[23]Compton et al., "Memorandum," p. 4 and enclosure, pp. 5–6. Thorkelson memo, 20 Oct. 1925, GEB 1.4 675.6990.

[24]Compton et al., "Memorandum." Thorkelson memos, 15 Aug. 1927, 16 Aug. 1928; G. H. Shull to Thorkelson, 19 Aug. 1927; Mason memo, 30 Oct. 1928; all in GEB 1.4 675.6991.

small part of a $20 million fund campaign, most of which was for residence halls, a chapel, and other collegiate things.[25] Hibben had his own pet projects, and endowment of scientific research was not one of them. He was also a good deal less canny than the Compton group in dealing with the GEB. He kept pressing Rose to consider his top priorities, namely, the undergraduate engineering school and a new chemistry building. He took little part in negotiating the GEB's project and did not seem to know or to care much about it.[26]

Hibben's approach made the GEB a little nervous. Including the science project in a collegiate fund drive, they feared, might suggest that the GEB was still interested in college endowment. Rose and Thorkelson also balked at the specific program in radiation research. Should they even appear to endorse a particular research agenda, they were sure to bring upon themselves a flood of similar proposals. (We can now see how wise Millikan had been to omit such specifics from Caltech's proposal.) Thorkelson suggested that the project be recast as a special campaign to raise $3 million of endowment for the physical and biological sciences, toward which the GEB would pledge $1 million. Compton acceded with alacrity and with Thorkelson quickly worked out the details: a research professorship for each department, research instructorships in physics, fluid research funds, and equipment—standard GEB format.[27]

Differences between Caltech and Princeton began to be apparent as fund raising began. Princeton and Caltech had quite different constituencies on which to draw. Hale and Millikan's chief benefactors were local civic and industrial boosters, who saw Caltech as a key element in developing a vibrant regional economy and culture.[28] Princeton, in contrast, had to rely on its college alumni, who, though wealthy and well connected, had little interest in scientific research and researchers. It was not that alumni were hostile to science, as Princeton's chief fund raiser Alexander Smith feared; they were just uninterested. Canvasing local alumni societies, Smith found little outright opposition to the GEB scheme, but got nowhere trying to raise matching funds from nickel-and-dime contributions. Smith did better cultivating the few alumni who were interested in science, but building a constituency

[25]Rockefeller, Jr., to Rose, 17 Apr. 1925; Rose diary, 30 Apr. 1925; Thorkelson memo, 22 May 1925; Thorkelson to Compton, 21 Oct. 1925; all in GEB 1.4 675.6990.

[26]Hibben to GEB, 21 Oct. 1925; Thorkelson to Hibben, 22 Oct. 1925; both in GEB 1.4 675.6990.

[27]Thorkelson memo, 28 Oct. 1925; Compton to Thorkelson, 22 Oct. 1925; Thorkelson to Compton, 26 Oct. 1925; "Budget for proposed research program," n.d. [Dec. 1925]; Brierley to Hibben, 21 Nov. 1925; all in GEB 1.4 675.6990.

[28]Kargon, "Temple to science."

from scratch was frustratingly show. Conklin complained that the administration was half-hearted and seemed to feel that the scientists themselves should take responsibility for raising the money.[29] However, the funds were finally raised, and the GEB's pledge was redeemed late in 1928.

While Compton and Conklin chafed, the Caltech troika moved swiftly ahead to the second and third stages of their grand scheme. In early 1927, with the last matching funds in sight for the first pledge, they rushed a second request to the GEB for an endowment of five departments: physics, mathematics, and geology ($250,000); organic chemistry ($625,000); and $1,250,000 as a start toward a division of biology.[30] This package was a compromise: Hale wanted to put all into a division of biology as a lure to T. H. Morgan, while Noyes wanted to develop organic and biochemistry as a first step. (Thorkelson thought both plans were too ambitious and suggested holding off on biology until Morgan said yes.)[31] But $50,000 a year was sufficiently sweet to tempt Morgan, and the whole of the GEB's grant of $1.05 million was diverted to biology. Within a year the matching million was secured.[32]

Thorkelson worried that the Caltech group would outreach themselves, but they always seemed able to make good on their promises of matching funds.[33] A third application was made and approved in May 1928, this one for $1 million toward $3 million for organic chemistry, geology, physics, and mathematics.[34] A crucial part of the Caltech group's success, clearly, was their ability to improvise and change plans to fit changing opportunities, without causing internal conflict. Noyes's plans for chemistry were deferred again and again, yet he never complained. Such altruism would probably have been impossible in an older institution with traditional departments and vested interests.

Princeton, in contrast, never managed to capitalize on the success of the GEB's initial grant. Hibben continued to press Rose on engineering and a chemistry laboratory, which in fact had already been built using

[29]Smith to Thorkelson, 26 Oct. 1925; Thorkelson memos, 6 Apr. 1926, 14 Nov. 1927; all in GEB 1.4 675.6990–6991. Conklin to Morgan, 30 May 1928, cited in Doel, "Department of Biology," p. 30.

[30]Brierley to Millikan, 9 June 1927, GEB 1.4 612.6476.

[31]Thorkelson memos, 24 Mar., 6, 8, 18 Apr. 1927; Thorkelson to Millikan, 25 Apr. 1927; Hale to Rose, 24 Apr. 1927; Millikan to Rose, 16 Feb. 1927; all in GEB 1.4 612.6476.

[32]Garland E. Allen, *Thomas Hunt Morgan* (Princeton: Princeton University Press, 1978), ch. 9. Hale to Rose, 17 Oct. 1927; Rose to A. Flexner, 3 Nov. 1927; both in GEB 1.4 612–6476. Millikan to Rose, 3, 16 Feb. 1928; Rose to Millikan, 7 Feb. 1928; Thorkelson memo, 23 Mar. 1928; all in GEB 1.4 612.6477.

[33]Thorkelson memos, 9 Apr., 26 Oct., GEB 1.4 612.6476.

[34]Rose diary, 8 Feb. 1928; Millikan to Rose, 3, 16 Feb. 1928; Thorkelson memo, 23 Mar. 1928; all in GEB 1.4 612.6477.

funds borrowed from general endowment when an alumnus reneged on a promise. Although the GEB usually declined to make good on a debt, in this case they agreed to a pledge of $1 million. Unwilling alumni, however, plus the deepening depression, conspired to defeat Hibben's efforts to raise the matching funds.[35]

Counterpoint: Stanford and Cornell

The dissonances that ruffled the Princeton project scuttled the projects at Stanford and Cornell. At neither of these institutions was there a core of scientists experienced in planning and working together. Nor did they have Princeton's wealthy alumni or Caltech's civic boosters. Thorkelson could coax and coach, but without strong scientific leadership, cooperative presidents, and sympathetic financial backers projects ripened slowly or died for lack of funds. Stanford president Ray Lyman Wilbur had been trying to transform an elite college into a research university by piecemeal improvisation, acquiring a few research stars to leaven teaching departments. L. G. M. Becking, a scion of the famous Dutch school of plant physiology, had been made a research professor in biology. (He had come to California on his honeymoon, run out of money, got temporary work at Stanford, and stayed on.) James McBain had been imported from England to start a program in the fashionable field of colloid chemistry. But Wilbur's European stars were not the sort to draw others into cooperative projects. Becking worked in splendid isolation at the marine station at Pacific Grove; McBain turned out to be temperamentally incapable of working with anyone.[36]

Wilbur, not the scientists, was the force behind the GEB project. He was active in the NRC and a trustee of the RF, and he had been instrumental in getting GEB funds to rebuild Stanford's medical school.[37] The medical connection, in fact, was the key to Wilbur's hopes to upgrade the natural sciences. Like Hale and Noyes, he saw biochemistry and biophysics as the strategic point where the momentum of medical reform could be imparted to the natural science departments. Wilbur made Becking and McBain the centerpiece of his appeal to the GEB. A few such stars, he assured Thorkelson, would transform traditional college departments. Thorkelson was not so sure; he had seen how little Wilbur's new men cooperated and how little progress had been made

[35]Thorkelson memo, 14 Nov. 1927; Rose diary, 14 Feb. 1928; Mason memo, 30 Oct., 27 Nov. 1928; all in GEB 1.4 675.6991. Mason diary, 31 Jan. 1929; Brierley to Hibben, 27 May 1929; both in GEB 675.6992.

[36]Thorkelson diary, 24 Feb., 5 Oct. 1926, 2 Nov. 1927, GEB 1.4 653.6809.

[37]Thorkelson memo, 5 Oct. 1926, GEB 1.4 653.6809.

toward improving the conservative Department of Biology. Wilbur's scheme, he thought, had the look of a last-minute improvisation.[38]

Thorkelson was eager to help Wilbur adapt his plans to GEB expectations. He noted how similar Stanford was to Princeton and thought that except for Caltech it was "the most promising private institution on the [west] coast for developing a center of university grade of instruction and research."[39] Within a year he and Wilbur had worked out a proposal that provided for new positions in mathematical physics, bacteriology, and biophysics (building on Stanford's strength in physics); also, a faculty research fund. The GEB pledged half of a total of $1,500,000.[40]

Despite liberal terms, Wilbur was not confident that he would be able to raise $750,000 for the sciences. (He had been trying to raise $4 million of general endowment, but was still far from his goal.) In his negotiations with Thorkelson, Wilbur tried every possible way to circumvent the GEB's requirement of matching funds, protesting that Stanford's collegiate alumni were not interested in research and graduate training. But Thorkelson insisted that the purpose of the GEB's program was to build a constituency for graduate research training.[41] Suggesting that Wilbur begin with a smaller project, Thorkelson only provoked a charge that the GEB was biased against California.[42] Wilbur was quite right to worry about matching a large grant. With much labor, he got pledges of $627,000 from alumni (most of it in amounts of $10–$15) but found it impossible to collect.[43] The GEB's pledge was never redeemed.

Cornell's president Livingston Farrand was in a similar position. Rural upstate New York offered little civic and commercial support, and Farrand had to rely on alumni whose loyalties were to the college. Ernest Merritt's school of experimental spectroscopy was the only important research group. The chemistry department turned out B.A.'s for industry; biology was old-fashioned and split between the college and the School of Agriculture (the real power at Cornell, as medicine was at Stanford).[44] Like Wilbur, Farrand hoped to leaven his

[38]Thorkelson memos, 24 Feb., 5 Oct. 1926, 7 Feb., 29 Mar. 1927, GEB 1.4 653.6809.

[39]Thorkelson diary, 29 Mar. 1927, p. 13; 5 Oct. 1926; both in GEB 1.4 653.6809.

[40]Thorkelson to Wilbur, 19 Oct., 18 Nov. 1927; Wilbur to Thorkelson, 24 Sept. 1927; both in GEB 1.4 653.6809.

[41]Thorkelson to Wilbur, 25 Apr. 1927; T. L. Hungate diary, 3 Mar. 1924; both in GEB 1.4 653.6807. Thorkelson to Wilbur, 24 May, 3 June, 19 Oct. 1927; Thorkelson diary, 29 Mar. 1927; Wilbur to Rose, 27 June 1927; all in GEB 1.4 653.6809.

[42]Wilbur to Thorkelson, 28 Apr. 1927, GEB 1.4 653.6809.

[43]Wilbur to Brierley, 4 Oct. 1928, GEB 1.4 653.6809.

[44]Thorkelson memos, 19 Mar., 12 Apr. 1925, 5 Mar. 1926, 9 Dec. 1927, 16, 20 Apr. 1928; all in GEB 1.4 649.6775. Merritt to Thorkelson, 14 May 1926, GEB 1.4 649.6777.

departments with a few European stars—hence his request that the GEB help to bring over Max Born and James Franck. Thorkelson encouraged Farrand to formulate a larger plan for all the science departments and assured him of the GEB's strong interest in Cornell.[45] When a year went by with no word from Farrand, Thorkelson gave him a nudge—Why had the expected proposal not arrived?

Several things inhibited Farrand from taking up Thorkelson's gambit. He doubted that he could ever raise the substantial sums required for the standard GEB grant. Also, he had not succeeded in devising a rationale for a large project. He had assembled a committee of the science departments, hoping that they would put together a plan, but only Merritt's physicists had shown any capacity to take an ecumenical view. The chemists and biologists were interested only in projects that would aid their parochial disciplinary concerns.[46] It was the problem of weak departments: improvement, for those in charge, spelled loss of status and authority.

Farrand proposed bringing in an experienced and disinterested outsider to do a survey and help formulate a plan: someone, say, like Karl Compton. (It was the physicists who came up with the idea.) Thorkelson suggested that Harlow Shapley would be an even better mediator, recalling, no doubt, Shapley's role in the IEB's surveys and planning.[47] In fact, the role of chief planner was filled by another veteran of the IEB's surveys, the energetic dean of the agricultural school, Albert R. Mann. Mann had the vision, experience, and relentless drive to get fractious individualists to agree on common goals. The biology departments in the agricultural college were strong, and the availability of state and federal funds for research had accustomed agricultural scientists to project-style research.[48] The GEB set aside $15,000 for a full-scale survey.

Mann needed no encouragement to think big. He wanted Cornell to become a national center of research in the natural and agricultural sciences, especially in modern specialties—plant and animal biochemistry, biophysics, general physiology—that would connect physics and chemistry to biology and agriculture. Farrand, too, saw general physiology as the nucleus of a special school, separate from the undergrad-

[45]Farrand to Rose, 28 Apr. 1926; Thorkelson to Farrand, 30 Apr. 1926; Thorkelson memos, 12 Apr., 17 June 1926; all in GEB 1.4 649.6777.

[46]Thorkelson to Farrand, 30 Apr. 1926, 29 Nov. 1927; Farrand to Thorkelson, 5 May 1926, 2 Dec. 1927; Thorkelson memo, 9 Dec. 1927; all in GEB 1.4 649.6777.

[47]Thorkelson memo, 9 Dec. 1927.

[48]Thorkelson memos, 9 Dec. 1927, 16, 20 Apr. 1928, GEB 1.4 649.6775. Mann to Thorkelson, 28 Apr. 1928; Thorkelson to Mann, 27 Apr. 1928; both in GEB 1.4 649.6777.

uate departments, and devoted entirely to research and graduate training.[49] At $9 million, Mann's grand scheme was obviously much too big to do all at once, and Thorkelson suggested that Mann and Farrand start by raising $3 million for a central building. Called a "center for general physiology," it would in fact bring together under one roof the graduate teaching and research of existing science departments.[50] Despite liberal terms, however, $1.5 million was far more than Farrand could hope to get from his limited constituency. No benefactors came forward, and the GEB's pledge was never redeemed.

The Caltech, Princeton, Stanford, and Cornell projects reveal the essential ingredients of institutional grant-making: scientists with the know-how to deal with foundations and manage disciplinary conflicts, sympathetic and flexible university presidents, loyal financial backers, and foundation managers good at tinkering together complex packages. It is also clear how this system of patronage gave the advantage to universities that already had strong science departments, like Caltech or Princeton; without a head start, institutions found it difficult to build a local constituency before the depression put all raising of endowment on ice.

The GEB's style of grant-making was more strictly institutional and less flexible than that of the IEB. There were no projects in the United States quite like the modest grants to specialized laboratories in Scandinavia or the Netherlands, and in hindsight perhaps there should have been. But Rose and Trowbridge's system of individual mentorship did not extend to American projects, and Thorkelson, lacking Trowbridge's scientific authority and freedom of initiative, was not able to make technical decisions about equipment and research agendas. Thus, the GEB's projects were uniformly large endowments for groups of departments, a form of patronage that was risky for institutions that were just beginning to acquire star-quality researchers and to build constituencies for scientific development.

View from the University of Chicago, Johns Hopkins, Harvard

For old elite universities, the problem was not getting to the top but staying there, and that almost inevitably meant doing in new ways what had worked for them in the past. In their heyday, Johns Hopkins and

[49]Thorkelson memo, 16 Apr. 1928, GEB 1.4 649.6775. Mann to Thorkelson, 21 Apr. 1928; Farrand to Thorkelson, 30 Apr. 1928; both in GEB 1.4 649.6777.

[50]Thorkelson memo, 20 Apr. 1928; Farrand to Thorkelson, 30 Apr. 1928; Farrand to GEB, 27 June 1928; Brierley to Farrand, 8 June 1928; all in GEB 1.4 649.6777.

the University of Chicago had enjoyed a real advantage in resembling most a comprehensive graduate university. It was a fleeting advantage, however, as the system grew larger and more competitive. To the jaundiced eye of Abraham Flexner, both Chicago and Johns Hopkins had been at their zenith at the start: there were still excellent departments in both, but in neither had the original ideal been upheld, that of being the best in every department.[51] Similar problems faced institutions like Columbia, Harvard, and Yale, where graduate programs had evolved within undergraduate departments. Abraham Flexner thought them "nondescript."[52] Like the cosmopolitan universities of Europe, they had grown through the efforts of individual entrepreneurs, who bequeathed to their universities a chaotic array of fiefdoms. "You know how individualistic Harvard has been throughout its history," Rose noted. "This extreme individualism has manifested itself in the field of science by the scattering of its science all over the place. . . . The university has had large resources in biology but they have been distributed in about five or six independent compartments."[53]

In the 1920s the old elite sought to regain their former eminence through bold reforms from the top down. The idea of a "real" university for research and graduate work was revived. Steps were taken at Johns Hopkins and Chicago to get rid of their undergraduate colleges, and at Harvard there were plans to consolidate the sciences into a few supradepartmental divisions. Usually such initiatives came from the top, from Chicago's new president Ernest D. Burton, Frank J. Goodnow of Johns Hopkins, and Harvard's A. Lawrence Lowell. These reforms were designed to restore the original ideals of a true university by stripping away the accommodations that earlier leaders had to make with the collegiate system. The idea was to build a new top story to the institutional pyramid, to elevate a few above the burgeoning middle ranks.[54]

Chicago was quicker off the mark than Johns Hopkins but less radical. President Burton proposed not to abolish the undergraduate college but to create a new and separate group of graduate departments.[55] Burton played on fears that the university was no longer the leader it once was in a region dominated by large public universities, whose budgets for research and graduate students were coupled to ex-

[51]A. Flexner, "A proposal to establish an American university," 27 Dec. 1925, GEB 1.4 588.6257.

[52]Ibid.

[53]Rose to Trowbridge, 29 Feb. 1928, IEB 1.1 18.266.

[54]Geiger, *To Advance Knowledge,* ch. 3.

[55]Burton, "The future of the University," 10 Feb. 1923; Burton and H. H. Swift to GEB, 5 May 1923; Burton to GEB, 2 Nov., 1923; all in GEB 1.4 657.6839.

ploding undergraduate enrollments and an expanding tax base. The only way for the University of College to keep the lead, he argued, was to capitalize on its special advantage as "an experiment station in the whole field of education."[56] Never having been dependent on its undergraduate college, Chicago was not hindered from taking bold steps by college alumni, as were older Eastern universities. But large new funds would be required.[57]

The leaders of Johns Hopkins stepped out even more boldly. Goodnow's plan, announced in January 1926, called for the abolition of the undergraduate college and the creation of the exclusively graduate university, of which Daniel Gilman had dreamed in 1876.[58] Goodnow was well aware that he faced a drastic loss of income from tuition fees and the State of Maryland's annual subsidy, to say nothing of the long-term loss of support from local benefactors and alumni.[59] The success of Goodnow's plan depended on his being able to raise an endowment of $10 million to offset such losses and raise all the departments to a uniform level of excellence. Naturally, Goodnow looked to the GEB.

The one enthusiast in the GEB for Goodnow's and Burton's schemes was Abraham Flexner, who had long cherished a dream of creating a "real university." He returned inspired from a visit to Burton, writing Wallace Buttrick: "We are going to have no end of fun together—you and Rose and the rest of us and Burton."[60] But his highest hopes were for Johns Hopkins, and he pulled every string he could to ensure that the Rockefeller boards would take quick action on Goodnow's plan. Raymond Fosdick received a characteristic admonition: "If you don't talk this thing over with Vincent and get some interest enlisted in the next few months, may the League of Nations go to the eternal demnition bowwows! There now!" On the same day, Vincent received a reminder to get after Fosdick: "These are the big things and for God's sake let us push them ahead and drop the little ones."[61]

Flexner was alone in his enthusiasm, however, George Vincent was ambivalent, as usual: a large project might force the Rockefeller boards

[56]Burton, "Future," pp. 2, 6–7.

[57]Ibid., pp. 6–8, 13–14.

[58]"From the minutes of the board of trustees of the Johns Hopkins University," 5 Jan. 1926, GEB 1.4 588.6257. F. J. Goodnow, "Address to the alumni," *Johns Hopkins Alumni Mag.* 13 (1925): 229–242.

[59]Goodnow to A. Flexner, 28 Oct. 1925; Rose memo, 26 Mar. 1926; both in GEB 1.4 588.6257. It was for just these reasons that Burton had resisted the counsel of some of his faculty to go whole-hog and abolish the college. Burton, "Future," pp. 9, 12–13, 18–19.

[60]A. Flexner to Buttrick, 30 Jan. 1923, GEB 1.4 657.6839.

[61]A. Flexner to Fosdick, 27 Dec. 1925; Flexner to Vincent, 27 Dec. 1925; both in GEB 1.4 588.6257.

to cooperate, but the price seemed very high.[62] Fosdick pronounced Goodnow's plan "very significant" but gave no sign of being deeply inspired. Rockefeller, Jr., favored a large capital grant as a way of speeding up the liquidation of the GEB—not exactly what Flexner had in mind.[63] Rose remained conspicuously quiet, his usual way of letting unwelcome ideas self-destruct. In November 1925 the GEB officers went on record against concentrating on a few large projects. Rose voiced the consensus view: that the GEB should aid individuals' research wherever opportunities, big or small, were found.[64]

It became clear within a few years that Burton and Goodnow's plans had no future. They were just too complicated and expensive. There is no evidence of enthusiasm from the faculties, and no progress was made in raising funds.[65] The postwar faith in large-scale organization was waning in the mid-1920s, and philanthropic zeal flagged as the public became jaded with constant appeals for funds. It is all quite reminiscent of the efforts to create one great university in Bloomsbury and to unify the science institutions of Paris.

A strategy of incremental improvement of specific departments proved more practicable, as it always had. Goodnow began to tout the advantages of an evolutionary strategy, much to Abraham Flexner's disgust: "[T]he presentation seems to me a very slight nibble at a very big affair. If it is called 'evolution,' perhaps it looks better than it would under any other title, but it will . . . be a geological process."[66] Geological perhaps: but American universities had always grown in that way, and ingrained habits and departmental interests made it almost impossible to engineer radical change from the top down, even if the GEB had wanted to do it.

The main result of Goodnow's and Burton's schemes was to stall for two years smaller but more easily realized projects initiated by Thorkelson in 1924 and 1925. When the process of negotiation finally revived, projects assumed the standard GEB form of grants to departments. But two crucial years had been lost: the last two years of prosperity before the crash, two years in which other groups had gotten ahead in the line for GEB largess. Success went to those who were

[62]Vincent to Fosdick, 5 Jan. 1926; Flexner to Fosdick, 27 Dec. 1925; both in GEB 1.4 588.6257.

[63]Fosdick to Flexner, 3 Mar. 1926; Flexner to Fosdick, 26 Mar. 1926; both in GEB 1.4 588.6257.

[64]Officers conference, 22 Nov. 1925, vol. 1, pp. 11–15, GEB.

[65]Rose diary, 26 Mar. 1925; Goodnow to Rose, 5 Apr. 1927; both in GEB 1.4 588.6257.

[66]A. Flexner, "Memorandum on President Goodnow's letter," 8 Apr. 1927; Goodnow to Rose 5 April 1927; both in GEB 1.4 588.6257.

quickest to retreat from radical restructuring to the more familiar and pragmatic mode of catch-as-catch-can.

The University of Chicago did rather better than Johns Hopkins.[67] The GEB's 1927 pledge ($1.5 million toward $2.79 million) was a typical grab bag, with hardly a trace of programmatic intent. It provided for completion of the chemistry building, an addition to the Ryerson Physics Laboratory, and greenhouses for the Department of Botany. In contrast, an ambitious scheme to unite the biological sciences and psychoneurology in one division never got to first base.[68] A similar scheme proposed by Johns Hopkins biologist, Herbert Spencer Jennings, similarly stalled. Jennings called for $3 million (half from the GEB) for a new laboratory building and research endowments for zoology, botany, plant physiology, and experimental psychology. These departments, in Jennings's scheme, would become institutes exclusively for graduate and postdoctoral training and for large research projects of the kind that could not be carried out in ordinary academic departments. (Jennings, e.g., had long wanted to do a long-term study of the effects of environment on successive generations of lower organisms.)[69] Jennings's idea of research departments would have been fine in the "real university scheme" and was probably proposed with that in mind, but it was quite out of tune with the GEB's standard departmental grant. Despite strong support for the natural sciences from the new president, Joseph Ames (a physicist), the ideal of a purely graduate university made it difficult to scale back proposals to a doable size. Thus, when the "real university" scheme failed, Jennings's plan went down with it. The GEB's science program went out of business without making a grant to Johns Hopkins.

A similar script was played out to a different ending at Harvard, where the biologists, aided by Harlow Shapley, asked the GEB for a new biology laboratory. At first, the idea was simply to consolidate scattered departments and research fiefdoms under one roof. However, Shap-

[67]Thorkelson memos, 11–12 June 1924, 6–8 July 1925, GEB 1.4 660.6862.

[68]A. Compton to Mason, 1 Dec. 1926; Thorkelson memos, 9 Dec. 1926, 4 Jan., 15 Feb., 29–30 June 1927, 31 Mar., 29 May 1928; Thorkelson to Mason, 5, 9 Jan. 1927, 20 Apr. 1928; Mason to Thorkelson, 11 Par. 1928; Brierley to Mason, 3 Mar. 1927; F. Woodward to Arnett, 10 Nov. 1928; all in GEB 1.4 660.6862. "The biological sciences and psycho-neurology," 10 Apr. 1929, GEB 1.4 660.6864. "A proposal for biological teaching at the divisional level," n.d. [Feb. 1931], GEB 1.4 660.6866 and 660.6862. A second pledge of $298,000 was made for additional staff in physics, mathematics, and chemistry.

[69]Thorkelson memos, 19 May, 15 June 1925, 27 Nov. 1926; H. S. Jennings, "A memorandum concerning the biological departments," July 1925; J. Ames to Rose, 10 July 1925; Jennings, "Biological sciences in the philosophical division," 22 Dec. 1926; Jennings to Thorkelson, 27 Nov. 1926; Thorkelson memo, 22 Dec. 1926; all in GEB 1.4 588.6257.

ley's committee went further, envisioning a set of specific, long-term research projects that groups of biologists could undertake in a cooperative way (e.g., oceanography, paleobiology, animal psychology, human biology, agricultural biology). These research projects would be organized as special research institutes, outside the system of teaching departments. At that, President Lowell balked: he wanted to reduce the number of independent departments and research enclaves, not create new ones. He refused flatly to have anything to do with raising funds for Shapley's scheme and made it clear that university development was the president's job, not that of departments or the GEB. Institutes and cooperative research were quickly dropped, and the GEB pledged $2 million for a new building, within which the biological sciences operated as more or less independent departments.[70]

Proposals to transform university departments into research "institutes" were quite common among leading universities in the late 1920s. In 1928, for example, Arthur Compton sounded out the GEB on an "institute" of physics at the University of Chicago. He did not get far.[71] A few years later Frank Lillie tried to sell the RF on an "Institute for Genetic Biology," with no greater success. After lengthy negotiations, the project ended up as a fluid research fund for the separate biological departments.[72] Making departments into institutes for research and graduate training, like creating purely graduate universities, was a strategy for escaping the inherent limitations of the American department system. They were an effort to undo the historical process by which research had been accommodated within the confines of the traditional college system.[73] Presidents dreamed of "real" universities or supradepartmental divisions, while professors dreamed of quasi-autonomous "institutes." All were equally out of tune with the reality of departmental control.

GEB leaders, except for Flexner, were not interested in any kind of radical restructuring. By turning a deaf ear to proposals for specific re-

[70]Rose memo, 6 May 1925; Shapley to Rose, 16 Feb. 1928 and enclosures; Rose memos, 25 Feb., 9 Mar. 1928; Thorkelson memo, 28–29 Feb. 1928; Rose to Trowbridge, 29 Feb. 1928; Lowell to Rose, 16 Mar., 15 May 1928; Osterhout to Rose, 10 May 1928; Brierley to Lowell, 15 June 1928; all in IEB 1.1 18.266.

[71]A. Compton, "An institute for research in the physical sciences," 9 Jan. 1928; Compton, "A proposed extension of the program of research in the physical sciences," 18 Jan. 1928; Thorkelson memo, 29–30 July 1927, 9 Jan., 31 Mar. 1928; all in GEB 1.4 660.6862.

[72]RF 216D 8.103 ff.

[73]Kohler, "The Ph.D. machine, a view from the shop floor." Some people expected that junior colleges, which were becoming popular in the 1920s, would enable universities to drop the first two years of college. Wilbur, "Further limitations of the lower division," 19 Apr. 1927, GEB 1.4 653.6807. Thorkelson diary, 20 Jan. 1928, GEB 1.1 76.667.

search projects and for "real university" schemes, the GEB added its institutional weight to the traditional system of mixed collegiate-graduate departments. The old elite discovered that a glorious past did not give them a competitive advantage over upstarts like Caltech. The GEB's projects at Johns Hopkins, Chicago, and Harvard turned out not very different from those at Princeton or Caltech, just as the IEB's projects at Paris and London turned out like those at the smaller provincial universities. Opportunistic, piecemeal improvement of departments was the best strategy at all levels of the university system. The GEB's institutional mode of patronage was an integral part of that system.

At the Grass Roots: State Universities

The process of patronage also goes a long way to explaining why the GEB made so few projects in public universities and in the scientifically less-developed regions of the United States. Certainly, it was not because of regional bias or elitism; Thorkelson's interest in the state universities of the Middle West was as keen as Trowbridge's in the universities of Latin Europe. Rose and Thorkelson took a similar interest in the American South. They tried hard to discover mechanisms for working with state universities, though they had not gotten very far when time ran out in 1928. What Spain, Italy, and Central Europe were to the IEB, the American South and Southwest were to the GEB. The GEB's experiments there reveal a good deal about the institutional system of patronage and its limits.

Strong personal feelings inspired Rose and Thorkelson's concern with the South and Midwest; both were deeply attached to their regions of birth. Rose spent his whole life working to improve education and health in the South. Thorkelson's feeling for the Midwest was deeply rooted in the egalitarian ideals of Norwegian-American culture, and in that special mix of populism and cultural striving that gives the upper Middle West its distinctive social character. Thorkelson's reports of visits to the universities of Wisconsin, Minnesota, Iowa, and so on betray a sense of being among old friends—which, indeed, he often was. Rose, too, seemed to expand on those rare occasions when he went "afield" in the South.

Opportunities for patronage in the state universities, as in other sectors of the American university system, were structured by their historical development. Every observer was struck by the extraordinary progress of the Midwestern state universities after World War I. It was the fruit of decades of active reform. Populated by immigrants from New England and transplanted Germans and Scandinavians, this region had long boasted more colleges per capita than any other, though standards of credentialing for academic careers lagged a generation behind those of New England. In the early 1900s, however, presidents of state universities discovered that graduate education and research, if packaged properly, had a strong appeal to state legislators. Graduate schools were reorganized on Eastern lines in the 1910s. Research expanded as states began to regulate agriculture and manufacturing, and connections with state bureaus of agriculture and geological surveys provided resources for research that few private universities enjoyed.[74] The regional habit of college going swelled enrollments and fueled a building boom, leaving many state universities with excellent laboratories. In the 1920s the region's cultural and material advantages were paying off—the center of gravity of the university system moved West.

Rose and Thorkelson wanted very much to be part of this great movement. The GEB had once prohibited projects in public institutions (Frederick Gates's doing), but this prohibition had been successfully challenged by Abraham Flexner just before Rose took charge. Over Gates's still very lively body, the trustees approved a grant to the University of Iowa Medical School and in early 1923 officially endorsed the principle of aiding public universities. In 1924 Thorkelson was authorized to survey state universities, and in the next few years he visited most of them.[75]

However, certain features of state universities made it difficult for the GEB to get involved. The very size of their graduate programs was an impediment. Rose and Thorkelson were never interested in graduate education as such. Swelling the output of Ph.D.s had no appeal, and large operations like the school of chemistry at Minnesota made Thorkelson nervous.[76] This may also be the reason why Thorkelson took no initiatives with the University of California, despite the impressive quality of its science departments, especially Gilbert N. Lewis's School of Chemistry.[77] Also, the history and egalitarian ethos of the land-grant system encouraged every university president to think that his institution had a right to an equal share of the resources. President George Norlin, for example, did not take kindly to Thorkelson's suggestion that the University of Colorado should be content to export its brightest talents to other centers.[78] Edward A. Birge, president of the University

[74] *Trans. Assoc. State Univ.* is a good source.

[75] GEB minutes, 24 May 1923; pp. 54–55, GEB 3. Officers conference, 22 Nov. 1924, vol. 1, pp. 11–13, GEB 5. E. Richard Brown, *Rockefeller Medical Men* (Berkeley: University of California Press, 1979), pp. 176–179.

[76] Thorkelson memo, 7 Dec. 1926, 1–2 June 1928, GEB 1.4 654.6819.

[77] Thorkelson memo, 10 Dec. 1924, 22 Mar. 1928, GEB 1.4 625.6593–6594.

[78] Thorkelson memo, 28 Nov. 1924, 19 Nov. 1926; Norlin to Thorkelson, 29 Nov., 6 Dec. 1924; Thorkelson to Norlin, 29 Nov. 1924; all in GEB 1.4 692.7102.

of Wisconsin, suggested that the GEB should give money equally to all state universities, so that some would not be discouraged by the good fortune of a few.[79] The experience of federal land grants and the Hatch and Adams Act was deeply rooted, and it created expectations that were exactly the opposite of those that were needed for making high peaks still higher.

Thorkelson's work was also complicated by the haphazard way in which graduate programs had evolved out of college departments, beginning with token individuals brought in from the outside and given reduced teaching loads and research funds. Such individuals symbolized an institution's intention to participate in research, but most departments that Thorkelson visited consisted largely of teachers who had never done much research and who resisted efforts to impose it as an official responsibility. Bootstrapping was not unique to state universities, but it was most pervasive there, and it made investment by the GEB an uncertain business.

Politics also intruded in the process of making projects. The Rockefellers and their works were not exactly popular in the Middle West, where railroad rebates and monopoly had been burning political issues only a few years earlier. Populist antipathy to private foundations remained strong among some university presidents and boards of regents. The most extreme case, to Thorkelson's great distress, as at his own alma mater, the University of Wisconsin. In 1925 the regents prohibited the university from receiving any funds from foundations, against strenuous opposition from Dean Charles S. Slichter and President Birge. Although this prohibition was instigated by one regent, the fact that eight others voted with him (against six) reveals how potent antifoundation sentiment was in the state's politics.[80]

While most university administrators were eager to do business with the GEB, Thorkelson did occasionally find one who was not. President Lloyd Morey of the University of Illinois told him frankly that he did not approve of foundations meddling in the affairs of public institutions. (It was not just foundations: Morey also resented the NRC's efforts to promote research in state universities.) Thorkelson regretfully concluded that it would not be possible to do business with the University of Illinois, although its excellent science departments

[79]Thorkelson memo, 31 May 1928, GEB 1.4 682.7048.

[80]Thorkelson memos, 30 Apr. 1926, 17 Mar. 1928, GEB 1.4 682.7048. M. H. Ingraham, *Charles Sumner Slichter: The Golden Vector* (Madison: University of Wisconsin Press, 1972), pp. 175–176. Hanson diary, 8 May 1936, RF 200D 164.2013. Merle Curti and Vernon Carstensen, *The University of Wisconsin: A History* (Madison: University of Wisconsin Press, 1929), vol. 2, pp. 223–232. W. R. Shepardson to Slichter, 29 Mar. 1923; Thorkelson to Slichter, 14 Feb. 1923. UW 6/1/1 box 37 New Philosophy. Slichter's views on education and business organization were much admired in the GEB circle.

(chemistry especially) made it a prime candidate. He also concluded that the best way for the GEB to help the state universities of the upper Middle West might be to make the University of Chicago a stronger role model.[81] For Thorkelson it was a bitter pill that Illinois and Wisconsin were out of bounds.

More commonly, university presidents wanted to deal with the GEB but did not know how. Often they used the same tactics on Thorkelson that they knew were effective in prying money out of state legislatures. Some, for example, presented lists of specific research projects, requesting that Thorkelson choose which ones the GEB would like to fund. (Legislators liked specific budget lines; the GEB did not.) Others pressed aggressively for their own favorite projects or, like good horse traders, asked for very large sums in the expectation of being bargained down to something reasonable. Being a "fighting type" was adaptive to the rough-and-tumble fights for state appropriations, but it complicated Thorkelson's efforts to teach potential clients how to frame a proposal that the GEB trustees would accept. Some were quick to learn new rules, others never did.

The greatest problem, however, was the GEB's requirement of matching funds. State legislatures were legally barred from appropriating money to universities for more than one or two years at a time, and volatile state politics and legislative logrolling made it impossible even for seasoned university presidents to be sure of support from one year to the next. Thus, there was no way for state universities to come up with funds to match a pledge of endowment, or even a five-year program of annual grants. Rose and Thorkelson were not inflexible and sometimes looked the other way when states reneged on promises to provide increased support. But that subverted the fundamental aim of institutional grants, which was to build local support systems. They tried out several mechanisms for aiding state universities, but none of these experiments worked very well.

Fluid Research Funds and Other Experiments

The first such experiment was a grant of $15,000 to the University of North Carolina for a grants-in-aid fund, to be distributed by a committee of the faculty. Matching funds were not required, though it was expected that the state legislature would be inspired to volunteer extra support for research. This kind of patronage was not new: many universities had set up such funds and committees in the early 1920s. Wisconsin was one of the first in 1919, and appropriations rose from

[81] Thorkelson diary, 12–13 Oct. 1925, 2–3 Nov. 1927, GEB 1.4 632.6644.

$23,000, to $50,000 by 1927, largely thanks to Charles Slichter's aggressive lobbying. Cornell's Heckscher Fund was yielding $50,000 by 1927, and California's Searle Fund about the same, to which the university added an additional $35,000 from general funds.[82] Beardsley Ruml used this method with some abandon to promote research in the social sciences.[83] In fact, one of Ruml's grant funds was operating at the University of North Carolina when the GEB began its project there.[84]

Thorkelson had high hopes for these faculty research funds. He made an informal survey and passed along information to administrators who came to him for advice on encouraging research. But he ceased to see such funds as the vehicle for GEB aid. The results of the North Carolina project fell far short of what Rose and Thorkelson had hoped for. The researches done were small and scattered among various disciplines, adding up to very little. Also, the grant did nothing to change the legislature's indifference to research. In his plea for renewal of the grant, President H. W. Chase had to fall back on its symbolic importance for Southern reform. But the GEB needed something more tangible than symbolism.[85] Thorkelson turned down requests for faculty research funds from the universities of Wisconsin, Iowa, and Minnesota.[86]

The GEB's other major experiment took the form of annual grants, starting with equal contributions and gradually shifting the burden to the state. In this way, legislators might be drawn into supporting research without having to face large initial costs. They might renege, of course, but Thorkelson could hope that they would gradually get used to the idea that support for research was a regular budget item. This method was formally adopted by the trustees in November 1926, though it was almost two years before the first grants were made, at the

[82]Thorkelson memo, 22 April 1925, GEB 1.4 682.7048. Thorkelson memos, 19 Mar. 1925, 16 Apr. 1928, GEB 1.4 649.6777. R. G. Sproul to Thorkelson, 26 Oct. 1927; Thorkelson memo, 22 Mar. 1928; A. O. Leuschner to W. W. Campbell, 1 July 1927. GEB 1.4 625.6593–6594. C. S. Slichter to Glenn Frank, 1 Dec. 1926, UW 6/1/1 box 13 Frank.

[83]M. Bulmer and J. Bulmer, "Philanthropy and social science in the 1920s: Beardsley Ruml and the Laura Spelman Rockefeller Memorial, 1922–29," *Minerva* 19 (1981): 347–407.

[84]Rose diary, 20 Mar. 1925; H. W. Chase to GEB, 27 May 1925; Chase to Thorkelson, 21 Apr., 6 June, 23 Nov. 1925; Brierley to Chase, 16 Jan. 1926; all in GEB 1.1 107.106.

[85]Chase to Rose, 9 Apr. 1928; Thorkelson diary, 27 Apr. 1928; both in GEB 1.1 107.106.

[86]Slichter to Thorkelson, 8 May 1925; Thorkelson memos 22 Apr. 1925, 31 Oct. 1927; all in GEB 1.4 682.7048. Carl Seashore to W. A. Jessup, 2 Feb. 1926; Thorkelson to Jessup, 17 Feb., 24 Mar. 1926; all in GEB 1.4 690.7109. Guy S. Ford to L. D. Coffman, 9 Dec. 1926; Ford to Thorkelson, 11 Dec. 1926; Thorkelson to Ford, 16 Dec. 1926; all in 654.6819.

universities of Virginia and Texas. Thorkelson and Rose moved cautiously, putting off other eager petitioners until they were sure the method worked.

It is no accident that the University of Virginia was selected as the GEB's demonstration project: among the Southern state universities it was the most eminent and the most like a private institution. Nor was it an accident that the funds came from the IEB and that Rose himself took the initiative in making it a showcase of regional development. In the fall of 1927, Rose personally laid before President Edwin A. Alderman a vision of the systematic development of science and higher education in the South, with Virginia as the paradigm. Alderman did not need to be persuaded of the potential of science for the cultural and economic uplift of the South. As Rose well knew, he had long been active in promoting that cause.[87]

Alderman had already made some progress in upgrading the university's science departments, bringing in one or two productive young scholars and giving them time and resources for research—the familiar bootstrap method but applied by Alderman with unusual skill. The new chairman of chemistry, a young and capable man with a Princeton Ph.D., was building a center of research on the mechanisms of catalytic and organic reactions.[88] The head of physics, Llewellyn G. Hoxton, had kept his hand in research as best he was able and had ambitious plans to start a modern program in spectroscopy and atomic structure. He and Alderman were wooing Jesse Beams, a young Virginia Ph.D. who had made good as an NRC fellow and instructor at Yale, but whose regional loyalties, they hoped, would draw him back to the South.[89]

The political situation also seemed auspicious. The new governor, Harry F. Bird, believed that higher education was vital to economic development, and he strongly supported Alderman's efforts to improve the university. Alderman found Bird open to his arguments that the science departments were especially deserving of support. A Southern business progressive, Bird impressed Rose with his business-like methods and lack of the populist anti-intellectualism that was so common in other state governments. Rose saw Virginia as an ideal opportunity for a demonstration of the sort that the IEB was doing in places like Spain. "If I were Czar in Virginia," Rose wrote, "I should begin by investing

[87]Rose to Alderman, 22 Sept. 1927, 8 Dec. 1927; Alderman to Rose, 9 Dec. 1927; Thorkelson memo, 2 Dec. 1927; all in IEB 1.1 21.311.

[88]Thorkelson memo, 2 Dec. 1927; A. F. Benton to Alderman, 15 Feb. 1928; both in IEB 1.1 21.311.

[89]G. L. Hoxton to Alderman, 15 Feb. 1928; Thorkelson memo 6 Mar. 1928; both in IEB 1.1 21.311.

funds in scientific research in the University." Alderman replied that he had been plying Bird with just that message, with good results.[90]

Alderman was also adroit in fashioning a project in the form that the IEB liked, proposing an eight-year grant of $15,000 per year each for chemistry, biology, and physics, with equal sums from the state. Rose did not try to impose a legal obligation on the state legislature, making the provision of matching funds an informal understanding.[91] Unlike the experiment at North Carolina, the Virginia project fully lived up to Rose's hopes. Jesse Beams did come home again, attracted in part by the IEB's grant, and made Virginia a center of research on the ultracentrifuge. The state did not always make good on its financial commitment in the depression years, but the effects of the annual IEB grants were evident in the growth of graduate programs and in the increased output of research.[92]

The circumstances that made Virginia a good demonstration were not easily duplicated elsewhere, however. Few presidents possessed Alderman's winning combination of vision and tact: certainly, President Walter Splawn of the University of Texas did not. Texas was of great interest to Thorkelson and Rose, as it was likely to become the largest university of the Southwest—and the richest. A lake of oil had been discovered under the land granted to the university by the state, and by 1926 it was producing an income of $1 million per year ($2 million in 1928). Oil had fueled a building spree, and Splawn aspired to make the university a center of graduate education and research.[93] Almost a caricature of the Texas booster, Splawn told Thorkelson there was no better place than Texas for the GEB to invest $5 million. When Thorkelson explained that was far too large a sum to be considered, Splawn handed him a long list of research projects and asked for $50,000 a year to support them. After some more negotiating, Splawn sent Thorkelson a trimmed-down list of projects, taking the GEB to task for lack of vision and challenging him to prove the GEB's goodwill with a token grant of $18,000.[94]

In fact, Thorkelson's interest in Texas focused on the one group of researchers who already possessed a national reputation. Assembled by John T. Patterson, the Texas school of cytogenetics included The-

[90]Rose to Alderman, 8 Dec. 1927; Alderman to Rose, 9 Dec. 1927; both in IEB 1.1 21.311.

[91]Brierley to Alderman, 4 June 1928, IEB 1.1 21.311.

[92]Hoxton, "A brief statement of the status in research in physics," 2 Mar. 1923, IEB 1.1 21.311. Mason diary, 23 Sept. 1929, GEB 1.1 179.1674.

[93]Thorkelson memos, 16 Mar., 29 Apr., 13 Nov. 1926, GEB 1.1 162.1517. Thorkelson memo, 9 Jan. 1928, GEB 1.1 162.1510.

[94]Splawn to Thorkelson, 13 Nov., 4, 24 Dec. 1926, 15 Jan. 1927; Thorkelson to Splawn, 29 Nov., 16 Dec. 1926; 7, 15 Jan. 1927; all in GEB 1.1 162.1517.

ophilus Painter and the future Nobelist, Hermann J. Muller. (Thorkelson saw no hope for the other science departments, a judgment confirmed years later by Warren Weaver.)[95] However, Splawn thought too big and was too accustomed to horse trading with legislators to learn how deals were made with the GEB, despite long and patient coaching by Thorkelson. His successor did better, simply authorizing Patterson to negotiate directly with the GEB. A grant for Muller's research on X-rays and mutation was soon worked out. Although the state reneged on its promise of increased funds after only two years, the GEB chose to interpret the requirement as a moral, not a legal, obligation and kept the money coming.[96]

The grant to the Texas geneticists was the only one to be made on the Virginia model, though others were in various stages of gestation when the GEB's science program came to a close. The one nearest to hatching was at the University of Michigan, where Harrison Randall had developed an eminent school of experimental spectroscopy and, by shrewd use of foundation patrons and international connections, was making Ann Arbor one of the centers of the new quantum physics in the United States.[97] Had Randall been as free as Patterson was to deal with Thorkelson, a project would undoubtedly have materialized by 1928. However, President Clarence C. Little had strong ideas of his own, which he insisted on pressing upon Thorkelson despite repeated rebuffs. Thorkelson mistrusted Little's unabashed self-interest and his habit of horning in on other people's plans. Refusing to take Randall's proposal to the legislature, Little recast it into a request for $100,000 for his own research on the genetics of cancer in mice.[98]

With much effort, Thorkelson managed to turn Little's grab-bag requests into a proposal that he hoped the GEB's trustees might take seriously. (His main problem was preventing Little from slipping in grants for himself and for miscellaneous needy departments of the university.) It was a very ambitious plan for physics and the biological

[95]Thorkelson memos, 16 Mar., 29 Apr. 1926, GEB 1.1 162.1517. Weaver diary, 30 Oct. 1933, RF 163.1518.

[96]Patterson to Thorkelson, 12 Dec. 1927; Thorkelson to Patterson, 20 Dec. 1927; Thorkelson diary, 9 Jan. 1928; Brierley to H. Y. Benedict, 4 June 1928; all in GEB 1.1 162.1517. Patterson to Thorkelson, 23 June 1931, GEB 1.1 163.1518.

[97]Stanley Coben, "The scientific establishment and the transmission of quantum mechanics to the United States, 1919–1932," *Amer. Hist. Rev.* 76 (1971): 442–466, see pp. 459–461. H. M. Randall to C. C. Little, 8 Dec. 1926; Randall to Thorkelson, 13 Dec. 1926; Rose to Little, 15 Nov. 1926; all in GEB 1.4 641.6712.

[98]E. M. East to W. H. Shepardson, 2 Nov., 8 Dec. 1925, IEB 1.1 1.6. Little to Shepardson, 21 Sept. 1926 and "Institute for biological research and research in forestry"; Rose to Little, 2 Oct., 15 Nov. 1926; Little to Rose, 29 Sept., 11 Nov. 1926; Thorkelson memo, 5 Mar. 1926; all in GEB 1.4 641.6712.

sciences: $740,000 from the GEB over eight years plus $840,000 from the state. It was perhaps too big, for in 1928 the board decided to defer until it was clear if the Virginia experiment would succeed. Delay that late in the game was, of course, fatal.[99]

Thorkelson's experiences in the South and Southwest were not unlike Trowbridge's in Eastern Europe and Spain. Both were eager to do business but succeeded only in exceptional cases like Virginia and Madrid. In the American South and Southwest, as in Spain and Eastern Europe, reactionary cultural politics threatened efforts to promote science and higher learning. In North Carolina and Texas, university administrators were harassed by the Ku Klux Klan and anti-intellectual populists, much as Castillejo was by Spain's church party.[100] It was just bad luck that the GEB's science program overlapped precisely with the high tide of the fundamentalist campaigns against science and cultural modernism. In Virginia, Harry Bird's political machine resembled Primo's dictatorship in Spain, with its mixture of economic progressivism and social reaction. The University of Texas was to Caltech or Princeton what Madrid was to Göttingen or Copenhagen; Splawn and Little were to Millikan and Karl Compton what Castillejo was to Hugh Anderson or Emile Borel. Every region had its scientific boosters, who dreamed of making universities and science the engine of regional development. But cultural and material factors conspired to frustrate Trowbridge and Thorkelson's intentions to channel funds to where funds were most needed.

The 1920s System of Patronage: Conclusions

How well, in hindsight, did the institutional system of patronage work? In the large cosmopolitan universities on both sides of the Atlantic the record was mixed. Few projects came up to expectations as well as at Cambridge; more commonly, grand visions were left undone or incomplete, as at Johns Hopkins, Chicago, London, and Paris. In smaller up-and-coming centers, the reality of grant-making was closer to expectations—in the small countries around the North Sea, the Boston-Washington corridor, and Southern California. The ideals and methods of Rose and his lieutenants were best adapted to mid-sized institutions like Caltech, Princeton, Uppsala, or Copenhagen. In both Europe and the United States, newcomers to the top rank of research universities had the advantage of small and tightly knit scientific lead-

[99]Thorkelson memos, 20 Mar. 1925, 16, 17 Feb., 16 Mar., 2 Oct. 1928; Little to Thorkelson, 19 July 1928; Thorkelson to Little, 1 Aug. 1928; all in GEB 1.4 641.6712.

[100]Thorkelson memo, 16 Mar. 1927, GEB 1.1 162.1517.

ership, understanding administrators and relatively uncomplicated internal politics, and access to civic or alumni constituencies. In the underdeveloped regions, where these elements were absent, even the best intentions on both sides led to meager results. It all depended on location.

These patterns reflect the basic fact that Rose's system of patronage was designed to aid educational institutions, not research as such, and to make strong centers stronger. These fundamental assumptions were embodied in the requirement of matching funds, and in the field practices of second-generation managers of science. Trowbridge, Thorkelson, and their colleagues were not designers of research projects but experts in university finance, laboratory design, the organization and social dynamics of academic departments, and educational politics. Disciplinary agendas and individual research achievement played a part indirectly in determining who got what, but they were less decisive than institutional factors, local support, and entrepreneurial know-how. Without matching funds, even the best projects failed. It was an advantage to learn the ropes of grant-getting in the Hale, Curie, and Hopkins circles and the research council networks.[101] But above all, it was location within university systems, and strategies for moving up the pecking order, that determined where and how quickly relations could be forged with the GEB and IEB.

The extension of the institutional mode of patronage from college development to science was a major discovery of the 1920s. It broke through the limitations of the prewar system of science patronage. (The CIW's program of grants-in-aid was a minor influence after 1920.) It completed the system of institutional patronage begun by the RF and CC grants to the NRC in 1919. Through this system of patronage unprecedented sums of money flowed into the natural sciences. Participation in the high culture of scientific research was expanded, and scores of scientists and their students gained experience in sponsored research. A dozen or so middling universities moved into the top rank, as further down the pecking order appetites were whetted, if not satisfied.

The experience of the 1920s also reveals the limitations of an institutional mode of patronage. The requirement of matching funds made it almost impossible for Thorkelson to do more than token projects in public universities and in regions that most needed help. Thorkelson chafed at the restrictions of matching endowments. He took a keen interest in individuals' research, especially in mathematical physics, and

[101] Access to elite networks was no guarantee of getting funds. Yale did little with the GEB, for example, despite President James R. Angell's insider's knowledge of foundations and the NRC (though Robert Yerkes's anthropoid research was supported by the Laura Spelman Rockefeller Memorial), GEB 1.4 634.6659, 636.6676.

was chagrined when the GEB's procedures kept him from helping talented young scientists in their careers. For example, he was moved by an appeal to aid a bright young physicist at the University of Iowa, Alexander Ellett, but could only hope that such obviously worthy cases would encourage the trustees to fund specific research projects.[102] So, too, Trowbridge lost no occasion to press Rose on the virtues of emergency aid to worthy individuals and institutions in places like Italy or Eastern Europe, but to no avail. The limitations of institutional patronage were more easily seen in the field than in foundation boardrooms.

Change was in the air in the late 1920s, but not because the institutional system of patronage was not working. If anything, Rose and Ruml's operations had become all too efficient spending machines. People like Rockefeller, Jr., George Vincent, and Raymond Fosdick were increasingly concerned about the irrationality of a system of patronage that had grown through individual initiative and was scattered over half-a-dozen separate organizations. As Rose and Ruml's programs reached a crescendo of spending in 1927–1928, efforts were already underway to rationalize the Rockefeller boards and bring spending under control. The result, not entirely anticipated by the architects of reorganization, was to transform the institutional system of science patronage. In the early 1920s, a new system of patronage emerged based on individual research projects, in which the partnership between scientists and foundation managers was more intimate than anything Rose and his lieutenants dreamed of.

[102]Thorkelson memos, 8–9 Oct. 1925, 14 Jan., 2, 17 Feb., 24 Mar. 1926; Thorkelson to Jessup, 1 Apr. 1926; all in GEB 1.4 690.7109. One exception was made to the rule against individual projects: T. H. Morgan got $10,000 to create a post in general physiology for Selig Hecht, GEB 1.4 681.7040.

Chapter Nine

The Rockefeller Foundation in Transition

It was already clear in 1926, when Rose put the IEB's grant-making machinery into high gear, that a major reshuffling of the Rockefeller boards was to take place. Rose could not have anticipated just how far-reaching the change would be—no one did, not even Raymond Fosdick who engineered the reorganization. But by 1930 a new system of patronage was emerging. Capital grants to university departments, the engine of large-scale patronage after 1923, became rarities after 1930. The mainstay of the new system was the individual project grant. These were not the occasional grants-in-aid of the pre-1919 era but systematic programs of investment, through individuals, in new fields of research, often fields that cut across disciplines.

Rose's system of patronage was an expression of the social mood and circumstances of 1919–1928—a buoyant economy and a belief in large-scale community development—and it did not long outlive these circumstances. The new system of individual grants was adaptive to the very different mood and conditions of the depression decade; no need then to expand capacity for scientific production. The linkage of research to expanding the production of researchers, so compelling in the postwar era of manpower shortages, lost its magic very quickly after 1929. Underemployment and excess capacity were the problems, and support for individual projects kept those people working who were already in the market. The language of relief again began to be heard, as industrial markets for researchers dried up and young scientists hung on desperately in graduate schools or in postdoctoral fellowships waiting for scarce academic jobs.

The Depression, however, only accelerated a cycle of change that had already begun to turn by 1928, the year of the Rockefeller reorganization. Declining income and concern with unemployment were not the cause of the Rockefeller Foundation's new system of patronage.

The key decisions had been made a year before the crash; the Depression only forced the RF's managers to be quicker about casting off old habits and putting new policies into practice. In fact, the reorganization of 1928 was remarkably little affected by any specific circumstances external to the Rockefeller boards. Fosdick was not inspired by specific events, as Vincent and Millikan were by the demobbing of the NRC and the postwar manpower crisis. Entrepreneurial scientists played no part in Fosdick's decision to devote the foundation to the advancement of scientific knowledge, as they did in the decisions of 1919. The sole aim of the reorganization was to rationalize the messy administrative structure of the boards. Fosdick's job was to take the five boards apart and put the pieces together in a more rational order, a final expression of the postwar zeal for system building.

Fosdick's biggest problem was what to do with the institutional programs in the natural and social sciences, which had grown wholly outside the RF and almost overshadowed its tradition in medical philanthropy. What Fosdick did, in brief, was to assimilate Ruml's and Rose's programs into the RF where, with the medical sciences, they replaced medical education as the foundation's raison d'être. The IEB was dissolved, and the Laura Spelman Rockefeller Memorial and GEB reverted to what they had been before Ruml and Rose co-opted them for their own ends. The vehicles improvised by these two entrepreneurs of patronage disappeared, but their programs lived on in a new guise.

The years 1929–1932 were transitional ones in the Rockefeller Foundation. Fosdick's creative solution to the problem of the Rockefeller boards contained implicit consequences, which were only revealed as his somewhat vague directives were translated into new programs. The momentum of projects inherited from Rose's operations dominated at first but gradually declined as the administrative machinery cleared. At the same time, foundation officers became more active in formulating new programs appropriate to their new role and the deepening social crisis. A program in the natural sciences did not really get underway until 1934.

The reorganization of 1928 set the process of change in motion and gave it a general direction. But the new system of patronage that eventually emerged was created in the field by a new generation of science managers. Thus much depended on the individuals chiefly involved—Max Mason, who succeeded Vincent, Raymond Fosdick, and especially the officers in charge of the divisions. In creating new philanthropic programs, these officers also created a new role for themselves. A system of project grants defined a larger space and a new authority for middle managers between their scientific constituents and foundation trustees. It made possible a relationship that brought this third generation of science managers far more intimately and actively into the research process than Rose, Trowbridge, or Thorkelson had ever been.

It was easy to miss this fundamental novelty. Edwin Embree thought that the Rockefeller Foundation, in abandoning institutional for individual grants, had thrown away its power to do a few big things and become just another place for people to get some money.[1] Embree was wrong: massed and concentrated in a few areas of science, individual project grants could and did have effects as large and lasting as institutional grants. Both systems of patronage aimed to develop communities of scientists. But whereas Rose and Trowbridge aspired to build regional or national scientific communities, the new generation of managers developed communities around specific tools, problems, and specialties.

In this chapter I will first examine the 1928 reorganization of the Rockefeller boards to see how such a surprising change in direction came to pass. I will then turn to what the reconstituted Rockefeller Foundation did in 1929–1932, to show how practical realities gradually transformed the older system of institutional grants into something new and unexpected.

Reorganization of the Rockefeller Boards, 1925–1928

Five years of uncoordinated entrepreneurship and growth inevitably resulted in duplication and conflicts among the various Rockefeller divisions and boards. Medical education was split between the foundation and the GEB. Richard Pearce, Abraham Flexner, and the IHB's Frederick R. Russell vied for control of public health. Nursing education was supported by four divisions, hygiene by three. Pearce tried to start a program in the basic premedical sciences, but the trustees dismissed the idea only to give it to Rose a few years later. Pearce, naturally, was put out.[2] Abraham Flexner resented Rose's diversion of the GEB from general education and became increasingly protective of his own turf and intolerant of his fellow officers. (Alan Gregg recalled one meeting when Flexner itemized the failings of each in turn, leaving everyone stunned.)[3] Rockefeller officers began to trip over each other in the field.

[1] Embree to Fosdick, 8 Jan. 1926, RF 903 1.2

[2] Vincent to Fosdick, 19 May, 1926, pp. 4–5, RF 900 17.121. Pearce to Fosdick, 16 Apr. 1926, p. 7, RF 900 22.165.

[3] Alan Gregg, oral history, pp. 38–40, CUL. Steven C. Wheatley, *The Politics of Philanthropy* (Madison: University of Wisconsin Press, 1988), pp. 112–115.

Embree recalled how, in the summer of 1926, no fewer than half-a-dozen officers converged on Cambridge University, each with his own agenda. It was, as Embree noted, "clearly not a good arrangement."[4] It was, in fact, a system of entrepreneurship gone haywire.

Some informal steps were taken in 1925 to encourage a more cooperative attitude within the RF, but they were ineffectual. A "Monday lunch" group intended to encourage cooperation among the five boards soon began to suffer from "parliamentary inefficiency," that is, it became just another forum for rival interests to stake out turf. Vincent wondered if a "benevolent despot" might be the answer and asked the trustees to clarify what official power he had over the heads of divisions and boards. A committee consisting of Fosdick, John G. Agar, and Simon Flexner was constituted to recommend structural changes.[5]

Fosdick concluded that "a pretty definite surgical operation" was needed.[6] The reforms that his committee proposed, however, were more on the order of administrative Band-Aids. The IHB was reconstituted as a division and told to stay out of education; the CMB's budding program in premedical sciences was transferred to Pearce's division. Both boards remained powerful and largely independent of Vincent's control. The only real casualty was Embree's fledgling Division of Studies.[7] It was clear even then that these stopgap measures did not touch the larger problem of competition among the five Rockefeller boards.[8] Endorsing the recommendations of Fosdick's committee, the trustees immediately appointed similar committees to investigate the IEB, GEB, and the memorial, plus an interboard committee, chaired by Fosdick, to plan a complete reorganization.[9]

As the scope of the reorganization widened, the IHB and CMB became less important issues than Ruml and Rose's enormously expanded operations in the social and natural sciences. In searching for a rationale for consolidating the boards, Fosdick had to balance and reconcile three highly successful but divergent philanthropic interests: general and medical education, basic science, and medicine and public health demonstration. The challenge to Fosdick was to invent a rational organization into which these varied programs would all fit. Should it

[4]Embree, untitled memoir, n.d. [c. 1930], p. 13, ERE box 1.

[5]Vincent, "The Rockefeller Foundation: A statement by the President," 24 Feb. 1926, pp. 3–4; Vincent to Fosdick, 5 Jan. 1926; Fosdick to Vincent, 24 Mar. 1926; Vincent, "The organization of the foundation," 19 May 1926, exhibit A; all in RF 900 17.121.

[6]Fosdick to S. Flexner, 26 Oct. 1926, RF 900 17.121.

[7]Fosdick, "Report of the Committee on Reorganization," 5 Nov. 1926, pp. 6–8, RF 900 19.136.

[8]Ibid., p. 4.

[9]Fosdick, memorandum, 5 Nov. 1926, RF 900 17.122. Vincent, "Memorandum on policy and organization," 29 Jan. 1927, RF 900 19.138.

be the various levels of the educational system, as in the GEB? Or disciplines, as in the IEB and the memorial? Should medicine and public health remain the hallmark of Rockefeller philanthropy? Should the IEB and GEB be preserved or absorbed into a single giant foundation?

Rose was the chief advocate of the "advancement of human knowledge." His plans for developing the basic sciences underlying agriculture and forestry, taking shape even as the interboard committee labored, were meant to ensure that his programs and organizational machinery would survive the reorganization. The RF, Rose argued, should become simply a holding company or endowment for these operating agencies. He was backed up by Simon Flexner, who went so far as to suggest that it should divest itself of educational work and transfer Pearce's division of medical education to the GEB.[10] The idea of making research the overarching mission of the RF had some appeal to Fosdick, but as advocated by Rose it was attached to an administrative scheme that perpetuated the worst features of the independent boards. Fosdick recognized that Rose wished above all to preserve the IEB.

Abraham Flexner was no less concerned to preserve the GEB (the old GEB, not Rose's) to carry on programs in general and professional education. He proposed that the knot between the GEB and IEB be untied, so that the true principles of an education foundation, subverted by Rose, could be revived. An educational foundation must, Flexner thought, deal with whole systems—primary, secondary, undergraduate, and postgraduate. He opposed the idea of disciplinary divisions.[11] Flexner was more than willing to let the RF take over the IEB's programs in science, since they could never be developed according to his, or any, system: "If you wish to develop physics, chemistry, biology, mathematics, etc., on a world-wide basis," he wrote, "you must practically disregard systems and select institutes or persons."[12]

In contrast to both Flexner and Rose, Vincent envisioned a single giant foundation, assimilating all the other boards into four divisions: (1) graduate education and research in the physical sciences, social sciences, and humanities; (2) general, secondary, and college education and teacher training; (3) professional education in medicine, public health, law, engineering, agriculture, and business; (4) applied social and economic science, government administration, city management, and criminology. "Does not the proposed combination," he wrote, "of-

[10]A. P. Stokes to Fosdick, 29 Jan. 1927; Fosdick, "Memorandum of a meeting, 5 May [1927]," RF 900 17.123.

[11]A. Flexner, "Memorandum regarding the GEB," 4 Jan. 1927, pp. 13–14; Flexner to C. Howland, 13 Apr. 1927; both in RF 900 17.123. On Flexner's views, see Wheatley, *Politics of Philanthropy*, ch. 4.

[12]Flexner, "Memorandum," p. 17.

fer at least a slender bridge between the 'promotion of science' idea and the 'co-operate with the educational system' theory?"[13] Vincent's personal agenda, of course, was to make sure that control of the Rockefeller philanthropies would finally come to rest in the office of the president.

By May 1927 Fosdick was at work trying to balance these conflicting visions. He rejected out-of-hand Rose's idea of a holding company—an invitation to more freewheeling. He also rejected Vincent's idea of four boards crammed into one—an administrative nightmare, he thought. Fosdick focused his efforts on devising a division of labor between the RF and GEB, but Rose and Ruml's programs in science stood in the way. It seemed impossible to transfer them to the GEB as branches of general education, but equally impossible to transfer them to the RF along with medicine and public health.[14] In May 1927 Fosdick arrived at a "sort of a compromise between what Flexner wanted, with more of a leaning to the Rose side:

> It leaves in the General Education Board responsibility for general education . . . the social sciences, the humanities and, perhaps, the fundamental sciences, but on this point I am not clear. Probably, the fundamental sciences ought to go with medicine in the RF. This idea would place in the RF responsibility for medical education, public health, agriculture, legal education (if it is ever undertaken), and probably the fundamental sciences.[15]

This left the basic sciences divided from education, however, and that gave Fosdick pause. After all, the 1920s system of science patronage was built on the premise that foundations aided science best by underwriting the education of scientists.

Fosdick was struggling with contradictions that were built into the compromises and improvisations made by the Rockefeller boards in the early 1920s. Vincent, by refusing to commit the RF to institutional grants in 1919, had left the way open for Rose and Ruml to do so. Rose, by improvising science programs within boards for general education, had created a system of patronage so successful that it outgrew the GEB's mission in general education. By confining the RF to medicine and public health, Vincent had made it difficult to accommodate large programs in the natural and social sciences. Science had become too big

[13]Vincent, "Memorandum on policy and organization," 29 Jan. 1927, p. 6. Vincent had previously favored two separate foundations, one for education and one for medicine, science, and public health. Vincent, "Memorandum on the organization of the GEB," 3 Jan. 1927, RF 918 1.3.

[14]Fosdick, "Memorandum on the reorganization of the boards," 13 May 1927, RF 900 17.123.

[15]Fosdick to Howland, 13 May 1927, RF 900 17.123.

to ignore but seemed to fit none of the traditional modes of Rockefeller philanthropy.

On the horns of this dilemma, Fosdick made little progress between May and October 1927. He even wondered if the issue might be left to Rose's successor to work out, but Rockefeller, Jr., urged him to decide on a policy independently of the personalities involved.[16]

A Foundation for the Advancement of Knowledge

Sometime in late October 1927 Fosdick hit upon a way of resolving the dilemma. He envisioned the RF as "a board for the advancement of knowledge" made up of four divisions: physical sciences (from the GEB and IEB), social sciences (from the memorial), humanities (from the GEB), and the arts. These divisions would include applications of basic knowledge to medicine, law, agriculture and forestry, and engineering. The IEB and the memorial would be dissolved, their programs reincarnated as divisions of the new RF, and the GEB would revert to general education.[17] But what about medicine and public health? Was the RF's corporate body to be taken over by an alien spirit, as the GEB had been by Rose? Fosdick made no mention of a division of medical education and implied that the IHB would be divested as a separate agency. (The way he expressed it was that the RF would not be an operating agency at all but delegate to outside agencies programs like public health and natural science fellowships.)[18] Fosdick had ceased to think of the RF as primarily a medical board and that, apparently, made it possible for him to conceive the idea of a foundation for the advancement of knowledge. His solution was more than a compromise between Rose, Flexner, and Vincent: it was a new invention, as pregnant of future change as Millikan's, when he split fellowships from institutional grants in 1919.

The germ of Fosdick's inspiration may have come from a group of officers who hitherto had not figured in his deliberations. On 17 October 1927 Fosdick met with Ruml, Embree, and Whitney Shepardson, who had asked for a hearing of their views on reorganization. What the three officers proposed bears a striking resemblance to Fosdick's subsequent plan. They, too, envisioned the RF as "a board for the advancement of knowledge," with four divisions in the physical sciences, social sciences, humanities, and "professional groups" (law, agriculture,

[16]Fosdick to Rockefeller, Jr., 6, 17 Oct. 1927; Rockefeller, Jr., to Fosdick, 13 Oct. 1927; all in RF 900 17.123.

[17]Fosdick to S. Flexner, 1 Nov. 1927; Fosdick to T. Debevoise, 21 Dec. 1927; both in RF 900 17.123.

[18]Fosdick to Flexner, 1 Nov. 1927.

and medicine). To the GEB they relegated all general education, and they proposed a third foundation to deal with the application of knowledge to mental hygiene, social welfare, and so on.[19] Most significant, in retrospect, was their insistence that the RF should divest itself of the IHB.

Fosdick was not greatly impressed at first. He did not like the idea of three separate foundations and thought it a mistake to abandon medicine and public health—the RF's hallmark. The idea of making the IHB an independent agency struck Fosdick as unrealistic, even a little ridiculous. As he wrote afterward: "The elimination of operating functions seems to these three gentlemen the key to the whole reorganization."[20] Later that day Fosdick dictated a plan that was virtually unchanged from the one he had entertained five months earlier. As before, he proposed that the RF emphasize medical education and public health, plus agriculture and physical science.[21] A week later he was still skeptical about the idea of divesting operating agencies but was wavering: it was "well worth considering," he admitted to Vincent, "although the difficulty of transforming the IHB into an independent agency, with its own Board of Trustees, seems almost insurmountable."[22]

Sometime between 17 October and 1 November, however, Fosdick realized that divesting the IHB would make it much easier to put the remaining pieces into a rational order.[23] So long as medicine and public health dominated the RF, the natural and social sciences had seemed not to fit. Without the IHB, however, everything fit comfortably under the umbrella of "the advancement of knowledge." Medicine became a field for the application of the natural sciences, along with agriculture—in principle, at least.

With these guiding principles Fosdick moved quickly toward a definite plan. The IHB was reconstituted as a separate operating agency, bankrolled by the RF, and the CMB was spun off with an endowment of $10 million. The memorial, with an endowment of $10 million and a new name (the Spelman Fund of New York), reverted to its original form of a charity for women and children. The GEB's remaining endowment ($39.5 million) was to be liquidated as quickly as possible, preferably by large grants in support of the RF's science programs. The IEB was dissolved. What remained of the endowments of the IEB and

[19]Fosdick, "Memorandum of a meeting at the Century Club," 17 Oct. 1927, in Fosdick to Vincent, 24 Oct. 1927; A. P. Stokes to Fosdick, 29 Jan. 1927; all in RF 900 17.123.

[20]Fosdick, "Memorandum of a meeting," p. 1.

[21]Fosdick memo, 17 Oct. 1927, RF 900 17.123.

[22]Fosdick to Vincent, 24 Oct. 1927.

[23]Fosdick memo, 1 Nov. 1927, RF 900 17.123.

the memorial ($19 million and $63 million) was transferred to the RF, making it a fund of $225 million devoted wholly to the production, diffusion, and application of basic scientific knowledge.[24] Though there would be some accommodation to the old RF's medical heritage, it was Ruml's program in the social sciences and Rose's in the natural sciences that became the core of the new Rockefeller Foundation. Thus did a dramatic change in policy emerge from a process that revolved largely on questions of internal structure and administration.

The idea of a foundation for the advancement of knowledge clearly was a victory for Rose—or seemed so. Less wary now of opposition from Rose, Fosdick drew him into the planning process only a few days after he conceived his plan. Rose was delighted, writing to Trowbridge in mid-November that things were taking shape rapidly and very satisfactorily for the IEB: "This whole reorganization scheme will tend to magnify the importance of science rather than otherwise." He was confident that agriculture and forestry would likewise have pride of place among the new foundation's programs.[25]

Vincent was naturally pleased by the consolidation of authority in the RF, though he worried that the IHB, now a division (the first of Fosdick's accommodations of principle to practical politics), might still be too independent. He worried, too, that Rose seemed to think that the IEB was going to live on.[26] He was quite right to worry: in fact, the International Health Division lost little of its old independence and power. Warren Weaver recalled that the IHD was the top of the pecking order in the 1930s, the only division with a capital "D." An aloof and favored elite with its own board of trustees and a vast field staff, it always secured the largest budget and in a lump sum once a year, with no constraints on how it was spent. No other division had that luxury.[27]

Abraham Flexner was the big loser in the reorganization, or so he felt. He was stung by the transfer of his beloved programs in medical education to Pearce, his old antagonist: "an irreparable blunder," he later called it. He did not respect Pearce's intellect and had never made any effort to hide his disdain, thwarting Pearce in the United States and stepping on his toes in Europe. And now Pearce had inherited all Flexner's programs. Flexner was gratified to see the IEB disappear along with his rival Rose, but he resigned from the GEB (in June 1928) in a

[24]Vincent, "Reorganization of the Rockefeller boards: An outline for discussion," 15 Dec. 1927; Fosdick to Debevoise, 21 Dec. 1927; Fosdick, "Memorandum on reorganization," 18 Jan. 1928; and "Memorandum of meeting," 19 Jan. 1928; all in RF 900 19.138.

[25]Rose to Trowbridge, 16 Nov. 1927, IEB 1.1 24.346. Fosdick to Rockefeller, Jr., 2 Nov. 1927; Fosdick to Debevoise, 21 Dec. 1927; both in RF 900 17.123.

[26]Vincent to Fosdick, 27 Feb. 1928, RF 900 17.124.

[27]Weaver, oral history, pp. 247–250.

bitter mood, convinced that "a tradition, which can never be restored or mended, was entirely destroyed."[28] Fosdick had qualms about dropping education from the RF, especially since it seemed that Vincent might well be succeeded by an educator.[29] But he did not look back.

Rose's joy, it turned out, was short-lived. Misled by the divesting of the IHB and CMB and by Fosdick's statements that the new RF would not be an operating agency, Rose assumed that Fosdick had adopted the idea of a holding company. When it finally dawned on him that the IEB would be dissolved, Rose balked. But his magic touch was gone. His skill in argument and debate, once irresistible, was no longer taken seriously. Indicative is a scene described by Fosdick:

> I had a hearty laugh over the reductio ad absurdum which [Ruml] made of Rose's phraseology. I believe we make a serious mistake when we speak of the RF as a "holding company." . . . I have never liked that description of the machinery we have in mind, and when Rose makes a premise out of it, it enables him by the sheer force of his irresistible logic to reach conclusions that from the standpoint of good organisation will not work. . . . We have had experience enough in the RF with subsidiary boards to know that they can easily become a tail that wags the dog.[30]

Finding no support among his colleagues, Rose appealed over their heads to Rockefeller, Jr. But Rockefeller realized that Rose wanted to scuttle the reorganization in order to preserve the IEB, and he told Fosdick to proceed whether Rose liked it or not.[31] Rose accepted defeat gracefully and bowed out. Two days later he asked Fosdick to handle the reorganization of the GEB and departed on a trip to California.[32]

Although the barons of the 1920s were no longer powers to be reckoned with, their programs were. Medicine remained the biggest problem. Pearce had ongoing projects in the medical sciences that could not simply be dropped. Medicine was too important for Rockefeller philanthropy to be set aside or lumped with agriculture. Its special claims had to be provided for. As with the IHD, Fosdick was obliged to bend ideal principles to the realities of history and practical administration. He entertained and rejected several different options: a third foundation for applications of the sciences seemed too complicated; a

28Wheatly, *Politics of Philanthropy*, ch. 5. A. Flexner, *Funds and Foundations* (New York: Harper, 1952), pp. 80–83. Alan Gregg, oral history, p. 140, CUL.

29Fosdick, "Memorandum of meeting," 19 Jan. 1928, RF 900 17.124.

30Fosdick to Vincent, 7 Mar. 1928, RF 900 17.124.

31Rose to Fosdick, 10 Feb., 1 Mar. 1928; Fosdick to Vincent, 7 Mar. 1928; Rockefeller, Jr., to Fosdick, 19 Mar. 1928; all in RF 900 17.124.

32Fosdick to Rockefeller, Jr., 19 Mar. 1928, RF 900 17.124.

division of biology, medicine, and agriculture combined research and application in a way that Fosdick liked but left the natural sciences divided.[33] In the end, Fosdick did the practical thing: he created a separate division for the medical sciences (plus some remnants of medical education), leaving biology in the natural sciences division.[34] That was the final scheme that Fosdick proposed to the trustees in May 1928—five divisions: IHD, medical sciences, natural sciences, social sciences, and humanities.[35]

In the transition years, more compromises were made between Fosdick's ideal policy and the real constraints of traditions and obligations. "The advancement of knowledge" was not a specific guide for deciding which of the numerous and diverse projects inherited from the boards should be continued. How Fosdick's principles were translated into projects and programs depended to a great extent on the officers whose job it was to lay suitable proposals before the board.

Changing the Guard

It was unclear, however, how much authority the officers should enjoy. Here again Fosdick's principles conflicted with historical practice. The new divisions were not to be operating agencies, according to Fosdick, yet the projects they inherited from Ruml and Rose obviously could not be administered without expert staffs. Fosdick seemed to assume that fellowships and individual grants, which involved a good deal of routine administration, would be delegated to outside agencies like the NRC. But the IEB's and the GEB's institutional projects could hardly be handled in the same way, and such important administrative responsibilities would—and did—give the officers de facto authority.

Vincent was the strongest advocate of restraining the officers' powers—no surprise. He clearly had Rose and the IEB in mind when he urged that the foundation eschew grandiose ideals like "the promotion of science internationally." The reality, he hoped, would be more modest: assisting the small number of exceptional scientists who were

[33]Fosdick, "Reorganization of the Rockefeller boards," n.d. [Jan. 1928], p. 5, RF 900 19.138. Pearce had earlier suggested a separate division of biology. Pearce to Fosdick, 16 Apr. 1926, RF 900 17.121.

[34]Fosdick to Debevoise, 21 Dec. 1927, RF 900 17.123. A separate division of agriculture was theoretically an option but had no strong advocate except Rose. Rose to Trowbridge, 16 Nov., 22 Dec. 1927, IEB 1.1 24.346–347. Shepardson to Rose, 27 Feb. 1928, IEB 1.1 24.349. Edsall to Fosdick, 19 Oct. 1928; "Summary of minutes," n.d. [c. 1933], RF 900 19.141. Vincent to Edsall, 14 June 1928, RF 900 17.125.

[35]"Report of the Interboard Committee on Reorganization," 22 May 1928, RF 900 19.139.

able to manage large research schemes. That, Vincent felt, would put control of grant money where it belonged, in the hands of the recipients, and keep it out of the grasp of ambitious officers.[36] Vincent, Pearce, and trustee David Edsall tried to persuade the board not the create formal divisions at all but simply to appoint officers in the various sciences. "The very name [division]," Vincent warned, "suggests a kind of autonomy from which we have suffered a good deal in the past."[37]

Divisions were organized, but with the proviso that policy decisions would be made by the president and trustees to keep officers from taking the bit in their teeth and imposing their own programmatic ideas on their scientific clientele.[38] Opportunism, not activism, was the message, but it was an ambiguous message: enjoining the officers to avoid "grandiose, comprehensive plans for world-wide propaganda," the trustees also urged them to "take the initiative in proposing certain developments of research."[39]

Vincent himself gave little direction from the top. Even as a lame duck he was more concerned with the symbols than the substance of power. His plans in 1928 were very much like those he had prepared ten years earlier: laundry lists of all conceivable methods of philanthropic work, fields of activity, and precedents from the Rockefeller boards.[40] Perhaps they should be read as symbolic demonstrations of the fact that the president, not the officers, now had the authority to set the agenda. Obsessed with preventing new baronies from taking root, Vincent never got around to deciding on a program of his own to counterbalance the momentum of inherited projects.

In the key divisions there was a distinct lack of experienced people to translate Fosdick's general directives into specific programs. Ruml and Embree had departed, as had Mann and Hutchison. Trowbridge returned to Princeton in 1928 to become dean of the graduate school. Thorkelson resigned to pursue a career in business, as assistant to Walter J. Kohler, president of the Kohler Co. and governor of Wisconsin. (It was the position he had declined in 1922 when he joined the GEB.) Only in the medical sciences was there real continuity of leadership: although Richard Pearce died early in 1930, he was succeeded by his experienced and capable associate, Alan Gregg.

[36]Vincent, "Memorandum on the reorganization of the boards," 20 Jan. 1927, pp. 4–5, RF 900 9.138. "Agenda for Conference," 19 Oct. 1928, RF 900 21.159. RF Board minutes, 3 Jan. 1929.

[37]Vincent to Fosdick, 14 June, 1928; Vincent to Gunn, 24 May 1928; both in RF 900 17.125.

[38]Vincent to Greene, 28 May 1928; Edsall to Fosdick, 29 May 1928; Fosdick to Edsall, 29 May 1928; all in RF 900 17.125.

[39]Agenda for conference, 19 Oct. 1928, RF 900 21.159.

[40]Vincent, "Memo on policy and organization," 19 Oct. 1928, RF 900 20.159.

The natural sciences division suffered most from the disruptions of the reorganization: it had five directors in three years. One would not have predicted that in May 1928, when Max Mason, president of the University of Chicago and a distinguished mathematical physicist, was appointed to take charge of the natural sciences. It seemed a brilliant coup. Trowbridge was delighted and relieved that the natural sciences would be in such capable hands.[41] But it was a makeshift, since Mason was already slated to succeed Vincent at the end of the year. Knowing he was temporary, Mason seems to have made little effort to bring some order to the chaotic roster of projects inherited from Rose. He did not even manage to find a director for the natural sciences by the time he moved into the president's office. Richard Pearce stood in briefly and was followed by his assistant, William S. Carter. Hermann A. Spoehr was appointed director in September 1930, but he stayed less than a year, departing some months before Warren Weaver arrived in February 1932.[42]

Spoehr was a distinguished organic chemist and a leading expert in the biochemistry of sugars (an appropriate choice of career for the son of a well-to-do manufacturer of candies). As a researcher in the CIW's Desert Laboratory since 1910, Spoehr had become one of the most productive and respected American plant physiologists. When the Desert Laboratory became part of Stanford in 1928, Spoehr was made chairman of the Division of Plant Sciences. A coauthor with Irving W. Bailey of the NRC's report on forestry science, Spoehr had been a favorite of Rose and Trowbridge long before Mason succeeded in luring him to New York.[43] However, he was never happy there. According to Weaver, Spoehr lacked a "good administrative conscience." He never felt satisfied that he knew enough to make informed judgments and, having made decisions, could not keep from worrying if he had done the right thing.[44] That may have been a kind way of saying that Spoehr was indecisive and lacked practical judgment. The paper trail suggests that he was slow to distinguish essential from nonessential issues. In any case, between the time that Trowbridge left and Weaver arrived there was no strong figure in the New York office to define a new program in natural sciences.

In the RF's Paris office, in contrast, there was a strong continuity in operating staff. Wilbur Tisdale and Harry Miller stayed on, both expe-

[41]Trowbridge to Mason, 14 May 1928, IEB 1.1 1.24.

[42]Norma Thompson to Mason, 31 Mar. 1930, RF 915 1.1. Mason diary, 29 Jan., 9 Dec. 1929, RF.

[43]James H. C. Smith and C. S. French, "Hermann A. Spoehr, 1885–1954," *Ann. Rev. Biochem.* 24 (1955): xi–xvi. Mason to Rose, 23 May 1928, IEB 1.1 24.347.

[44]Weaver, oral history, pp. 121–122.

rienced IEB hands. Rose hoped that Harlow Shapley could be persuaded to succeed Trowbridge, but to no avail. Finally, the Princeton chemist Lauder W. Jones, who had done surveys of chemistry for the IEB, was appointed associate director. Jones welcomed the change. Following the death of his wife and the tragic death of his daughter (swept over a waterfall before his eyes), his professional work had become his whole life. Jones was a capable manager. He knew the routines of negotiating projects and obviously enjoyed it. He became a gourmet and bon vivant and relished traveling around the cultural centers of Europe. But he lacked Trowbridge's boldness, imagination, and independence. And he had little interest in filling the gap in top management in New York. He was less a successor to Trowbridge than a caretaker of the Rose legacy. Wilbur Tisdale, too, was a reliable manager but not a leader, it was felt in New York.[45] During the period of transition, the Paris office kept the machinery set up by Trowbridge and Rose well oiled and running. Insulation from the political upheavals in the New York office helped preserve good organization and morale in the absence of direction from the center. It was a highly efficient staff, but a staff without a chief.

Entre'Acte: Policy and Practice, 1929–1933

The transition years, including Weaver's first year, were largely a continuation of IEB and GEB programs. In the absence of new leadership, it was the administrative machinery that largely determined how funds were spent, and the administrative machinery was set up to make institutional projects. Some projects Jones simply picked up where Trowbridge had left them. Tangled projects that had been deferred were resolved. About $12 million was spent, in effect, on completing what Rose had begun.[46] Most grants were for equipping laboratories and endowing research in leading institutions rather than for particular research lines—testimony to the power of Rose's spirit, hovering over the RF's counsels. Postdoctoral training was still important, and matching local support continued to be a key issue in negotiations. In the absence of a plan of concentration, physical scientists continued to enjoy their natural advantages of prestige and experience in grant-getting. IEB advisors like Harlow Shapley and James Franck remained influential.

[45]Trowbridge to Rose, 8 Nov. 1927; Mason to Rose, 16, 23 May 1928; Rose to Trowbridge, 7 June 1928; all in IEB 1.1 24.347. Trowbridge to Rose, 21 Feb. 1928; Mason to Tisdale, 3 Jan. 1929; both in IEB 1.1 24.349. Mason diary, 5 Oct. 1928, 29 Jan. 1929, RF. Weaver, oral history pp. 304–308.

[46]Agenda for special meeting, 11 Apr. 1933, pp. 36–37, RF 900 22.168.

There were some significant changes, however. Recurring annual grants became more common, as the collapse of financial markets made it impossible for the RF to spend capital. Jones and Mason were more willing to invest in good research schools in provincial universities that would never have qualified as "centers" for Trowbridge and Rose. There were more projects in Germany, especially in Berlin. A few grants were made to individuals for research on specific problems. Such changes came about less because of policy than because the momentum of Trowbridge's initiatives spent itself and because the depression forced Mason to invent substitutes for institutional projects. Practice preceded policy in these years.

The most striking new feature of the transition years was the large and systematic investment in oceanography and marine biology. The $2.5 million grant to the new Woods Hole Oceanographic Institute was grand on the scale of the Mt. Palomar telescope. The chief marine biology stations of the United States and Europe also received substantial grants for facilities and endowment (table 9.1). Yet what seems a major new initiative was in fact Rose's legacy. One of the GEB's last acts in 1928 was to give $75,000 to the National Research Council for a survey of oceanography, modeled on the earlier survey of forestry that had so impressed Rose. Written by the Harvard oceanographer Henry Bigelow, the huge report was delivered to the NRC and RF late in 1929. It was, as Rose intended, a blueprint for developing the basic sciences of oceanography on an international scale. Mason and the officers adopted it almost without discussion. It was taken for granted, apparently, that Rose's sponsorship of the survey was an implicit guarantee that the RF would underwrite the program.[47] Rose had hoped that the reports on forestry and agriculture would likewise form the basis of RF programs, but it was only in oceanography that his hopes were fulfilled.

Almost all the RF's projects in oceanography either revived projects deferred by Rose (Banyuls and Roscoff, e.g.), or carried through recommendations in the NRC report, like the grand central institute at Woods Hole, the deep sea station at Bermuda, and the chain of oceanographic stations on the Pacific Coast of the United States—the Scripps Institute, Stanford's Hopkins Marine Laboratory, and Friday Harbor in Puget Sound.[48] Over $3 million was spent with minimal discussion and effort.

[47]Mason diary, 2 Sept., 13 Oct. 1929; Mason to Lillie, 25 Nov. 1929; Lillie to Mason, 14 Oct. 1929; all in RF 200D 129.1588. Bigelow's report is in RF 200D 153.1882. Harold L. Burstyn, "Reviving American oceanography: Frank Lillie, Wickliffe Rose, and the founding of the Woods Hole Oceanographic Institution," in M. Sears and D. Merriman, eds., *Oceanography: The Past* (Berlin: Springer, 1980), pp. 57–66.

[48]On Bermuda, RF 430D 1.1. Banyuls and Roscoff, RF 500D 11.119. Scripps, RF 1.2 200D 225.2170. Hopkins, RF 200D 157.1925. Friday Harbor, 253D 1.89.

TABLE 9-1
ROCKEFELLER FOUNDATION EXPENDITURES IN MARINE BIOLOGY (IN $1,000S)

Institution	*1929*	*1930*	*1931*	*1932*	*1933*
Roscoff, Banyuls	24				
Bermuda	245		12		
Scripps		40			
Pacific Grove		20			
Friday Harbor		250			
Woods Hole Oceanographic		2,500	10		
Plymouth			23		
Geneva			40		
Cold Spring Harbor			20	20	
Amoy			3		3
Tihany (Hungary)			13		
Naples				18	
Total	269	2,810	121	38	3

This is not surprising, perhaps, given the lack of strong leadership in the natural sciences (NS) division and Mason's aversion to administrative routine. The momentum of Rose's program and the intellectual leadership of the NRC's oceanographic experts and lobbyists were decisive where there was no comparable programmatic vision in the RF's New York office.

A similar interpretation can be put upon Mason's delegation of authority to NRC committees to manage foundation programs: the Committee for Research on Problems of Sex (inherited from the Bureau of Social Hygiene), a new committee for cooperative research in radiation biology, and a program of small grants-in-aid (table 9.2). Confronted in 1929 with the problem of administering the diverse projects, fellowships, and grants inherited from four different Rockefeller boards, Mason and his staff were more than willing to delegate as much as possible of the retail business of patronage to committees of experts.[49]

49R. A. Lambert memo, 10 Nov. 1930; Tisdale to Jones, 13 Jan. 1932; both in RF 200E 170.2070 and 170.2072. On the grants-in-aid, RF 200 37.418 ff.

TABLE 9-2
ROCKEFELLER FOUNDATION EXPENDITURES IN NATURAL SCIENCE (GENERAL) (IN $1,000S)

Institution	*1929*	*1930*	*1931*	*1932*	*1933*
Fluid research funds:					
N. Carolina	15[a]				
Washington Univ.		100[a]			
Pennsylvania		40[a]			
Minnesota			150[a]		
MIT			170		
Iowa State			30		
Fellowships:					
NRC	197	245	321	292	275
RF program	23	89	160	149	132
Grants-in-aid:					
NRC (U.S.)	25	75	75	75	50
RF program (Europe)	2	7	10	32	31
Publication:					
NRC	67	82	98	102	53
Other				2	8
NRC research committees:			17	75	70
Miscellaneous				24	14
Total	329	638	1,031	751	633

[a]Jointly with the medical sciences division.

EUROPEAN PROJECTS, 1929–1933

The momentum of Rose's program was strongest in Europe, thanks to the rich field experience of the old IEB hands. After a brief pause in 1929 it was business as usual in the Paris office (table 9.3). The IEB's appropriation to zoology at the University of London was redrawn and scaled back when a fund drive collapsed. A grant was made for astronomy at Leiden when Jones managed to untangle a complex negotiation

TABLE 9-3

Rockefeller Foundation Expenditures in Natural Science (Europe) (in $1,000s)

Institution	*Field*	*1929*	*1930*	*1931*	*1932*	*1933*
Kaiser Wilhelm Gesellschaft	Physics, physiol.		655[a]			
Bristol	Physics		250			
Royal Institution	Physics		113			
Munich	Biol., chem.		372			
Leiden	Astron.		110			
Vienna	Physics		13			
Utrecht	Physiol.		100			1
Kaiser Wilhelm	Chem.		8		13	
Oslo	Geophys.			90		
Szeged	Gen.			155[a]		
Stockholm	Physiol.			28		
London	Zool., anat.			370		
Warsaw	Physics			50		
Freiburg	Chem.			25		
Hannover Techn.	Chem.			20		
Göttingen	Chem.				50	
Jungfrauhoch	Geophys.					36
Other			10	10		
Total			1,631	748	63	37

[a]Jointly with the medical sciences division.

that Trowbridge had begun.[50] Renovation of the inorganic chemistry institute at Göttingen was an extension of the IEB's grants there to mathematics and physics. It rounded out the natural sciences in a key scientific center, bringing up to par one that, because it was not fashionable, had not received adequate support for decades. The style of the project was vintage IEB. It was the physicists and mathematicians, es-

[50]On London University, RF 401D 46.589–590. On Leiden, RF 650D 3.42.

pecially Richard Courant and James Franck, who knew from their experience with the IEB that a grant by the RF was the right lever to pry funds out of a reluctant state bureaucracy. These veteran grant-getters saw the opportunity to fulfill their vision, which they shared with Rose and Jones, of an international center of all the physical sciences.[51]

A similar development was begun at Munich. Rose and Trowbridge had seen the University of Munich as a European center, and they were prevented only by last-minute errors in building estimates from launching a big project with an appropriation to zoologist Karl von Frisch. (When time ran out in 1928, Trowbridge and Rose were planning a survey of all the sciences at Munich.) Frisch took the first opportunity to renew his suit to Lauder Jones.[52] With Jones's encouragement, the physical chemist Herbert Fajans also presented a plan to build an institute for a discipline whose growth in Germany had been stunted by the long shadow of organic chemistry. The idea was to make these two requests into a comprehensive plan for all the natural science institutes, just as Rose would have done. Jones knew the script and, stepping into Trowbridge's role, invited Mason to take the role of Rose. But the roles were in fact reversed: where once Rose pressed Trowbridge to think big, now it was Jones who pressed Mason to approve comprehensive planning. Mason was quite content for Jones to run the show, and only hoped that the physicists too would be ready soon with a proposal.[53] Individual grants-in-aid to Arnold Sommerfeld and Richard Willstätter kept future options warm, but further development was cut short by the Nazi seizure of power.

The largest European project was the grant of $655,000 to the Kaiser Wilhelm Gesellschaft (KWG) for two new institutes at Dahlem: in cell physiology for Otto Warburg, and in physics for Albert Einstein and Max von Laue. Warburg's modest laboratory in the Institute of Biology (one of its five subdivisions) was in 1930 one of the two world centers of research in general physiology. The fame of Warburg's researches on the enzymology of respiration in normal and cancerous cells clearly, in Alan Gregg's opinion, merited a separate institute. A so-called Institute of Physics had existed since 1917, but when funds earmarked for a laboratory evaporated in the postwar inflation it became a grants-in-aid fund, with Einstein as figurehead. The spectacular development of quantum and atomic physics made it intolerable, Jones

[51]Jones diary, 4, 5 July 1931; Franck to Jones, 3 Nov. 1931; Jones to Tisdale, 18 Nov. 1931; Jones to Weaver, 24 Mar. 1932; all in RF 717D 13.121.

[52]K. von Frisch to Jones, 2 May 1929; Jones diary, 12–15 May 1929; all in RF 717D 14.133. Rose to Trowbridge, 18 June 1928, IEB 1.2 35.133.

[53]Jones diary, 13 May 1929; Jones to Mason, 18 June, 7 Aug., 2 Sept. 1929, 10 Mar. 1930; Mason to Jones, 16 Aug. 1929; all in RF 717D 14.133.

thought, that the KWG had no proper institute in a field largely invented by German physicists and mathematicians.[54]

Jones and Gregg's support for these two institutes was part of a general pattern of interest in the KWG. The foundation made large grants to the institutes of brain research and psychiatry, and smaller ones to Herbert Freundlich and Fritz Haber's chemistry and physical chemistry institutes, and Eugen Fischer's institute of anthropology. An appropriation of $750,000 was authorized for an institute in international relations even before the RF was asked for it (the plan fell through). All in all, the RF channeled over $2 million to the network of Kaiser Wilhelm institutes in the years just before the Nazi takeover.[55] The IEB, in contrast, had seemed to avoid Berlin. Why the change? The question is only made more pointed by the controversy that engulfed the RF in 1934, when it decided to honor its pledge for the physics institute, even though Einstein and all Jewish scientists were now nonpersons in the Third Reich.[56]

The documents reveal no simple answer to the question. No conscious decision was made to rebuild science in the old Prussian capital. Indeed, Gregg and Richard Pearce worried that the growing system of research institutes was a threat to the German universities, especially in Berlin, and opposed a grant to Warburg's institute when the idea was first brought up, despite their keen desire to assist Warburg.[57] For Jones, too, the danger of injuring the University of Berlin at first outweighed his eagerness to aid the most famous physicist in the world. Just before the proposal went to the RF board, Jones sought reassurance from James Franck that aiding the Kaiser Wilhelm institutes would not simply drain off the best researchers from universities and cause the Prussian government to shift support from the University of Berlin to the institutes—robbing Peter to pay Paul.[58]

The nationalistic politics of the KWG also put Jones off at first. He deplored the tendency to put resources into creating new institutes

[54]Forman, "The financial support and political alignment of physicists in Weimar Germany," pp. 39–66. Board minutes 30099; Jones, "Report on the Institute of Physics," 5 Mar. 1930; Gregg, "Report on the Institute for Cell Physiology," 20 Feb. 1930; all in RF 717 2.9.

[55]Board minutes 30099; Tisdale to Jones, 22 Jan. 1932; both in RF 717 2.9. Board minutes 30253, 32260, RF 717D 13.110. On international law, RF 717S 20.184. On anthropology, RF 717A 10.63. On brain research, RF 717A 10.64. Weindling, "The Rockefeller Foundation and German biomedical sciences, 1920–1940," in Rupke, ed., *Science, Politics and the Public Good,* pp. 119–140.

[56]Kristie Macrakis, "Wissenschaftsförderung durch die Rockefeller-Stiftung im 'Dritten Reich,'" *Geschichte und Gesellschaft* 12 (1986): 348–379.

[57]Officers conference, 18 Jan. 1930; Pearce to Gregg, 24 Oct. 1929; both in RF 717 2.9.

[58]Jones diary, 25 Mar. 1930, RF 717 2.9.

rather than helping established institutes to expand their research programs. (There were no less than thirty-one institutes in 1932.) This policy was a reflection of the nationalistic cultural politics of the Weimar years.[59] The institutes were showpieces of German cultural superiority, and every region and scientific interest lobbied to have one. The result was, to Jones's eye at least, a mindless overbuilding and waste of scarce resources. The acting director of the KWG, Friedrich Glum, hoped that outside support from the RF would free his hands from regional politics and logrolling. Gregg was not so sure: more likely, he thought, a grant to one new institute would simply result in requests to underwrite others. It was not an unreasonable fear: Glum did in fact sound out Jones about possible aid to Max Bergmann's Institute für Lederforschung at Dresden, and for three proposed new institutes for the chemistry of silicates, fibers, and proteins.[60] Given these difficulties, why then did Jones and his fellow officers change their minds about Berlin and the Kaiser Wilhelm institutes?

The immediate cause, it appears, was a seemingly small technical matter. Jones learned by accident that the KWG had an option to buy one of the last large parcels of land in Dahlem, large enough for three institutes and strategically placed near the institutes of biology and chemistry. The society had funds to buy the site but not enough for both site and buildings. Since the option was soon to expire, and since the price of land in Dahlem was rising fast, it was now or never for the RF and for the KWG. If the RF provided funds for the site and two buildings, Jones perceived, the KWG's fund could be earmarked instead for an endowment, thus technically satisfying the RF's requirement for matching increases from the state. The political and financial problems seemed to evaporate, and the project was concluded with remarkable ease.[61]

It appears also that Jones had a different mental picture of the scientific map of Europe than Trowbridge did. He took for granted that Berlin was still the center of German science.[62] He shared the Germanophilia of most American organic chemists, uncomplicated by Trowbridge's idiosyncratic love of Latin culture. Jones wanted to do projects in Berlin and found a way around the difficulties. Also, the magnitude of the new complex at Dahlem was appealing. In his report

[59]Forman, "Scientific internationalism." Kurt Zierold, *Forschungsförderung in drei Epochen* (Wiesbaden: Franz Steiner Verlag, 1968).

[60]Officers conference, 18 Jan. 1929; Jones diary, 11–12, 17–18 Feb. 1930; all in RF 717 2.9. Adolf Morsbach, "The Kaiser Wilhelm Gesellschaft," 1932, RF 717A 10.64.

[61]Jones diary, 11–12 Feb. 1930; Gregg report, 20 Feb. 1930; Jones report, 5 Mar. 1930; all in RF 717 2.9.

[62]Jones to Mason, 2 Sept. 1929, RF 717D 14.133.

to Mason, Jones emphasized the concentration of institutes at Dahlem and the prospect of cooperation among the new and existing laboratories.[63] He depicted the institutes not as scattered and isolated facilities but as a comprehensive center of research in all the natural sciences. In short, Dahlem began to figure in Jones's plans much as the capital universities had in Rose and Trowbridge's map of European science, as a successor to the University of Berlin.

The prosperity of the Kaiser Wilhelm institutes also allayed Jones's misgivings. Since the war the German government had been putting what money there was for research into these institutes, and the RF officers could not help but be swayed by the sight of commitment and affluence in a general scene of indifference and dearth. Gregg's lieutenant, Richard Lambert, admitted that he had felt impelled to endorse a request in full by his "surprise and bewilderment at finding in Germany institutes of such magnificence as those of the KWG group."[64] It is a revealing remark: revealing of the Paris office's frustration in not being able to do more to aid science in Germany, and their sense that the KWG's grand plan was, finally, a sign of general recovery.

Germany was the last of the European scientific powers to recover from World War I: from the devastating postwar inflation, official ostracism by the International Research Council, and, not least, from the chauvinistic isolation that many German scientists imposed upon themselves. Things began to improve after 1926, with currency stabilization, Locarno, and the end of official shunning. But it was not until almost 1930 that real development seemed possible. Jones seemed always to be hearing about prewar plans still not completed after twenty years.[65] Despite the growing political crisis of the late 1920s, many hoped that long-delayed plans might finally be realized. Alan Gregg had always been more Germanophilic than Trowbridge, and the merging of the IEB and RF programs gave Gregg a greater influence.[66] For all these reasons, he and Jones began to see opportunities where Rose and Trowbridge had seen only reasons to wait and see. The RF's German projects in 1929—1933 are best understood as a delayed working out of the IEB policies and field practices.

A similar interpretation can be put on Jones and Mason's willingness to invest in research schools in provincial universities that were unlikely ever to become centers in Rose's sense. Take, for example, A. M. Tyn-

[63]Jones report, 5 Mar. 1930, RF 717 2.19.

[64]Lambert to Gregg, 5 Apr. 1932; 4 Apr. 1932; RF 717A 10.64.

[65]Jones to Mason, 18 June 1929; Wilkens to Jones, 8 Mar. 1929; both in RF 717D 14.133. Schröder-Gudehus, *Deutsche Wissenschaft und Internationale Zusammenarbeit 1914–1928*. Forman, "Scientific internationalism."

[66]Weindling, "Rockefeller Foundation and German biomedical sciences."

dall and J. E. Lennard-Jones's small but excellent school of physics at the University of Bristol. Trowbridge had declined their overtures in 1927 because Bristol was not a university of the first rank; but what deterred Trowbridge appealed to Jones and Mason as a "counter-irritant to Oxford and Cambridge."[67] Jones also arranged a grant to the "Norwegian school" of mathematicians and earth scientists at the University of Oslo, led by Vilhelm Bjerknes, who were pioneering in the application of fluid dynamics to the atmosphere and oceans.[68] And to S. Peinkowski's Institute of Physics at the University of Warsaw, which had produced several IEB fellows and was the most important center for advanced training in physics in Poland.[69] And to the newly created University of Szeged in Hungary, which provided scarce employment for some very talented alumni of the IEB and RF fellowship programs, among them the future Nobel biochemist, Albert Szent-Györgyi. Rose would probably have put such a project on ice. Jones went out of his way to push for a grant to a provincial university that was not a "center" and could not boast a group of "distinguished older scientists" but which was vital to the development of science in Hungary.[70]

What was happening in these provincial centers was in fact precisely what Rose had forecast in his original plan for the IEB. Practices of the leading universities were being emulated by new ones, carried by alumni of the IEB and RF fellowship programs from the centers of European science to the scientifically developing margins. Rose had lost sight of that ideal in his final binge of big projects. But much of what Jones saw and tried to nourish at places like Bristol and Szeged were in fact the first fruits of Rose's original vision, long in coming and soon to be cut short.

Projects in the United States, 1929–1933

The momentum (or inertia) of Rose's administrative machinery is also apparent in the RF's American projects (table 9.4). Thus the Caltech

[67]Tyndall to Tisdale, 8 Jan. 1929 and "Developments in physics in the University of Bristol"; staff conference, 8 Jan. 1930; Jones diary, 9–10 Apr., 22, 30 Oct. 1929, 14 Mar. 1930; Jones to Spoehr, 25 Nov. 1930; all in RF 401D 41.533. S. T. Keith and Paul K. Hoch, "Formation of a research school: Theoretical solid state physics at Bristol 1930–54," *Brit. J. Hist. Sci.* 19 (1986): 19–44.

[68]Staff conference, 20 Mar. 1931; Jones to Spoehr, 4 Mar. 1931; both in RF 767D 2.19. Robert M. Friedman, "Constituting the polar front, 1919–1920," *Isis* 73 (1982): 343–362. Friedman, *Appropriating the Weather: Vilhelm Bjerknes and the Construction of Modern Meteorology* (Ithaca: Cornell University Press, 1989).

[69]Jones to S. Peinkowski, 6 Feb. 1931; Jones to Spoehr, 6 Mar. 1931; both in RF 789D 4.45.

[70]Jones to Spoehr, 11 Mar. 1931, RF 750 2.12.

TABLE 9-4

ROCKEFELLER FOUNDATION EXPENDITURES IN NATURAL SCIENCE (U.S.) (IN $1,000S)

Institution	*Field*	*1929*	*1930*	*1931*	*1932*	*1933*
Harvard	Astron.	500				
Chicago	Biology	150				
Chicago	Physics	15				
Princeton	Geology	100				
Minnesota	Geology	15				
Alaska Agricult.	Geophys.	10				
Field Museum	Botany	15				
Johns Hopkins	Chem.	40				
Johns Hopkins	Biology		388[a]			
Caltech	Nat. sci.		500			50
Harvard	Geophys.			50		
MIT	Geophys.				6	8
Ohio Wesleyan	Astron.				20	
Princeton	Physics				16	
Caltech	Chem., physics				40	10
Harvard	Chem.				45	
Jackson Lab.	Genetics					11
Total		845	888	50	127	79

[a]Not including a pledge of 500 (i.e., 500,000) payable in 1940).

troika easily got yet another pledge of $1 million of endowment (though Spoehr objected, for reasons that hint of a preference for Stanford, his own academic base).[71] An appropriation to the Harvard Observatory, deferred in 1927, was easily revived—Harlow Shapley had only to inquire.[72] A grant for geology at Princeton, engineered by

[71]Board minutes 30227; Millikan to Spoehr, 8 Nov. 1930; staff conference, 28 Nov. 1930; all in RF 205D 5.66.

[72]"Historical record"; Shapley to Mason, 21 Mar. 1929; staff conference, 11 Oct. 1929; all in RF 200D 139.1718.

Trowbridge (now graduate dean), rounded out prior GEB grants to the other natural sciences. Vincent would have been glad to give more had Trowbridge only been so bold as to ask for it.[73] GEB connections and precedents made it possible to consummate projects with almost unseemly ease.

Rather more interesting was the expansion of Rose's and Thorkelson's experiments with fluid research funds (table 9.2). The grants to the universities of North Carolina and Virginia were renewed, and new ones, previously turned down by Thorkelson, were inaugurated at Minnesota and Iowa State. These grants introduced into state universities a system of patronage different from the established system of state and federal project grants. Though smaller, foundation grants were more conducive to change because they channeled funds to younger researchers working in newer fields. Project grants from states and the U.S. Department of Agriculture, in contrast, tended to go preferentially to established researchers in a few (mainly agricultural) departments. The political process made it so: it was politically difficult to get support from legislators for new projects but easy to continue projects once started. This system of funding perpetuated research programs, whether or not they remained productive, and excluded younger faculty with new and often less obviously utilitarian research interests. The system of agricultural patronage, in its early years, had worked extremely well. It empowered a whole generation of scientists to make productive careers in state universities and to discover how basic research could flourish around agricultural problem solving. But like any system, this one was generation-specific, and its procedures made it inflexible. It could not make funds available to attract the best young scientists to the state universities and to participate in the newest and most productive research lines. Reformers like Iowa State's president, R. M. Hughes, and the geneticist Edward W. Lindstrom hoped that foundation funds would enable them to break the generational deadlock without disrupting the agricultural support system on which they would always depend.[74]

With his own roots in Midwestern academic culture, Mason was well disposed toward such reformers. He believed that the RF should encourage research everywhere by making small sums broadly available. Besides, fluid research funds were a convenient and affordable substitute for capital grants—a practical necessity in the 1930s. Mason rebutted those who felt that RF funds would be better spent in a few

[73]J. Hibben to Mason, 9 May 1929; Vincent diary, 10 May 1929; both in RF 200D 154.1898.

[74]Lindstrom to Mason, 14 Oct., 8 Nov. 1929; staff conference, 13 May 1931; all in RF 218D 1.6.

centers, pointing to the golden age of German science, when research flourished even in minor provincial universities.[75] Fluid funds began to be more than a practical makeshift for aiding public universities.

Mason extended fluid research funds to private institutions like the University of Pennsylvania and Washington University. These urban institutions, like the state universities, were having difficulty moving into the research elite. Their basic science departments were overshadowed by powerful medical schools, which had got millions from foundations and were exploiting opportunities for research in much the same way that agricultural scientists were in the land-grant system of patronage. The basic sciences were cut off from the medical patronage system, as they were also from the largesse of college alumni. Reformers like Penn's able provost, Josiah H. Penniman, were trying hard in the late 1920s to bring their institutions into the modern world of organized, sponsored research by organizing local research committees and surveys of resources for research.[76] This middle tier of private universities was only about five years behind the earlier group (Caltech, Princeton), which had been the first to take advantage of the postwar boom and mood of uplift. Had the latecomers come along just a little earlier they could well have caught Thorkelson's eye. In the early 1930s, they intersected with Mason's growing interest in new modes of patronage to replace the 1920s system of capital grants.

On the whole, Mason's experiment with fluid grants-in-aid funds was not a great success. In principle, faculty committees would channel support to the best young researchers without getting the RF entangled in deciding who was best. In practice, of course, faculty committees were entangled in academic politics and found it easier to spread funds around than to concentrate on a few productive individuals. Reviewing the results of the grants to North Carolina, Penn, and Washington University, Warren Weaver felt that the RF's money had been largely dissipated in unexceptional projects.[77] Where there was strong leadership, however, fluid research funds could be used to develop specific research schools. In a few places such funds began to be used as a replacement for institutional endowments of specific departments.

At MIT, for example, money was not spread around to rank-and-file faculty but was used strategically to strengthen the programs in chemistry and physics. In fact, MIT's grants-in-aid fund was a surrogate for

[75]Staff conference, 13 May 1931.

[76]J. Penniman to Mason, 17 June 1930, and "Report of the research committee," 23 May 1930; Mason diary, 26 Nov., 1 Dec. 1930; all in RF 241 1.5. On Washington University, RF 228 1.1.

[77]Weaver, "Appraisal," May 1939, RF 224D 4.41. "Appraisal," June 1939, RF 218D 1.6. "Appraisal," Nov. 1939, RF 228 1.3. Weaver diary, 2, 3 Mar., 28 Oct. 1932, all in RF.

an endowment that MIT never received. President Samuel W. Stratton had been trying for years, without success, to get a capital grant from the GEB to revive MIT's once distinguished tradition of research in biology and chemistry.[78] Mason, to whom Stratton renewed his suit in 1929, was more receptive than Rose, but Stratton seemed incapable of doing anything to accomplish the reforms that he himself preached. Powerful members of MIT's board agreed, like AT&T's Frank Jewett and GE's Gerard Swope. When Mason asked them how he should respond to Stratton's request for endowment, Jewett and Swope asked him to stall until Stratton could be eased out and a more active president installed.[79] Stratton's successor was Karl T. Compton, an experienced and successful academic entrepreneur and fund raiser and a paragon of the post-Hale generation of science organizers.[80] Compton lost no time in reviving Stratton's application for $1 million to modernize the basic science departments.

Mason was eager to aid Compton's reforms, but he declined to consider a grant of endowment on the grounds that aiding a school of engineering did not fall within the RF's program. Compton protested in vain that MIT was as much a school of science as of engineering and that the RF had not shrunk from giving a large sum to Caltech. How was MIT different? he wanted to know.[81] In fact, Compton had misunderstood the problem: it was not that the RF arbitrarily ruled out grants to schools of technology but that a grant of endowment implied a commitment to the institution as a whole and might be mistaken as support of education, not research. A fluid research fund, in contrast, was unambiguously for individuals' research projects. It is not clear who first suggested this compromise: Mason, probably, but Compton went along when he realized it was that or nothing.[82] The episode recalls Millikan and Vincent in 1919, trying to discover a way of developing institutions without setting a precedent of direct investment, and in so doing inventing a new system of patronage. So too Mason and Compton invented a form of grant that provided institutional support in the guise of individual grants—a harbinger of things to come.

In practice, Compton used the fluid research fund to woo fresh tal-

[78]S. W. Stratton to GEB, 23 Mar. 1923; Rose diary, 25 Mar. 1927; both in RF 224D 4.38.

[79]Stratton to GEB, 21 May 1929; Mason diary, 11 Apr., [?] Sept., 24 Oct. 1929; all in RF 224D 4.38.

[80]Robert Kargon and Elizabeth Hodes, "Karl Compton, Isiah Bowman, and the politics of science in the Great Depression," *Isis* 76 (1985): 301–318.

[81]Compton to Mason, 7, 11 Apr. 1930; "Application for support of fundamental sciences," 11 Nov. 1930; Compton to Mason, 21 Jan. 1931; all in RF 224D 4.39–40.

[82]Spoehr diary, 27 Jan., 5 Feb. 1931; Compton to Spoehr, 30 Jan. 1931, 12 Feb. 1931; staff conference, 16 Mar. 1931; all in RF 224D 4.40.

ent to the departments of physics and chemistry and support projects in productive new lines of research. It was the only one of the RF's fluid grants that Weaver regarded as a success—a showpiece, in fact.[83] By using the MIT grants fund strategically, Compton transformed a vehicle designed to aid isolated individual talents into an engine for realizing Compton's vision of a Caltech on the Charles. The patronage relation was evolving into something new as the era of large capital grants came to a close.

Other hybrids between individual and institutional grants also appeared in the early 1930s. At Johns Hopkins, for example, the expiring grant to Raymond Pearl's Institute for Biological Research (a white elephant, in Mason's view, inherited from Embree's program in human biology) was transformed into a tapering grants-in-aid fund, partly for Pearl and partly for Herbert S. Jennings's group of biologists. Similarly at the University of Chicago, Frank Lillie's scheme for an Institute of Genetic Biology became a fluid research fund for the biological departments and Lillie's school of reproductive physiologists.[84] (Both these grants were eventually capitalized.) At Caltech a stalled drive for endowment was tided over with individual grants for Linus Pauling's researches on the physics of chemical bonding and for a project in the physics department on the molecular structure of solids.[85] It had been customary for the RF to pay interest on pledges, but this was the first time that prepayments had been earmarked for specific individuals and research lines. Unable to disavow a prior pledge of capital, Mason did what he could to make it look like an individual research grant.

Toward a New System of Patronage

These improvisations signaled a fundamental change in the system of foundation patronage, from institution building to aiding individual projects in specific research fields. It was a change not so much in policy as in practice. Fosdick's conception of a foundation for the advancement of knowledge did not rule out institutional grants. The choice of philanthropic methods had been left completely open, and the momentum of IEB and GEB programs ensured that for a few years institutional grants would dominate. It was a recapitulation, in a way, of the events that followed the creation of the Rockefeller Foundation in

[83]Weaver, "Appraisal," May 1939, RF 224D 4.41.

[84]On Johns Hopkins, RF 200D 145.1791–1794. On Chicago, RF 216D 89.103.

[85]Weaver to Noyes, 9 May 1932; A. Goetz and F. Zwicky, "Program on the new physics of solids," 8 Apr. 1932; Pauling, "A program of research in structural chemistry," April 1932; staff conference, 20 April 1932; Mason diary, 24 Oct. 1932; all in RF 205D 5.70.

1913, when Rose's public health operation dominated the larger organization of which it was a part. In 1913, the newborn foundation had had no program of its own to compete with the hookworm program, and so, again in 1929, Rose's ongoing program in the natural sciences dominated a reborn foundation, which had no programs except those from which it had been cobbled together out of the pieces of the IEB, GEB, and the Spelman memorial.

The indeterminate potential of the new Rockefeller Foundation was a reflection of the way it was created by Fosdick. It was not, as we have seen, a calculated response to specific concerns about society or science, as Ruml's and Rose's programs were responses to postwar agendas of social reconstruction. Resolving the administrative disharmonies among the boards was an in-house problem, which neither required nor encouraged specific directives, only a reshuffling. Fosdick did not intend to set in motion a process that would result in far-reaching changes in the system of patronage. Yet he did, and the potential for change clearly was a reflection of Fosdick's unusual character and vision. He probably could have solved the problem by tinkering with the machinery. But Fosdick was not a tinkerer. Everything he ever did reveals a person who did not act without a compelling intellectual vision. The idea of a foundation for the advancement of knowledge was such a vision, pregnant of future development, but one whose potential would be realized only in the process of putting it to work.

Policy in the transition years was conflicted and uncertain. Thus the zoologist Caswell Grave got a mixed message from Mason when he came in 1929 to seek endowment for his department. Mason blew hot and cold, recalling in one breath the RF's longstanding interest in Washington University and, in the next, telling Grave how endowment of the zoologists at Washington University would only discourage those in other equally deserving places, like the University of Wisconsin. Grave reported that Mason expressed interest only in research that was interdisciplinary: "institutes of research in which several departments combine—Mathematics + Physics + Biological and Physical Chemistry + Biology."[86] Grave also left thinking that Mason was interested only in major centers, like Chicago or Harvard. More likely, Mason hoped that grants to multidisciplinary projects would enable the RF to aid particular universities without setting a precedent for aiding all of them. Unused to dealing with foundations, Grave could not have known how much Mason was improvising policies to fit an uncertain and changing context.

[86]Caswell Grave to George R. Throop, 6 Sept. 1929, Throop Chancellor Records, ser. 2, box 1, Washington University Archives, St. Louis, Mo. I am grateful to Glenn Bugos for showing me this document.

It was financial pressure, ultimately, that forced fundamental changes in practice and policy. The era of endowment came to an abrupt end as a Wall Street panic turned into the Great Depression. Within months of the Crash, Mason was under pressure from Rockefeller, Jr., who was alarmed by plummeting stock prices and wanted to terminate large capital expenditures as quickly as possible.[87] Rockefeller thought begging scientists always exaggerated their needs, and he did not feel any moral commitment to complete projects just because negotiations had been started. (He was scrupulous about honoring contracts but was insensitive to personal or moral obligations.)[88] Anyway, it was crystal clear that the RF would no longer be able to make large appropriations from its deflating capital. In late 1930, Mason took steps to engineer consent by the officers to a strict policy of concentration. "Restriction within fields and co-operation between fields" was to be the new watchword. Yet everyone seemed to assume that fields of concentration would be developed through institutional grants to university centers of research and training, in the style of Rose. It was still possible, Mason thought, to "build a permanent machine" in the major universities.[89] A new policy was in the making, but old practices were hard to cast off.

The conflict of old habits and new circumstances resulted, as we have seen, in compromise. Capital gifts virtually ceased by 1930. On-going projects were tinkered into new forms. Institutional grants took a more personal form, and individual grants served institutional purposes. Late-blooming projects of the IEB and GEB were metamorphosing into a new system of patronage. Smaller grants for individual projects were clearly the form most appropriate for hard times. But policy changed more slowly. It was not clear, for the natural sciences at least, what the appropriate fields of concentration were, nor what the role of officers should be in fostering specific fields. There was no enthusiasm for a return to the pre-World War I system of small grants-in-aid, but no one saw very clearly how individual grants could systematically be given institutional purpose. These questions would be thrashed out in the mid-1930s. As with so many aspects of economic and social life of the 1920s, the old system of patronage expired before a new one was fully formed.

[87]Mason diary, 27 Dec. 1929, 9 Jan. 1930, both in RF.

[88]Embree, "Rockefeller Foundation," n.d. [c. 1930], p. 3, ERE box 1.

[89]Officers conference, 2 Oct. 1930, RF 900 21.160. "Report of the committee of appraisal," 11 Dec. 1934, pp. 51–54, RF 900 22.166.

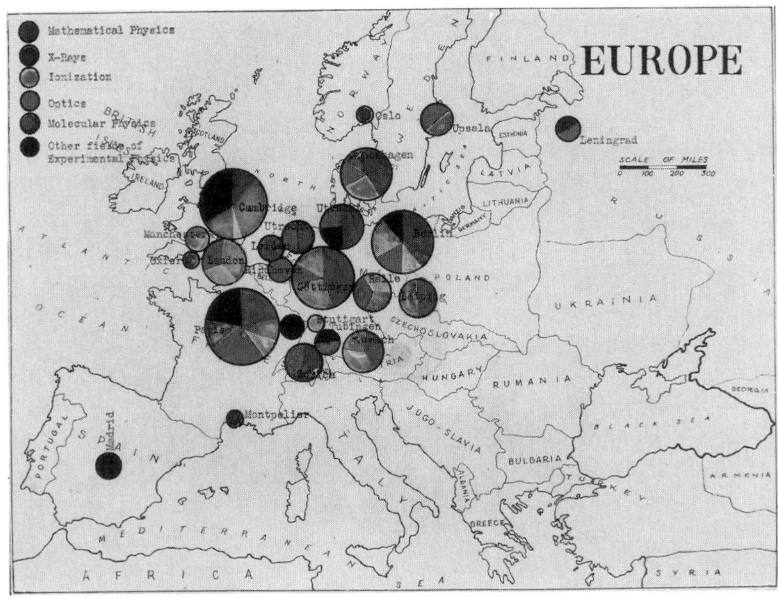

13. IEB map of European centers of physics, 1926. Courtesy of the Rockefeller Archive Center.

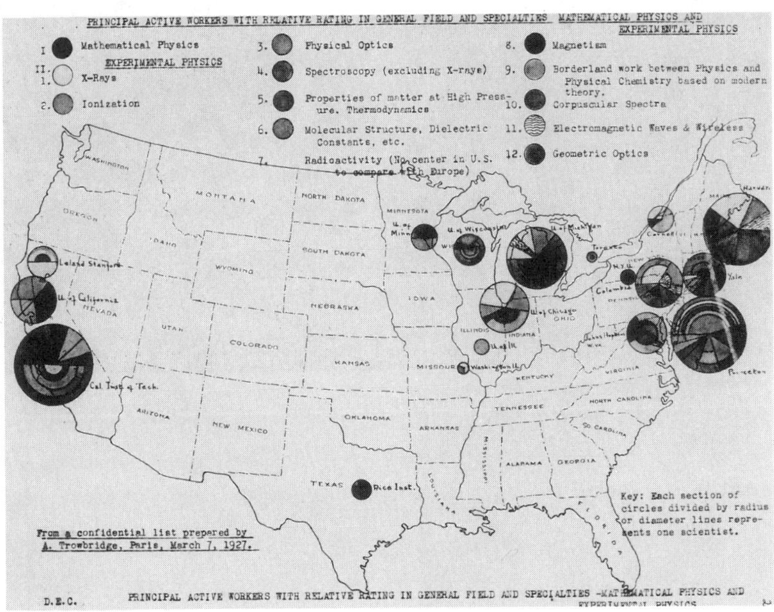

14. IEB map of U.S. centers of physics, 1926. Courtesy of the Rockefeller Archive Center.

15. Raymond B. Fosdick, mid-1930s. Courtesy of the Rockefeller Archive Center.

16. Max Mason, early 1930s. Courtesy of the Rockefeller Archive Center.

17. Kaiser Wilhelm Institute for Cell Physiology, c. 1932. Courtesy of the Rockefeller Archive Center.

18. Warren Weaver, University of Wisconsin, c. 1921. Courtesy of Helen Weaver.

19. Warren Weaver in his office at the Rockefeller Foundation, 1932. Courtesy of Helen Weaver.

20. Warren Weaver in retirement, c. 1970. Photograph by and courtesy of Helen Weaver.

21. Grasshopper culture in Joseph Bodine's laboratory, University of Iowa, 1935. Courtesy of the Rockefeller Archive Center.

22. Joseph Bodine's laboratory of cell physiology, University of Iowa, 1935. Courtesy of the Rockefeller Archive Center.

23. Frances O. Schmitt, c. 1940. Courtesy of the MIT Museum.

24. John Runnström, 1938. Courtesy of Dr. Vera Runnström-Reio.

PART III

Disciplinary Relations, the 1930s

Chapter Ten

Warren Weaver and His Program

The new system of science patronage had three key elements: first, the belief that it was necessary and proper for sponsors to identify specific disciplines or research specialties in which to invest. There was no precedent for such an interventionist kind of sponsorship, and when it had been tried it did not work very well (recall Embree's disaster with "human biology"). The second and most crucial element was the activist program manager, who knew enough science to devise a program, select fields of concentration, and make informed judgments about individuals and their projects. Again, there was little precedent in the history of science patronage for officers' taking such an activist role. Woodward got burned when he tried, and Trowbridge and Thorkelson steered clear of such intervention. Indeed, the system of institutional grants was designed in part to avoid the problems of a more intimate and demanding partnership. Finally, there was the project grant, a vehicle of funding that made an activist, programmatic style of patronage practicable, and that gave individual grants a larger institutional purpose.

The success of this new system depended absolutely on program managers who had the technical training and authority to select among alternative programs and manage projects without offending their scientific clientele. It was a rare combination of talents that was required, and without the right person an activist system of sponsorship could be an invitation to disaster. Mason's tendency to tinker and improvise reflected the basic fact that in the natural sciences there was no officer able to design and manage a coherent program. Alan Gregg was such a person in the medical sciences, and his program in psychiatry got off to a strong start. In the natural sciences it was a different story, as we have seen.

Eventually, of course, Mason found the person he was looking for in

Warren Weaver, who proved to have the ambition and vision to make the new system of patronage work. Even then, however, the process of construction was a difficult one. Weaver had simultaneously to invent a program that was politically and intellectually defensible, and to create a role between his constituents and the board that would give him the authority to carry out his program. That was no easy task. When Weaver arrived in February 1932, the RF was still uncertain of its new mission and was in financially straitened circumstances. The public health and medical science programs enjoyed continuity in practice and unquestioning acceptance by the trustees. In the natural sciences none of Weaver's predecessors had set clear precedents. Fosdick's 1928 design implied that division heads should be active parties in making projects, but the trustees were afraid to let officers get the bit in their teeth. With a budget crisis looming, it was not clear that Weaver would be able to put a new program into effect, even if he had one.

In short, Weaver had simultaneously to learn the ropes of grant-giving, plan a program, and win the confidence of the trustees. Every decision was tricky, because it had implications for the division of labor between divisions and for his own authority. And his efforts were not unopposed. Some important people would have been happy if Weaver had been less active in defining a program and a less vigorous participant in scientists' projects. Simon Flexner resented mathematicians intervening in biologists' business, and he was still a powerful mover behind the scenes. Flexner's feelings were shared by some members of the board, especially the biomedical scientists. There is no evidence that Weaver's scientific clients ever objected to his activist style; but scientists, when invited to review and advise, reflexively favored a more laissez-faire relation between grant-giver and grant-getter. Weaver did have the constant support of Raymond Fosdick, who believed that officers should be empowered, but he had to tread carefully and keep his fences mended. As a newcomer to the RF he made some political mistakes that, but for Fosdick's backing, could have been fatal.

Weaver became an activist manager of science in the process of constructing his program in "vital processes," and how this process worked is the subject of this chapter. We will begin with Mason's efforts to find a director for the natural sciences and to define a program; then probe Weaver's academic background for experiences that, I will argue, shaped his unusual philanthropic vision. Finally, we will follow Weaver's apprenticeship as a manager of science and the development of his conception of "vital processes" through the vicissitudes of RF politics in the 1930s. We will see how his vision of a realm of "vital processes," in which natural scientists of various sorts could engage cooperatively, was well suited to a system of sponsorship in which officers played an active personal role. Disengaging "vital processes" from established biological disciplines enabled Weaver to sponsor particular individuals or modes of practice, without running afoul (as Woodward did and Trowbridge feared he might) of disciplinary interests and logrolling. "Vital processes" not only specified an area of research, it also entailed a system of philanthropic practice.

Attempts to Define a Program, 1930–1932

"Restriction within fields and co-operation between fields," that was Mason's watchword as Vincent's successor. But what fields? That was not so clear, especially in the natural sciences. Biology was an obvious choice, given the RF's tradition in medicine and the trustees' lack of enthusiasm for perpetuating Rose's program in mathematics, physics, and chemistry. The physical sciences were already well supported by universities, Fosdick felt, and they were beginning to be talked about as causes of technological unemployment.[1] There was a growing sense in 1930–1932 that biology was more in tune with the times than physics. But if biology, then biology in what connection? Biophysics and biochemistry? Medicine and "psychology"? Agriculture? All had precedents in the programs of the 1920s, but the departmental grants to Jennings and Lillie were the only two left on which to build a new program.[2] Mason did not hide his personal interest in earth science but seemed inclined to leave decisions to whomever was in charge of the natural sciences division.

It is hard to imagine that Mason hired the plant physiologist Hermann Spoehr without some thought that he might develop biology in its connection with agriculture and forestry. The Bailey-Spoehr report was a readymade blueprint for a program of research in the basic sciences underlying forestry. Mason and Tisdale agreed that the program in agriculture, though "quiescent," could be revived.[3] Egged on by Mason, Spoehr waxed eloquent at his first meeting with the trustees on the opportunities in plant physiology and forestry.[4] The head of the NRC's agricultural division, Lewis R. Jones, was a regular visitor in Mason's office, obviously hoping that the RF would do for agriculture and forestry what it had done so spectacularly for oceanography. Although Mason showed little personal interest, he probably would have sup-

[1]Fosdick to Greene, 25 Mar. 1937, RF 915 1.2 C. W. Pursell, Jr., "'A savage struck by lightning': The idea of a research moratorium, 1927–1937," *Lex et Scientia* 10 (1974): 146–161.

[2]"Agenda for special meeting," April 11, 1933, p. 33, RF 900 22.168.

[3]Mason diary, 15–16 Oct. 1929, RF.

[4]"Conference of trustees and officers," 29 Oct. 1930, pp. 90–95, RF 900 22.167.

ported such a program had Spoehr pushed for it.[5] But Spoehr did not push, despite the continual urging of hopeful colleagues like Henry S. Graves, dean of Yale's forestry school, who pressed him to say what the RF intended to do about the NRC report and suggested that Spoehr set up a small planning group to formulate a program in forestry. Spoehr replied bluntly that the RF had no program in forestry and that he had no opinion about what should be done.[6]

We should not assume that Mason meant to concentrate in biology. Scouting Harvard and MIT in April 1931 for a successor to Spoehr, he seemed bent upon getting a physicist or chemist. Possible candidates included John Johnston, physical chemist and director of research at U.S. Steel; Harvard physicist George W. Pierce; Floyd K. Richtmyer, physicist from Cornell; Harvey N. Davis, physicist and president of Stevens Institute of Technology; and several other MIT physicists and chemists. None had experience in biology. Nor did the most likely prospect, the Harvard chemist Arthur B. Lamb. Mason was impressed by Lamb's administrative ability (he was director of Harvard's chemistry laboratories and editor of a leading journal), and by his pleasing and tactful personality—"a safe and a fine cooperator," Mason noted.[7] Mason's actions remind us that his vision of a multidisciplinary program was not limited to physicochemical biology. Mason had an active interest in geophysics, as we have seen, and the grants to Pauling and Zwicky suggest that he was also contemplating a program in the nascent field of chemical physics. This conjecture is supported by a grant in 1932 to the Harvard physical chemist George Kistiakowsky for his research on chemical thermodynamics.[8] Mason's problem was to limit claims by university departments on the RF, and multidisciplinary projects did that. The choice of disciplines was less important, and Mason obviously meant to leave that to his divisional directors. When it became clear that Lamb was not prepared to give up a scientific career, however, Mason was again left with no prospects for a director of the natural sciences.

Weaver and the Wisconsin School of Mathematics

In the fall of 1931 Mason telephoned Warren Weaver, his former student and colleague, and asked him to take the next train to New York

[5]Mason diary, 26 Nov. 1928, 28 May, 15–16 Oct., 22, 27 Nov. 1929; all in RF.

[6]Spoehr diary, 2, 7 Mar. 1931, 21 Oct. 1930, all in RF. "Conference of trustees and officers," 29 Oct. 1930, pp. 91–95, RF 900 22.167.

[7]Mason diary, 22 Apr. 1931, RF 915 1.1.

[8]J. B. Conant to Mason, 7 May 1932; Weaver diary, 22 Nov. 1932; G. Kistiakowsky to Weaver, 20 Dec. 1934; Weaver to Kistiakowsky, 17 Jan. 1935; all in RF 200D 139.1720.

to talk about a career in philanthropy. Weaver was taken aback by Mason's suggestion. It had never crossed his mind to leave the academic world or the Middle West (the trip East would be his first). But he felt that a request from Max could not be disregarded. The two had become close friends during their years together on the faculty of the University of Wisconsin (1920–1925), Weaver as a starting assistant professor in mathematics, Mason as a brilliant and popular professor of mathematical physics. They shared a common interest in applied mathematics, and Weaver treasured the hours they spent talking and dreaming about a fundamentally new mathematical approach to quantum phenomena. They spent several hours a day for years writing a textbook of electrodynamics. Their talents were complementary: Mason did not like hard work and could not stick to a regular schedule; Weaver loved to work and was well organized. Weaver drafted, Mason criticized. It is characteristic of their relationship that Weaver later gave Mason credit for "at least 90 percent of the brains of the combination and probably 99 percent of the imagination."[9] Although the difference in their ages was not great (thirteen years), they related to each other as master and disciple. In every respect but their interest in applied mathematics they were completely different. Weaver was a small-town boy who made good but never lost the plain style of his roots. Practical and hardworking, he stood in awe of Mason's brilliant, articulate, and cosmopolitan style.

Mason was undoubtedly counting on Weaver's devotion when he called him to New York. He must have been feeling desperate. Spoehr had left some months before, and negotiations with Lamb had been broken off when Lamb began to talk about a halftime research position at the Rockefeller Institute.[10] It would be consistent with Mason's character to suppose that he called Weaver with the thought that it would be easy to talk Weaver into following him East and getting him out of a jam. Shortly after accepting Mason's offer, Weaver wrote to a friend: "Six weeks or two months ago I would have laughed heartily in the face of anyone who would have suggested that any job could blast me out of our snug situation here at Wisconsin."[11] What kind of person was Weaver, and why was he persuaded to follow Max into the unknown?

Weaver came from a small town in central Wisconsin, where his father owned a drugstore, and he grew up in Madison. It was a place where a boy's nascent interest in science led to a career in engineering

[9]Weaver, *Scene of Change* (New York: Scribners, 1970), pp. 54–57, 28–31. Weaver, oral history, pp. 110–113, 129–146. CUL Weaver, "Max Mason," *Biog. Mem. Nat. Acad. Sci.* 37 (1963): 205–236.

[10]Mason diary, 10 June 1931, RF 915 1.1.

[11]Weaver to Chancey Leake, 6 Jan. 1932, UW 7/22/18 box 1.

(no one in his family knew there was such a thing as physics). Weaver was several years into an engineering course at the University of Wisconsin before he realized that he really wanted to do mathematical physics. To become a professor at Wisconsin was the height of his ambition. A wartime stint at the National Bureau of Standards and a year as lecturer at Caltech taught him to admire Robert Millikan's cosmopolitan stylishness, but did not much change the plain style of his Midwestern roots. A boyhood friend of the LaFollette family, which produced two generations of populist political leaders, Weaver never lost his region's egalitarian values, belief in social service, and intolerance of social snobbery.[12] He never shrank from exercising what authority he had, but never exercised power merely for the pleasure of doing so.

Weaver was deeply influenced by the academic culture of the University of Wisconsin, where he spent all but two of his adult years before going to New York. Nowhere was the land-grant ideal of knowledge and public service more deeply rooted—the "Wisconsin Idea" was the quintessence of high progressivism.[13] The most important influence on Weaver (and Mason) was Charles S. Slichter, the founder of the Wisconsin school of applied mathematics. Of Swiss-German descent, Slichter represented the generation of Midwestern academic reformers who, in the early twentieth century, integrated Eastern ideals of high scholarship with land-grant ideals of practical service. Slichter combined an academic career with an extensive consulting practice, working with hydrological surveys and irrigation projects, and applying mathematics to model the flow of water in aquifers. He took applied mathematics from engineering and made it a dynamic basic discipline quite unlike the abstract academic style that prevailed in most universities. He was an imposing figure: charming, energetic, unorthodox, and interested in everything, especially things scientific—an intellectual in the distinctive Midwestern style.[14] Mason and Weaver emulated different aspects of Slichter's style: Mason, his intellectual flamboyance; Weaver, his practicality and service values.

Slichter, Mason, and Weaver were very close in the 1920s and shared a commitment to cross-disciplinary research. They worked together on mathematical modeling of geological processes of sedimentation

[12]Weaver, oral history, pp. 2–30, 54–70.

[13]Curti and Carstensen, *The University of Wisconsin.*

[14]Weaver, oral history, pp. 104–109. Weaver, *Scene of Change,* pp. 24–27. Ingraham, *Charles Sumner Slichter,* pp. 40–101. John Servos, "Mathematics and the physical sciences in America, 1880–1930," *Isis* 77 1986: 611–629.

(Weaver's best research, he always thought).[15] Mason applied his wartime experience with acoustical submarine detection to geophysics and organized a consulting firm for geophysical exploration, in which Slichter's geophysicist son Louis was a partner. Slichter was always promoting sponsored interdisciplinary research. As dean of the graduate school in the 1920s, he took charge of the faculty research fund and used it to develop strategic lines of research. He was instrumental in creating the Wisconsin Alumni Research Fund and actively cultivated connections with the Rockefeller boards. Though frustrated there by the university's injunction against foundation gifts, he was ever watchful of opportunities to promote programmatic, interdisciplinary research projects. Mason and Weaver's subsequent careers as managers of science make clear just how much they learned at Slichter's knee.

Weaver was most closely involved with Slichter's schemes for planned research in 1930–1931, not long before he left for New York. In the summer of 1930, the two conceived a plan to make the university a center for interdisciplinary research in geophysics, by pulling together in one place research lines that had been established in several departments of science and engineering. Weaver formed a committee which called for a new professorship of geophysics (for Louis Slichter) and for two full-time codirectors of a program of cross-departmental research—no names were mentioned, but Weaver would undoubtedly have been one of them. Dean Slichter hoped to put together a funding package from the university, the Wisconsin Alumni Research Fund, and—most important—the RF, now headed by Max Mason, who expressed a keen interest in the project.[16] Optimism about the RF may not have been misplaced, since it had just made several grants to similar interdisciplinary projects in chemical physics and biophysics, and Mason was very interested in a nascent project in geophysics at Harvard.

Although Weaver's project collapsed in 1931, when Louis Slichter went to MIT, his efforts were hardly wasted.[17] He learned from Slichter about entrepreneurship and organizing interdisciplinary science. And he came to Mason's attention at a moment when Mason needed someone with just such interests and skills. This was the context in which Weaver received a call from Mason to become not his client, as seemed most likely, but his colleague.

[15]Weaver, oral history, p. 32. Jones and Weaver diary, 8 May 1933, RF 800D 7.75.

[16]Weaver to Slichter, 11 Dec. 1930, and "Memorandum," both in UW 6/1/1 box 21 Miscellaneous. Curti and Carstenson, *University of Wisconsin,* vol. 2, pp. 230–231. The prohibition against foundation grants had just been lifted, but the new governor, Philip LaFollette, was threatening to reimpose it.

[17]Weaver to Slichter, 23 Mar. 1931, UW 6/1/1 box 21 Miscellaneous.

Weaver's confessed lack of sympathy with quantum theory has reinforced a suspicion that philanthropy might have been for Weaver a way out of a flagging career in science.[18] That was hardly the case. Weaver had been full professor and chairman of his department since 1928, when he was only thirty-four years old. Trowbridge ranked him among the top fifteen Americans in mathematical physics.[19] In 1928 Weaver was one of two mathematicians considered by Arthur Compton to lead a large project in mathematics and physics at the University of Chicago.[20] He had good prospects in 1931 for fruitful research in mathematical geology. It is quite true that Weaver published nothing of fundamental importance in pure mathematics. His talent lay in putting basic mathematics to work in other fields. Compton's proposal to the GEB noted that Weaver "has a personality remarkably well fitted to the development of inter-departmental fields."[21] His personality and training suited him well for the new life that Mason had in mind for him. But if anything, Weaver was overqualified for the job.

Weaver would not have known that, of course. He was a modest man and tended to underrate his talents. Looking back on his career, he described himself as "extremely unimaginative" and "not fundamentally creative."[22] He was thinking, apparently, of his inability to make fundamental contributions to pure mathematics, and it may be that Weaver's knack for mediating between disciplines deprived him of the self-assurance that comes from being in the mainstream of a well-defined and prestigious scientific community. But it was also the source of what he himself identified as his special talent: an ability to assimilate varied knowledge easily, fit it into a larger framework, and make other nonspecialists understand it. He was a natural teacher and interpreter, and that gift served him well in his dealings with a great variety of scientific specialists and personalities. Weaver was a highly creative and resourceful person, but his creativity came less from brilliant intellect or cosmopolitan sophistication than from a taste for practical problem solving. He liked to work and had a "good administrative conscience." He was ambitious, but his ambition was always tempered by an instinct for service and by his awareness of his own limitations. He brought to philanthropy the habits of the Wisconsin school of applied mathematics and the values of a man who had come up from nowhere through his ability to learn and teach.

[18]Weaver, oral history, pp. 135–136, 146–155.

[19]Trowbridge, "Mathematical physics and experimental physics in the United States," 7 Mar. 1927, IEB 1.1 10.143.

[20]Ibid.

[21]"A research program for pure and applied physics," 18 Jan. 1928, GEB 1.4 660.6862.

[22]Weaver, oral history, pp. 30–32.

We know very little of what was talked about in Mason's office on 10 December 1931. Weaver apologized to Mason for being too concerned with the practicalities of living in New York City—salary, housing, social and family life. (Could anything make up for leaving the intimate and informal circle of family and friends at Madison, he wondered.) But that was only, he assured Mason, because there was so little question about the appeal of the job. He was flattered and grateful: "I have never been able to understand in the past why you did so many things for me," he wrote Mason, "and this situation does not help clear up the mystery."[23] Weaver had been snowed ("stunned," "dazed," he wrote), and it is clear that Mason had made sure he was snowed. Mason painted a picture of a job with much freedom (a five-day week, two months or more free time in summers), and lots of travel on a very liberal allowance. (The last, at least, was true.) Weaver was given a view not just of the natural sciences division but of the RF's vast operations: a staff of 150 in the New York office (including a complete travel bureau) and sixty more in Paris; representatives in every major city of Europe and many in Africa and Asia; the variety of programs—laboratories, research projects, salary supplements, public health bureaus, six to eight kinds of fellowships and exchange, medical schools, universities, all on the grand scale. The prospect was held out of being received by European ministers of education as a kind of "scientific diplomat."[24] No wonder Weaver was dazzled with the magnitude and importance of what Mason was offering. It was impossible to refuse, even though it was a wrenching experience for him and his wife to pull up their roots in Madison.[25] On 1 February 1932 he began to learn what it was really like to be a manager of science.

Learning the Craft of Philanthropy

What Weaver found at the Rockefeller Foundation did not quite match the rosy picture that Mason had painted for him. There was no one in the national science division to show him the ropes and tell him what to do with the knotty administrative problems that appeared mysteriously on his desk. Nor did he get much help from colleagues in other divisions. It was not the custom in the New York office to hold general staff

[23]Weaver to Mason, 8, 24 Dec. 1931; Weaver to H. B. Helstrom, 30 Jan. 1931; all in UW 7/22/18 box 1.

[24]Weaver to R. N. Hunt, 29 Dec. 1931; Weaver to T. C. Fry, 18 Dec. 1931; both in UW 7/22/18 box 1.

[25]Weaver to Leake, 6 Jan. 1932; Weaver to P. LaFollette, 29 Jan. 1932. UW 7/22/18 box 1. Not wanting to burn all his bridges, Weaver arranged to be on leave from the university for 18 months.

conferences, and the divisions made little effort to work together. The only opportunity Weaver had to learn from his colleagues was watching their semi-annual performances before the trustees.[26] Obviously, the foundation's officers had not quite recovered from the trauma of reorganization.

Max Mason was one of the reasons for the disarray in the New York office, and Weaver soon began to see his intellectual idol in a new light. Despite his quick and brilliant intellect (or because of it), Mason was not a good administrator, simply leaving everyone to do as they pleased. As Weaver later recalled, Mason was "in reality awfully lazy." Everything came so easily to him that he was never forced to acquire habits of hard work and self-discipline. He liked to take up odd problems that intrigued him but lost interest as soon as he saw that they could be solved. "He was so clever, so brilliant, so able—and so lazy," Weaver recalled. "He simply could not devote himself, in any sustained way, to any task. He could not write a good memorandum on any subject. . . . He hated to write." Charming and fascinating as a colleague, Mason never succeeded in any major undertaking. He had stopped doing research at an early age, after some brilliant work in the calculus of variations. Weaver felt that he never did an honest day's work as president of the University of Chicago. He began to remind Weaver of the stereotypical brilliant person who never produced and so went about with his original store of brilliance undiminished. He had been a superb and inspiring teacher; perhaps, as Weaver suggested, he never should have been tempted into administration.[27]

Weaver also felt some strain in his relations with his fellow officers, who saw him as Mason's pet: "[W]henever Max and any one of the other major officers would collide, I was always on the cow-catcher, right in between, whenever two of them would bump."[28] It was a bumpy few years. Lauder Jones bristled when Weaver wrote to him that the "shotgun methods" of the past would have to give way to a more "imaginative" and selective approach. Weaver thought he was just passing along Mason's policy; Jones took it as a condemnation of the IEB's legacy and lectured Weaver on the dangers of substituting fancy theories for tried and true practices.[29]

Weaver knew that the Paris office was a valuable repository of experience. It was there, indeed, that he learned the ropes of field practice during an extended visit in the summer of 1932. The Paris office was as

[26]Weaver, oral history, pp. 262–280, 295–296. Wheatly, *Politics of Philanthropy*, pp. 176–180.

[27]Ibid., pp. 110–118, 122–134, 240. Weaver, "Max Mason," pp. 224–225.

[28]Weaver, oral history, pp. 125–127, 251–253, 305.

[29]Jones to Weaver, 14 Dec. 1932; Weaver to Jones, 19 Nov. 1932; both in RF 915 1.1.

different as it could be from the New York office. Its staff was active and well organized, and there was excellent communication and cooperation among the divisions. In weekly meetings all the officers exposed their projects to informal and often vigorous criticism. Weaver later described his visit in 1932 as "an extraordinary piece of luck." It was, but it was also Mason's cunning: he had arranged this apprenticeship even before Weaver arrived.[30] The following year Weaver asked Jones to be his guide on a grand tour of the foundation's projects from Scandinavia to Eastern Europe.[31] In New York Weaver learned about foundation politics; at the Paris office he learned the craft of field work.

The honeymoon, such as it was, lasted about a year. The April 1933 board meeting was to be the occasion for Weaver to unveil his program for the board's approval—his maiden appearance as a manager of science.

"Vital Processes": Inventing a Program

Weaver put his first tentative thoughts on paper in late 1932 and early 1933 (table 10.1).[32] It is clear from parts I and II of his scheme that Weaver had been a diligent student of the history of Rockefeller programs. He spent a good deal of effort at first classifying and reshuffling the different kinds of grants. It was all moot, however, since institutional and general programs were slated by the trustees for elimination (too expensive, too open-ended). The really critical issues, therefore, had to do with the "specific program." How many of the natural sciences should it include? How much emphasis should be put on biology? It was Weaver who had finally to resolve the contest between Rose's legacy in the physical sciences and the RF's biomedical tradition.

The policy that was eventually worked out was to concentrate on biology but not on biologists. The natural sciences (NS) program was officially "experimental biology," but many or even most projects involved chemists and physicists. There has always been some confusion over exactly what Weaver was doing. The trustees and some scientific advisors worried that he was bootlegging physical science into a biology program. Some historians have imputed to Weaver a desire to reduce biology to chemistry and physics.[33] In fact, Weaver was neither simple-

[30]Weaver to R. N. Hunt, 29 Dec. 1931; Weaver to C. Leake, 6 Jan. 1932; both in UW 7/22/18 box 1.

[31]Weaver, oral history, pp. 295–296, 320–324. Weaver to Jones, 19 Jan. 1933, RF 915.1.1 Jones and Weaver diary, May–July 1933, RF.

[32]Weaver memo, 27 Jan. 1933, RF 915 1.6. For earlier versions of this program see Weaver to Jones, 19 Nov. 1932, and Weaver memo, 18 Oct. 1932, both in RF 915 1.1.

[33]Pnina Abir-Am, "The discourse of physical power and biological knowledge in the

TABLE 10-1

Weaver's Program for Natural Sciences Division (January 1933)

I. Institutional Program
  A. *Departmental development*
  B. *Fluid research funds*
II. General Program
  A. *Fellowships*
  B. *Grants-in-aid*
  C. *Publication*
III. Specific Program
  A. *Major*
    1. Mathematics, physics, and chemistry of vital processes
    2. Genetic biology
    3. Mathematics, physics, and chemistry of the earth, sea and air
  B. *Minor*
    1. Fundamental construction problems
    2. Physical and colloidal chemistry
    3. Theory of probability and statistics
    4. Cooperative and unassigned

minded nor Machiavellian; he was just not thinking about science in the usual compartmentalized way. The scope of the NS division, as Weaver saw it, was all the natural sciences; but the concentration was, to use his favorite term, in "vital processes."

Look again at Weaver's first formulation of a program: it was from the start a program in all the natural sciences. Biology was important, but it was an integral part of a program that took in the whole range of disciplines from mathematics and physics to psychology. Clearly, too, Weaver was doing his best to avoid traditional disciplinary rubrics. Familiar labels were modified: psychology was "quantitative," physical chemistry was "colloidal," biology was "genetic." Unusual terms like "vital processes"—Weaver's own invention—or "fundamental construction problems" (atomic physics and astrophysics) reveal how Weaver was thinking in terms of problem solving, not of disciplines. Multidisciplinary rubrics like "mathematics, physics, and chemistry of X or Y" were the result of Weaver's effort to invent rationales for collaborative projects that cut across disciplines.

As he later recalled, Weaver invented the term "vital processes" precisely "to get away from any ancient limitations adhering to the word 'Biology.'" In other words, he meant to avoid any implications that

1930s: A reappraisal of the Rockefeller Foundation's 'policy' in molecular biology," *Soc. Stud. Sci.* 12 (1982): 341–382.

whatever biologists did was in program, or that any work by mathematicians, physicists, and chemists was out of program: "I wanted to say that this was a science of vital processes . . . in which you bring to bear the whole battery of science to the study and investigation of the phenomena of living things."[34] Rose and Trowbridge might have used the term "general physiology," but Weaver wanted his program to be open to any kind of natural scientist. Likewise, his expression "mathematics, physics, and chemistry of earth, sea and air" was not just a complicated name for "geophysics." Rather it indicated "a single unified package in which one would seize upon great techniques of mathematics, physics and chemistry . . . and would apply them to the central problems of our physical environment."[35] Weaver's aim, in both biology and earth science, was not to build disciplines but to break down the barriers between them. The time would come, he believed, when conventional disciplinary labels would be so far from the reality of scientific practice as to seem quite artificial and illusory. Such ideas were far from uncommon at the time; few took them as far as Weaver did, however, or developed them so systematically. In Weaver's hands they became the basis for a new system of patronage.

It is clear how much Weaver's ideas were indebted to his prior experience with Charles Slichter's cross-disciplinary research projects. His plan for the "mathematics, physics, and chemistry of the earth, sea and air" is the most obvious link: it came directly out of the ill-fated project in geophysics that Slichter and Weaver had planned in 1930. Weaver's vision of "the mathematics, physics, and chemistry of vital processes" was the very same idea extended to biology. The accident of history that kept Weaver from ever doing anything in the earth sciences has obscured the roots of Weaver's vision of "vital processes." Those roots were in the University of Wisconsin school of applied mathematicians and geologists.

Weaver's ideas for "vital processes" may have had other antecedents, too. The geophysical project was only one of several such projects that Charles Slichter was trying to get started in the early 1930s. Another one, which apparently never got very far, was in experimental biology. Slichter envisioned a program of planned and extramurally sponsored research in the rapidly developing fields of hormones, vitamins, and filterable viruses. It was not a random selection. Eminent researchers at the university had already staked out a position in these fields, and the three were sufficiently alike to form an integrated whole: "[T]hey converge," Slichter wrote, "in a great and important field of infracellular

[34]Weaver, oral history, pp. 332–333.

[35]Ibid., p. 413. Weaver to Jones, 26 Jan. 1933, RF 915 1.1.

Chemistry."[36] All three fields required close collaboration between organic or biochemists and biologists, with physical chemistry as "an important suburb of the work."[37] Slichter's strategy was to build on this nucleus by raising large funds from foundations; undoubtedly the RF was in his mind.[38] Slichter's "infracellular chemistry" and Weaver's "vital processes" were both meant to give intellectual legitimacy to a new system of sponsorship: they were two sides of one coin.

The parallels are striking: Weaver, too, had a program in hormones and vitamins and filterable viruses. He, too, looked to cell physiology as the key area in which the physical and biological sciences intersected. Both foresaw chemists and biologists converging on common problems. It is unlikely that Weaver knew of Slichter's biological projects—he was surprised in 1934 to discover how alike their thoughts were. The point is, rather, that there was a common context of ideas about programmatic, transdisciplinary science upon which Slichter, Mason, and Weaver all drew. Mason was perhaps being a little chauvinistic when he asserted that the University of Wisconsin was five years ahead of other universities in knowing how to organize multidisciplinary research.[39] But through Weaver and the RF's projects in "vital processes," this distinctive subculture would be spread far and wide.

Weaver never really succeeded in making clear to the trustees or to his fellow officers what he meant by terms like "vital processes"—not surprisingly, perhaps, since they lacked Weaver's peculiar experience. Some of the trustees shared his belief that disciplinary rubrics were merely artefacts of academic politics: Ray Lyman Wilbur, for example. "What impression do you get of the artificial division we have made of physics, chemistry, bacteriology and pathology?" he demanded of Alan Gregg. "Are they in the way now or not? Isn't it about time we forgot those names and scrambled the whole thing to see if we cannot get some new terms? Haven't we interfered with our development?"[40] The trustees feared getting entangled with academic disciplines; but terms like "vital processes" seemed even more dangerously open-ended invitations to spending money. For Lauder Jones, Weaver's novel terms were incomprehensible and threatening. Since "vital processes" included studies by biologists, where was the name "biology?" he wanted to know. Where was astronomy? What was "quantitative psychology," and why

[36]Slichter to H. L. Russell, 12 Mar. 1930, UW 6/1/1 box 16 WARF Research Projects.

[37]Ibid.

[38]Slichter was working also on a project in the chemistry and biology of molds. Slichter to WARF, n.d. [1929]; Russell to Slichter, 11 Nov. 1929; both in UW 6/1/1 box 5 Mould Project.

[39]Mason diary, 7 Jan. 1931, RF.

[40]"Conference of trustees and officers," 29 Oct. 1930, pp. 64–65, RF 900 22.167.

was colloidal chemistry a separate category when it clearly was part of the chemistry of vital processes?[41]

Weaver had to admit that the term "fundamental construction problems" was "totally incomprehensible." But he tried to make Jones see that he was using a different principle of classification: "You may well ask where is physics, where is chemistry, where is mathematics, just as you have already asked, 'Where is astronomy?' The answer is that I have been thinking in terms of problems rather than in terms of the classical divisions."[42] Jones did not see, and how could he, when everything in his experience as an academic chemist and as a designer of grants to university departments reinforced an orthodox view of disciplines.

One reason why Weaver found it hard to explain what he meant by "vital processes" was that he did not know exactly what special fields of biology that rubric should include. He knew he wanted to exclude taxonomy, and probably morphological and ecological fields as well.[43] But his ignorance of experimental biology made it harder to see what should be included. None of his early plans, right up to his statement to the trustees in April 1933, say anything specifically about the content of "vital processes." He wrote in detail of the merits and demerits of institutional grants, fellowships, and grants-in-aid, but when the point came to talk about scientific particulars, Weaver broke off, saying that he was working on that.[44] He had begun to read systematically in biology, starting with genetics, which he hoped would be more accessible to a mathematician because it involved statistics.[45] But the Balkanized world of experimental biology was difficult for an outsider to penetrate. Weaver first talked at length about the technical side of biology at the board meeting of December 1933, and that was a joint presentation with Alan Gregg, on whom Weaver leaned heavily.[46]

Weaver's ignorance of biology would have been less of a problem if biology had remained one part of a much broader program in all the natural sciences. But in 1933 and 1934 that grand vision was relentlessly whittled away until there was nothing left but biology.

It was not principle but financial necessity that was the knife. The year 1934 was the nadir financially for the RF. Payments were coming

[41]Jones to Weaver, 14 Dec. 1932, RF 915 1.1.

[42]Weaver to Jones, 26 Jan. 1933, RF 915 1.1.

[43]Weaver to Lillie, 6 Oct. 1933, RF 216D 8.105.

[44]Weaver draft, 27 Jan. 1933, p. 16, RF 915 1.6.

[45]Weaver, *Scene of Change,* p. 69.

[46]"The medical and the natural sciences," 13 Dec. 1933, RF 915 1.7. Weaver included some technical detail in "Agenda for special meeting, 11 Apr. 1933," pp. 78–81, RF 900 22.168.

due on prior pledges, the NRC fellowships and NRC grants-in-aid projects were due for renewal, and income from endowment was sharply down. Unless old and nonessential programs were eliminated, there would be literally no money for any new projects in the fields of concentration.[47] In 1932 Mason warned Weaver that the Paris office should present no new projects, for fear of being turned down by the board. Weaver instructed Jones that the board would approve only projects in specific fields of concentration.[48]

The first cuts of the budget axe were made by Mason and Weaver themselves, hoping that would prevent the trustees from doing far more drastic surgery on the natural sciences program. Mason was not entirely displeased. He had been trying for two years to get the officers to concentrate, especially the guardians of the Rose tradition in the Paris office, and necessity would ensure that Weaver did not fall into the same bad habits. "It will be anything but bad for us," he wrote to his protégé. "It is a chance for a remaking of program, and gives you just the time I would hope for to determine significant lines of endeavor that you are willing to back and back hard."[49]

The first casualties were the institutional and general parts of Weaver's grand scheme (table 10.1, parts I and II). Fluid research funds were terminated, and other kinds of grants were permitted only as adjuncts to projects in specific fields of concentration. The NRC fellowships and grants-in-aid were sharply reduced.[50] These were no great loss, since Weaver did not think much of the fluid grants and wanted to get control of the fellowships back from the NRC. New fellowship and grants-in-aid programs became part of "vital processes."

Less welcome was the elimination of the "minor" specific programs (table 10.1 part III*b*)—the physical science part of Weaver's program. "Fundamental construction problems" and statistics were quietly dropped before Weaver's plan got to the board. (He and Mason knew the trustees were determined finally to end Rose's programs in physics.) Physical and colloidal chemistry and psychology were folded into "vital processes." The program in earth science survived the first cut in April 1933, downgraded from a major to a minor field, but in De-

[47]Weaver to Jones, 14 Mar., 19 Jan. 1933, RF 915 1.1.

[48]Weaver to Jones, 19 Nov. 1932, 19 Jan., 14 Mar. 1933; Jones to Weaver, 14 Dec. 1932; all in RF 915 1.1.

[49]Mason to Weaver, 9 June 1932; Weaver to Mason, 5 July 1932; both in RF 100 65.530.

[50]Staff conference, 14 Mar. 1933, RF 900 21.160. "Agenda for special meeting," 11 Apr. 1933, pp. 76–87, RF 900 22.168. Weaver, "Progress report of the N.S.," 14 Feb. 1934, RF 915 1.7.

cember 1934 the trustees ruled it out as "interesting but not vital."[51] It had never really gotten started: in early 1934 Weaver wrote to Jones that earth science had "left the earth and is up in the air for the present" and directed him not to enter into any negotiations.[52] The fact that he saw it coming did not make the loss of the earth sciences a less bitter blow. Weaver never forgot and never quite forgave it.[53]

Thus, within two years, a program embracing all the natural sciences had been contracted to one in biology. Chemistry, physics, and mathematics were included only insofar as they were relevant to biological problems. The balance had swung decisively from Rose's legacy to the RF's tradition in biology and medicine. It was not clear what Weaver had been led to expect. Mason had made it plain even in Weaver's interview in December 1931 that biology should be a prime field for the RF, and Weaver heartily agreed. But he would not have devised a program in all the natural sciences, surely, had Mason meant to restrict him to biology. Much later, Weaver gave Mason credit for perceiving "with brilliant clarity" that the RF should concentrate on biology.[54] It could hardly have seemed an unmixed blessing at the time.

In some ways Weaver did benefit from the radical surgery done to his program. The elimination of the institutional and general programs cleared away precisely those kinds of patronage over which officers had least control, indeed, which were designed to keep officers from getting too involved in their clients' work. Weaver was glad to recover control of fellowships and grants-in-aid from the NRC, and individual project grants entailed an active relation with his clients. In "maturing" applications for the board, Weaver could not help but make judgments about the significance of individual scientists and their research problems.[55] Most important, perhaps, concentration on "vital processes," by denying Weaver the luxury of aiding pure chemistry and physics, forced him to devise new ways of combining them with the biological disciplines. Concentration spurred the invention of a new system of patronage and a greatly expanded authority for managers of science.

These were long-term consequences, however, foreseen dimly if at all in 1933. In the short term, the disadvantages of concentration were

[51] "Agenda for special meeting," pp. 76–87. "Report of committee of appraisal and plan," 11 Dec. 1934, p. 61, RF 900 22.166.

[52] Weaver to Jones, 16 Feb. 1934; Weaver to Fosdick, 14 Nov. 1934; both in RF 915 1.1.

[53] Weaver, oral history, pp. 413–419.

[54] Weaver, *Scene of Change,* pp. 59–63. Weaver, oral history, pp. 126–128, 241–244, 280.

[55] Weaver, "The natural science budget," 11 Dec. 1935, pp. 16–19, RF 900 23.171. Weaver and Gregg, "The medical and the natural sciences," 13 Dec. 1933, RF 900 22.168.

more obvious. Weaver had been largely barred from disciplines in which he had an unassailable authority and was left with those in which he was far from expert. One result was that Weaver became more dependent on Alan Gregg, who had a broad knowledge of biomedical science, a rich experience in field practice with the IHB, and enjoyed the trustees' absolute confidence.[56] Programmatically, there seemed to be a good basis for cooperation, since Gregg's plans for developing psychiatry called for building up its foundations in the basic biomedical sciences. "Gregg feels strongly that psychiatry cannot lift itself by its own bootstraps," Weaver wrote, [and] "definitely wishes . . . to dip down into those broader fields of research, such as neurology . . . genetics, [and] nutritional studies . . . which can contribute invigorating and illuminating ideas to psychiatry."[57] But whereas Weaver came almost to idolize Gregg, Gregg was not so sure about the new boy in the New York office. He resented Weaver's favored status as Mason's protégé, and was one of those who objected to mathematicians being put in charge of a program in the biomedical sciences. Gregg may also have misread Weaver's plain Midwestern style as a lack of intellect.[58] In any case, the increasing pressure on all division officers to concentrate and cooperate threw the two men together.

Under pressure to be more specific about what he meant by "vital processes," Weaver borrowed heavily on Gregg and his program in "psychobiology." In December 1933, the two presented a single combined program in psychiatry and "the sciences underlying the behavior of man" (table 10.2).[59] Weaver would take responsibility for basic research in each of their nine subfields, leaving the more clinical aspects to Gregg. The balance of basic and clinical work in each field would thus determine who got what fields. In practice, of course, that division of labor was not quite so clear-cut, but neither Weaver nor Gregg were at all territorial. Both thought it more important to get something done than to be in control. When disputable cases did arise, then, a tête-à-tête sufficed to avoid conflicts over turf. Office politics notwithstanding, in the field Weaver and Gregg appear to have had an excellent working relationship, as did also Tisdale and Daniel P. O'Brien, their lieutenants in the Paris office.[60]

[56]"Report of the committee of appraisal and plan," 11 Dec. 1934, pp. 51–54, RF 900 22.166. Wheatley, *Politics of Patronage,* ch. 6.

[57]Weaver to Tisdale, 27 Mar. 1935, RF 915 1.2. "Agenda for special meeting," 11 April 1933, pp. 66–75, RF 900 22.168.

[58]Wheatley, *Politics of Patronage,* pp. 176–185.

[59]"The medical and the natural sciences," 13 Dec. 1933, RF 900 22.168.

[60]Ibid. Weaver, "The program in experimental biology, relationship of MS and NS," 3 Dec. 1934; Weaver, "Relation between medical and natural science programs," 2 Oct. 1935; both in RF 915 1.7. Weaver to Tisdale, 28 Oct. 1935; Tisdale to Weaver, 13 Nov.

TABLE 10-2

PROPOSED PROGRAM IN NATURAL AND MEDICAL SCIENCES (DECEMBER 1933)

1. Psychobiology (psychiatry, psychology, neurophysiology)
2. Internal secretions (hormones, enzymes)
3. Nutrition (vitamins)
4. Radiation effects (ultraviolet, X-rays, cosmic rays, mitogenetic rays)
5. Biology of sex (physiology of sex, fertility)
6. Experimental and chemical embryology (fertilization, sex determination, transplantation, regeneration, organizers)
7. Genetics (chromosomes, genes, cytology)
8. General physiology (cell physiology, nerve conduction, electrical effects, osmosis, permeability)
9. Biophysics and biochemistry (spectroscopy, microchemistry, basic studies)

RELATIONS BETWEEN OFFICERS AND TRUSTEES

Relations between officers and trustees were less harmonious. The board fretted over possible duplications in Weaver and Gregg's divisions and were uneasy about the powers they had in effect delegated to the officers when they insisted on a policy of concentration. The trustees had been comfortable making judgments of institutional and capital grants. Now, decisions about projects in biophysics or genetics required a level of scientific expertise to which they could never aspire. The important decisions, they found, had already been made by the officers before proposals ever reached the board. The trustees felt reduced to wielding the rubber stamp. Unable to participate as equals, they naturally grew suspicious of the officers and questioned the propriety of intervening in scientists' decision making. Someone quipped at the time that "the trustees used to come out of meetings rubbing their hands; but that [more recently] they came out . . . scratching their heads."[61]

As Weaver recalled, the officers dreaded appearing before the board and were enormously relieved when the inquisitions were over. Afterward they would have a few more drinks than they should, each one "rejoicing in whatever triumphs he had had and hoping to drown out the memory of whatever defeats he had had. Because . . . practically all of the things that happened to him, he would be inclined to classify as

1935; O'Brien to Gregg, 24 June 1935; Gregg to O'Brien, 19 July, 17, 29 Oct. 1935; Weaver diary, 10 Dec. 1935; all in RF 915 1.2.

61Weaver, "Program and administration," 1 Oct. 1937, p. 18, RF 915 1.8.

either a victory or a defeat." There was an atmosphere, he recalled, "of—well, conflict is too strong a word, but contest is not—between the officers and the trustees."[62] The trustees fell into the role of questioning and resisting any new initiative by the officers, who naturally felt defensive:

> Any officer, when he appeared before the trustees, had the distinct impression that he was faced by a somewhat hostile audience. He had to prove himself. He had to prove his case. It was not a question of a group of colleagues inquiring together as to what was the wisest and most constructive thing to do. It was not a procedure of adding together the wisdom and the insight of the two groups, but rather a procedure of using one group as a stone and the other as the flint iron, and striking them together to see whether in this rather violent, and sometimes rather unhappy and unpleasant contact some sparks would be thrown off.[63]

This relationship was symbolized in the physical setting of board meetings: the trustees at a table in the center, the officers seated around the walls of the room.[64]

This adversarial relationship was exacerbated by Max Mason's way of dealing with it. All of his worst characteristics came out. While the officers would work themselves "into a blue funk" mastering their material and polishing their presentations, Mason never prepared a thing. He treated the meetings as a game of wits and delighted in outthinking and outwitting the whole group. A brilliant extemporaneous speaker, Mason was in fact able to skate circles around the trustees. For a time they thought Mason was wonderful, but soon they grew disillusioned: "This was a spectacular performance. The man was incredible. But when would he start to really get down to solid, substantial business? Well, the answer is he never did."[65] The symbolic message of Mason's improvised performances was that he did not take the trustees seriously. Mason's relation with Rockefeller, Jr., also soured. A cold, formal man, Rockefeller had a passion for careful documentation and analysis. He was not one to take well to someone "who was just going to whirl about him." Much taken at first by Mason's public presence, Rockefeller soon changed his mind. As Weaver noted, their characters were so different that conflict was in the cards from the start.[66] The trustee's growing mistrust of Mason spilled over into an adversarial at-

[62]Weaver, oral history, pp. 297–298.
[63]Ibid., p. 296.
[64]Ibid., p. 304.
[65]Ibid., pp. 123–124.
[66]Ibid., pp. 124–127.

titude to the officers and their programs. The relation between the officers and trustees did not begin to improve until Fosdick succeeded Mason as president in 1937.

The Depression also impeded a more cooperative and trusting relationship. As the crisis deepened, some trustees began to wonder if money spent on advancing knowledge would not be better spent on immediate relief. These sentiments were most clearly expressed by Ernest M. Hopkins, president of Dartmouth College. What was the point, he wrote Fosdick, of developing "a scientific Utopia for remote descendants whom we may never have unless crime, misery, and want can be greatly reduced."[67] Hopkins recalled an anecdote of World War I, when a Dartmouth graduate, fresh from artillery school and in charge of a field battery, was unceremoniously enjoined by a French officer to throw away his mathematics and precision instruments and start shooting. Hopkins wanted the RF to do the same: "I should like to take and begin to shoot, without books of logarithms, without too much concern from celestial mechanics, and without formulas or rigidity or preconceived plans." Hopkins's criticism of "academic disciplines, research enthusiasms, and brain trust characteristics" was obviously aimed at the officers, including his old friend Mason.[68]

Hopkins's frustration with planning was a sign of the times. In Washington the New Deal was abandoning the planning strategy of the National Recovery Administration and the "brains trust" for pragmatic economic and social programs.[69] Public attitudes to science and technology took a pessimistic turn in the years of the "science holiday."[70] In the prosperous 1920s, the notion that science and technology changed more rapidly than other social institutions had seemed grounds for optimism; in the depths of the depression, the same argument made scientists and engineers seem socially irresponsible and complicit in the economic collapse. Fosdick later recalled that one of the reasons why the trustees excluded the physical sciences from Weaver's program was their feeling that the good they had done for society was also mixed with much that was harmful.[71] These were commonplace sentiments in the mid-1930s.

Weaver's response to the trustees' flagging enthusiasm for basic science was to educate and inspire them. He adopted a more popular language, which emphasized the relevance of science to social and

[67]Hopkins to Fosdick, 16 Nov. 1934, RF 900 21.160.

[68]Hopkins to Mason, 16 Nov. 1934, RF 900 21.160.

[69]Otis L. Graham, Jr., *Toward a Planned Society* (New York: Oxford University Press, 1976), ch. 1.

[70]Pursell, Jr., "'A savage struck by lightning.'"

[71]Fosdick to Greene, 25 Mar. 1937, RF 915 1.2.

human issues. Weaver wrapped "vital processes" and "psychobiology" in a fulsome rhetoric, which reached its highest pitch in a report to the trustees in February 1934:

> As one views the present state of the world, with its terrific tension, its paradoxical confusion of abundance, and its almost uncontrollable mechanical expertness, one is tempted to charge the physical sciences with having helped to produce a situation that man has neither the wits to manage nor the nerves to endure. . . . The challenge of this situation is obvious. Can man gain an intelligent control of his own power? Can we develop so sound and extensive a genetics that we can hope to breed, in the future, superior men? Can we obtain enough knowledge of the physiology and psychobiology of sex so that man can bring this pervasive, highly important, and dangerous aspect of life under control? Can we unravel the tangled problem of the endocrine glands, and develop, before it is too late, a therapy for the whole hideous range of mental and physical disorders which result from glandular disturbances? . . . Can we release psychology from its present confusion and ineffectiveness and shape it into a tool which every man can use every day? Can man acquire enough knowledge of his own vital processes so that we can hope to rationalize human behavior? Can we, in short, create a new science of Man?[72]

If the past century saw the supremacy of physics and chemistry, Weaver wrote, the next should see the creation of "a new biology and a new psychology." Mason used similar though less purple language.[73]

Weaver never meant such statements to be taken literally. They were intended to make technical matters accessible and inspiring to the trustees. It inspired them all right: not with enthusiasm for science but with concern that an officer would display such an excess of enthusiasm over cool judgment. Fosdick realized that Weaver's efforts to inspire the trustees were making matters worse. Wishing Weaver well, he called him to an evening tête-à-tête at his home, to warn him that his reports and presentations to the board were causing alarm. The juicy rhetoric was one problem; another was "bizarre" projects (like spectroscopic analysis of biological materials) that seemed more reflections of Weaver's personal enthusiasm than of a program in biology. Weaver's overblown rhetoric just heightened the trustees' fears of losing control to officers whose enthusiasms could not be trusted or challenged.

Feeling very deflated, Weaver explained to Fosdick that he had been

[72]Weaver, "Progress report of the N.S.," 14 Feb. 1934, RF 915 1.7.

[73]Staff conference, 14 Mar. 1933, RF 900 21.160.

impelled to use a more literary and promotional style by the trustees' lukewarm response to his soberly scientific descriptions of opportunities, especially in the earth sciences (a sore point).

> [I]s it not natural for one to try to go behind a cold, factual, scientifically reserved picture of the opportunity and to attempt to give a warmer, less formal, less technical description . . . ? The result is an inevitable upturn in the percentage and potency of the adjectives and adverbs used in our documents and our comments. It probably was too sharp and too great an upturn; but is there not some point to the fact that circumstances seem to require a semi-popular kind of explanation which would emphasize values, purposes, and . . . applications?[74]

Weaver's interest in popular science writing was also a sign of the times, which saw a general revival of science journalism, deliberately encouraged by professional organizations of both scientists and journalists to regain public approval of science.[75] Fosdick was sympathetic. Himself the author of one of the many popular statements of the "culture lag" idea, he had no patience with talk of a moratorium on research.[76] But although his own faith in science was unshaken, he realized that the trustees' doubts could only be laid to rest by being openly vetted.

Fosdick was the prime mover behind the series of investigations of the officers and their programs that began in December 1933 with the appointment of a Committee of Appraisal and Plan. A smaller committee was appointed in December 1935 to look into allegations by Jerome Greene that the officers were mismanaging the programs. A five-year review was carried out in 1938–1939. The officers thoroughly disliked these investigations, though in every instance the trustees ended up reaffirming Fosdick's policies and endorsing the officers' new powers. In hindsight, they were a way of working out the new power relation between officers and trustees through carefully orchestrated conflict. In this fashion, the shift of authority from board to operating staff, implicit in Fosdick's policy of concentration, became an accepted part of the new system of patronage.

Committee on Appraisal and Plan, 1933–1934

The charge to the Committee on Appraisal and Plan was to ask themselves, If the RF were starting in 1933 with a clean slate, would it choose

[74]Weaver to Fosdick, 22 Mar. 1934, RF 915 4.38.

[75]Hillier Kriegbaum, *Science and the Mass Media* (New York: New York University Press, 1967). John C. Burnham, *How Superstition Won and Science Lost* (New Brunswick: Rutgers University Press, 1987).

[76]Fosdick, *The Old Savage in the New Civilization* (New York: Doubleday Doran, 1928).

the same policies it had in 1928?[77] It sounded like an invitation to drop research for more practical activities, but Fosdick, who chaired the committee, certainly did not expect to repudiate his own policy. It is unlikely that many of the trustees subscribed to the pessimistic reading of the "culture lag" concept anymore than Fosdick did. After all, their lives and livelihoods as educators and professional men were based on producing and using knowledge. As Jerome Greene put it a few years later: "I . . . hope that the Trustees . . . are not under the influence of the dangerous notion that because science has been enlisted in creating the horrors of war, therefore science should take a holiday. The logical consequences of that conclusion would have a devastating effect on all education."[78]

For Fosdick the appraisal was an occasion to remind both the trustees and the officers that "the advancement of knowledge" did not mean research for its own sake. To him the phrase meant promoting the production, diffusion, and application of knowledge that would ultimately help solve social problems—a conception rooted in Fosdick's experience as a Progressive social reformer. Fosdick's report in 1934 had much to say about this broader meaning of research. He clearly shared the trustees' feeling that Weaver had strayed some distance from this conception toward a view of science for science's sake.[79] But Fosdick's homilies were aimed primarily at the social sciences and humanities, which were closest to his own heart and to the social problems of the depression.

The Committee's report did contain some passages that, read out of context, would appear to be quite radical. For example:

> The limitations of research in influencing public thinking on social problems are, in times like these, vividly demonstrated. Research leads to publication, and publication as a device has at best restricted possibilities. The mere accumulation of facts, untested by practical application, is in danger of becoming a substitute rather than a basis for collective action.[80]

The issue was not research as such, however, but the different conceptions of research held by natural and social scientists. The issue turned

[77]"Report of the Committee on Appraisal and Plan," 11 Dec. 1934, p. 1, RF 900 22.166.

[78]Greene to Fosdick, 29 Mar. 1937, RF 915 1.2.

[79]"Report of the Committee on Appraisal and Plan," 11 Dec. 1934, pp. 22–24, 35–39, 40–45, RF 900 22.166.

[80]Ibid., p. 43. Weaver was relieved that a still stronger statement had been omitted, to the effect that the RF was "not interested in promoting scientific research as an end in itself." In fact, those words do appear in the report (at p. 45), in a quotation from Beardsley Ruml's program for the Laura Spelman Rockefeller Memorial! Weaver to Tisdale, 27 Dec. 1934, RF 915 1.1.

up again and again in the New Deal period, when political and social troubles strengthened the claims of social scientists to a share of research dollars.[81]

Not surprisingly, the foundation's programs in the social sciences and humanities were most affected by the pressure for social relevance. The natural sciences were relatively untouched. Fosdick's committee did make it clear that the trustees would not take seriously arguments that appealed to purely scientific opportunities, such as Weaver had made for the earth sciences. In more important ways, however, Weaver's authority was again strengthened. The policy of concentration was reaffirmed, and with it the right of officers to actively define and develop fields for concentration. However much the trustees feared officers' authority, they had no choice: concentration was the only practical way to economize, and concentration gave authority to program managers. "I think it can be honestly said," Weaver wrote Tisdale, "that the work of the N.S. division came in for a relatively small amount of negative criticism, and also that our program requires a relatively mild amount of readjustment."[82]

In "vital processes" the only change the trustees insisted upon was a change of name to "experimental biology." As Weaver recalled, the trustees were suspicious of the unfamiliar phrase and "never quite understood whether they were being sold a bill of goods or not." In fact, "experimental biology" described Weaver's program less well, since it did not cover projects in organic, physical, or biochemistry and biophysics. Weaver was not expected to give up projects in these fields, but he was obliged to accommodate to disciplinary conventions.[83] The idea behind "vital processes" survived intact, though the name did not.

The review gave the trustees a chance to vent some steam, but it did not reconcile them to their lessened control over what the officers were doing. The tension between Weaver and the trustees is apparent, for example, in skirmishes over the issue of "small projects." The trustees did not like Weaver's propensity to do "retail business."[84] Weaver's active role in defining programs and selecting projects was displayed most clearly in fellowships and grants-in-aid. They were a constant reminder that an enthusiastic officer could commit the trustees to projects, the scientific value of which they were in no position to judge.

The volume of retail business brought before the trustees was in fact unusually large in the mid-1930s, because they had enjoined the of-

[81]Carroll W. Pursell, Jr., "The anatomy of a failure: The Science Advisory Board, 1933–1935," *Proc. Amer. Phil. Soc.* 109 (1965): 342–351.

[82]Weaver to Tisdale, 27 Dec. 1935, RF 915 1.1.

[83]Weaver, oral history, pp. 332–333.

[84]Weaver to Tisdale, 19 Nov. 1936, RF 915 1.2.

ficers to make no commitments beyond a single year, even in cases of multiyear projects. The result was that board meetings were snowed under with extensions, renewals, and reviews. Each one of these technical decisions about a few thousand dollars brought home to the trustees their inability to understand, much less control what the officers were doing.[85] Weaver tried to avoid irritating the trustees by transferring one-year grants to his grants-in-aid budget, from which he could spend without the approval of the executive committee.[86] However, both Gregg and Fosdick warned Weaver that eventually he would have to face the issue of small projects openly.[87]

Jerome Greene was especially critical, complaining that the officers did not reveal their principles of selection and did not tell the board about projects they had rejected. He also alleged that the executive committee was dominated by an exclusive inner circle close to the Rockefellers and their "employees" (i.e., officers).[88] A committee appointed to look into Greene's allegations reported that the foundation's organization and procedures were basically sound. They simply enjoined the officers to keep the board more fully informed.[89]

Weaver sympathized with the trustees' desire to be better informed about applications and the results of projects, and he worked hard to improve communications. His annual reports became essays in popular science writing, aimed in part at the trustees. In 1937 he started the system of monthly "confidential reports" for the trustees, popular and inspirational accounts of projects, written by a professional science journalist, George W. Gray.[90] Weaver became quite good at getting the trustees excited about science, at least about such large and dramatic projects as Ernest Lawrence's 184-inch cyclotron.[91] Sharing of information, he hoped, would give the trustees a greater sense of participation in technical decisions. The fact remained, however, that the officers decided what the trustees would see.

[85]Gregg to Fosdick, 12 Nov. 1940; Weaver to Fosdick, 27 Nov. 1940; both in RF 900 21.160.

[86]Weaver memo, 20 Apr. 1936, RF 915 1.2.

[87]Gregg to Weaver, 10 Nov. 1936; Tisdale to Weaver, 18 Dec. 1936; both in RF 915 1.2. Weaver, "A case for small projects," 5 Nov. 1936, RF 915 1.8.

[88]Greene, "Memo on the relation of the trustees to the policy and control of the RF," 28 Oct. 1935, RF 900 19.141.

[89]"Report of a special committee," 10 Dec. 1935, RF 900 19.141.

[90]Weaver to Tisdale, 26 May 1936, RF 915 1.2. Bruce V. Lewenstein, "Public understanding of science" (Ph.D. diss., University of Pennsylvania, 1987), pp. 329–357.

[91]Weaver, oral history, pp. 391–393.

Scientist-Trustees and Scientific Advisors

Better communication did seem to relieve tensions, but it could not dispel the underlying conflict. In November 1937 Greene, together with physiologist Herbert Gasser (Flexner's successor at the Rockefeller Institute) and the eminent pathologist George H. Whipple, "referred with some show of disapproval to the large number of small projects, particularly in the NS division."[92] Implicitly, they were questioning Weaver's authority to make grants for research on specific problems. Such signs of disaffection worried Fosdick, especially when they came from trustees who were themselves scientists.

Fosdick had deliberately appointed more scientists to the board, hoping that they would help bridge the gap between officers and trustees. He urged the officers to discuss proposals privately with the scientist-trustees in advance of board meetings, to ensure a more informed and harmonious discussion. Weaver was not so sure it would work that way. Eminent scientists might not make any greater effort than their lay colleagues to understand what the officers were doing, but their prestige as scientists would be sure to have great authority with the other trustees. "It is my own belief," Weaver wrote, "that technical competency must . . . rest with the officers and upon the advice which the officers can get from the scholars of the world."[93] The best trustees, he thought, were experienced business executives who could provide guidance on overall policy but would not be inclined to challenge the officers on their technical competence. Weaver argued that if scientists were to be appointed to the board, there should be more than a few.[94]

Undoubtedly, the few that Weaver was thinking about were Simon Flexner and his biomedical friends. Flexner had been a persistent critic of Weaver and Mason, and although he had not been a trustee since 1930 he remained a force to be reckoned with. In 1933, for example, he was empowered to assemble a panel of scientists to assess the scientific worth of Weaver's program for the Committee of Appraisal. Flexner used this panel to attack planned research in general and Weaver in particular. The gambit failed, but it is revealing of the basic tension that existed between Flexner's laissez-faire values and Weaver's belief that patrons should have programs.

The membership of the scientific panel was divided into two distinct groups, the one picked by Mason, the other by Flexner. Mason's men

[92]"Special trustees' meeting," 30 Nov. 1937, RF 900 23.172.

[93]Weaver, "Program and administration," 1 Oct. 1937, p. 23, RF 915 1.8.

[94]Ibid., p. 25.

were biologists with experience in organized, programmatic research: Frank R. Lillie and the Harvard physiologist Walter B. Cannon. Flexner's, in contrast, were older men from his own circle who shared his skepticism about planning in science. Henry Dakin, an old-style individualist, operated a private biochemical research laboratory. The physiologist William H. Howell, though still active at seventy-four (he was chairman of the NRC), was very much a scientist of the old school.[95]

The scientific panel gave surprisingly little attention to the scientific content of Weaver's program. No one took issue with Weaver's choice of subfields or suggested better ones. Cannon and Lillie were enthusiastic, of course, psychobiology being their major interest. Flexner grumbled that Weaver's program was hardly as novel as was claimed, but he did not suggest that Weaver had been unwise in choosing what he did.[96] The panel contented itself with disapproving of the name "vital processes." Lillie suggested "experimental biology," and the other panelists agreed that the "more established" term was best.[97] Clearly, disciplinary jurisdictions were being reasserted—jurisdictions that Weaver had hoped to subvert by avoiding the "established" terminology.

The panel spent most of its energy, however, on the old controversies about individual and managed research. They took issue with the idea that foundation officers could select fields and projects, and they offered various suggestions for shifting control back to scientists themselves. Howell and Lillie suggested that the foundation might spend its money better by endowing research institutes in a few strategic universities. Flexner as always favored strict separation of research and teaching. Cannon favored project grants for productive academic researchers.[98]

Organized research came in for a scourging from Flexner, Dakin, and Howell, who worried that the RF's programs might corrupt scientists into tailoring their research priorities just to get grants. Dakin bluntly put the case for laissez-faire: "less plan, less emphasis on the future coordination of scientific knowledge and its human implications, and more scientific opportunism."[99] Dakin and Howell also worried that project grants might tempt academic scientists to adopt the narrowly focused style of industrial research. Such fears were not uncommon among older scientists, and Flexner played upon them,

[95]Mason to Fosdick, 28 Mar. 1934; Fosdick to panel, 9 Oct. 1934; both in RF 915 4.41.

[96]Cannon to Flexner, 21 Nov. 1934; Lillie to Flexner, 10–12 Nov. 1934; all in RF 915 4.41.

[97]Howell to Flexner, 10 Nov. 1934; Lillie to Flexner, 10–12 Nov. 1935; Flexner to panel, 19 Nov. 1934; all in RF 915 4.41.

[98]Howell to Flexner, pp. 3–4; Lillie to Flexner, p. 8; Flexner to panel, pp. 4–5. Cannon to Flexner, 21 Nov. 1934, p. 7, RF 915 4.41.

[99]Dakin to Flexner, 16 Nov. 1934, p. 3, RF 915 4.41.

professing to see no difference between routine factory research and the organized research that Weaver and Mason were trying to encourage—differences that Cannon and Lillie saw perfectly clearly.[100]

Flexner's animus against Mason and Weaver was revealed more forthrightly in a confidential letter to Fosdick. Was Weaver's program being "wisely and discriminatingly pursued?" Fosdick had asked. Flexner was not sure that the officers, "captivated by their own notions, may not have imposed their ideas on individual laboratories." Flexner also found "something anomalous in mathematicians and physicists dominating in a wide way research in biology and medicine." Was the panel satisfied with the qualifications of the officers? A delicate question, Flexner thought: "A disturbing element is that the chief men in charge are so completely 'sold' to the program." Flexner thought it unlikely that such active officers could avoid exerting improper influence on their clients: the resources and prestige of the Rockefeller Foundation were just too great.[101]

Lillie and Cannon did not share Flexner's antipathy to Weaver's activist style of patronage. They ran organized research programs and had no objection to mathematicians sending funds their way. Lillie summed up the case for a definite program and activist officers: "Seeing that influence [by the RF] is unavoidable, it should be intentional."[102] There is no more pithy statement of what Fosdick and Weaver aspired to do.

Flexner's attempt to undermine Weaver's position backfired. Fosdick was decidedly unimpressed by the panel's report. "Frankly," he wrote, "I got very little from their reports except, perhaps, their general feeling that if properly limited the program was good. Their irrelevancies seem to center about two points: (1) The old row between university research and institute research . . . [and] (2) Planned research versus general research."[103] Fosdick took for granted that foundation officers would be active participants in selecting fields and projects; the only question was, Were they making a good job of it?

Fosdick got more substantive advice from David Edsall, Harvard's medical dean and a key member of the board's executive committee. Edsall was skeptical about the programs in hormone research and psychobiology. Weaver was much too optimistic about the practical results to be expected, he thought, and should proceed more cautiously. At the

100 Ibid., p. 2. Howell to Flexner, p. 2; Flexner to panel, pp. 2–3; Cannon to Flexner, pp. 1–2.

101 Flexner to Fosdick, 20 Nov. 1934; Fosdick to Flexner, 19 Nov. 1934; both in RF 915 4.41.

102 Lillie to Flexner, pp. 5–6.

103 Fosdick to W. W. Stewart, 25 Nov. 1934, RF 915 4.41.

same time, Edsall strongly supported the basic ideas behind Weaver's program. Important discoveries in biology and medicine would come from the physical sciences, he thought, and RF officers ought to take an active part in setting scientific agendas.[104] Fosdick was impressed, all the more so because he had always believed that Edsall liked the program in psychobiology.[105]

In his report for the Committee on Appraisal, Fosdick enjoined Weaver (without mentioning names, of course) to cool the rhetoric about "new" sciences, avoid grandiose projects in psychobiology, and keep his eyes open for opportunities in basic physical science. At the same time, he affirmed that officers were expected to take an active role in formulating programs and choosing projects, and he neatly parried Flexner's private personal criticisms of Weaver by pointing to his close cooperation with Gregg. Without such cooperation, he wrote, "we might be placed in the position of asking experts in the physical sciences to assume responsibility for technical competence in biology and medicine."[106]

Flexner continued to snipe at Weaver when he got the chance, but Weaver was on to his tricks and could give as good as he got.[107] Flexner's prejudice against active officers was shared by Herbert Gasser, his successor at the institute and on the RF board.[108] When Weaver called on him to discuss a pending application, Gasser bluntly told him that he did not like "projects" and programs. The RF would do well, in his view, to abandon programs and return to fluid grants-in-aid or unrestricted grants to exceptional individuals without regard to field. Weaver was taken aback by Gasser's ignorance of foundation practices and was alarmed by his imperious tone, almost the tone that an employer would take with an employee. "A process which, to the slightest degree, involves knuckling under to a succession of individual trustees," Weaver warned Fosdick, was "a process which no self-respecting person would tolerate."[109] Little wonder that Weaver did not share Fosdick's hopes for scientist-trustees.

Relations between officers and trustees did improve in the late 1930s, however, for a lot of reasons: improving economic conditions, Fosdick's presence in the president's office, and Weaver's excellent track record. Productive projects did more than anything else to dispel the trustees'

[104]Edsall to Fosdick, 23 Nov. 1934, RF 915 4.41.

[105]Fosdick to Stewart, 25 Nov. 1934, RF 915 4.41.

[106]"Report of the committee on appraisal and plan," 11 Dec. 1934, pp. 55–60, quote on pp. 55–56, RF 900 223.166.

[107]Weaver diary, 11 Jan. 1936, RF 200 170.2076.

[108]Weaver diary, 24 Sept. 1935, RF.

[109]Weaver to Fosdick, 10 Nov. 1937; Weaver diary, 6 Nov. 1937; both in RF 205D 5.78.

doubts. More self-confidant, Weaver felt better able to stand up to the trustees. In 1937, for example, he asserted that the officers had been "too long on the defensive," too willing to give in to continual demands by the trustees to explain and justify and modify their programs.[110] His improved relations with the trustees was apparent in the five-year review of 1938–1939, which could hardly have been more different from the appraisal of 1933–1934. Weaver produced a vast report and analyses of his projects (demonstrating, if nothing else, the defensive value of mountains of technical information).[111] The scientist-trustees were generally supportive, and the board pronounced themselves delighted with what seemed to them a well-chosen, well-managed, and productive program.[112]

Weaver handled a panel of scientific experts adroitly. "I don't want to drive this committee with too tight a rein," he wrote Fosdick, "but I am determined to use every proper effort to assure that we get somewhere."[113] Members were selected for their expert knowledge and experience in programmatic, sponsored research; there was no one of Flexner's generation or persuasion. Although biologists were in the majority, the panel also included a chemist and a physicist. Unlike the 1934 panel, whose deliberations were kept secret from the officers, this one worked closely with Weaver and not as adversaries. They did not ride personal hobbyhorses and did not get sidetracked by the pros and cons of laissez-faire. Everyone, including Weaver, enjoyed the experience of working together as colleagues with a common aim.[114] The panel put an enthusiastic stamp of approval on Weaver's procedures and his program, objecting only—but of course—to the name "experimental biology"![115]

Most important, the scientific panel gave implicit approval to Weaver's role as an active partner in setting research agendas. As the biologist Ross Harrison gingerly put it in his report, several members of the panel thought that officers could encourage applications that, from their own studies, they thought worthy. Weaver was more bold. In a footnote added to explain points that Fosdick found ambiguous, he interpreted Harrison's statement as "an expression of [the panel's] confidence in the ability of the RF officers to participate rather actively in the strategy of general research planning."[116] This was an overstatement of

[110]Weaver, "Program and administration," 1 Oct. 1937, RF 915 1.2.
[111]Weaver, "Activities in the natural sciences . . . 1933–1938," RF 915 2.10.
[112]"Report of the committee of review, appraisal, and advice," May 1939, RF 915 2.12.
[113]Weaver to Fosdick, 2 Dec. 1938, and other correspondence, all in RF 915 3.26–27.
[114]Weaver to Fosdick, 2 Apr. 1939, RF 915 3.26.
[115]"Report of the committee of review, appraisal, and advice," May 1939, RF 915 2.12.
[116]Ibid., p. 14.

an ambiguous endorsement; however, Weaver's gloss accurately reflected the reality of his relations with his now numerous scientific clients. Weaver's role as an activist manager of science was secure.

Molecular Biology

With security in his role came a growing confidence in his program in vital processes. Weaver became more independent of Gregg after 1935. References to human applications gradually disappeared from his reports, and appeals to "psychobiology" seemed more ritual gestures than descriptions of what he was actually doing. At the same time, terms appeared like "physicochemical biology," or "chemistry in its relation to biology," or "molecular biology," which emphasized the connections among the natural sciences. The knot between the natural and medical sciences divisions was officially cut by Fosdick in 1937, when he assigned all the psychobiological fields to Gregg. This left Weaver free to cultivate experimental biology in its connection to physics and chemistry. Many projects in "vital processes" already involved physicists and chemists. In 1939, Weaver divided the program under three heads: experimental biology, chemistry in relation to biology, and physics in relation to biology.[117] It was "vital processes" as Weaver had originally understood the term before his alliance of convenience with Gregg.

This trend obviously suited Weaver. He had always wanted a broad program in the natural sciences, and despite his use of psychobiological jargon he never seemed comfortable with the behavioral sciences. Psychology, for example, proved an intractable field for him. A planned survey of the field bogged down when Weaver discovered that not even the experts could agree on what the field was.[118] Sex biology and endocrinology lost their appeal for him as practitioners moved closer to clinical applications.[119] Combining the physical and biological sciences was much more to his taste and closer to his own experience in Charles Slichter's circle.

The shift from psychobiology to natural science was a gradual process. The 1934 appraisal loosened the knot that Fosdick untied three years later. Weaver had worried that the committee, in striving for practical relevance, might want every project to have psychobiological

[117]RF *Ann. Rep.* 1939.

[118]Weaver, "Progress report," 14 Feb. 1934, p. 9; Weaver to Tisdale, 8 Feb. 1935; both in RF 915 1.7 and 1.2.

[119]Weaver, "Progress report," 16 May 1936, pp. 31–32; "Progress report," 16 May 1938, p. 47; both in RF 915 1.8. "Report of the Committee of Review," Nov. 1938, p. 29, RF 915 1.12.

applications. (It would be analogous, he warned, "to a holding company which too heavily milks its subsidiaries in order to meet current dividend requirements.")[120] But the committee's message was that the natural sciences division could best contribute to "understanding and rationalizing human behavior" by strengthening basic biology, chemistry, and physics.[121]

Weaver and Tisdale were less certain about the programmatic unity of their eight subfields: if not psychobiology, then what? Tisdale tended to think of separate fields, each of which was to be developed in general but with a leaning toward the human side. (In genetics, e.g., this meant a leaning toward mammalian projects.)[122] Weaver saw more clearly the importance of problems and experimental methods as unifying themes: "I hope . . . that it is not true that we have eight or more separate objectives," he wrote. "I would say, rather . . . that the direction of our activity results from the fact that we are attempting to sponsor the application of experimental procedures to the study of the organization and reactions of living matter."[123] Loosening of the apron strings to psychobiology, in short, sharpened Weaver's conception of a unified transdisciplinary field of "vital processes."

Scientific interest groups within the Rockefeller circle also encouraged Weaver's retreat from psychobiology. Alan Gregg told Weaver that physics and chemistry were the real basis for progress in biology.[124] Simon Flexner advised Fosdick that Weaver should extend his program into basic physics and chemistry.[125] (Probably he meant to force Weaver to be less aggressively programmatic.) Edsall's warnings about the less reputable side of psychobiology was an invitation to Weaver to engage more physical scientists in his program of vital processes. The trustees hoped that opportunistic projects in pure physics would prevent officers from becoming too attached to their own programs.[126] Their fears of duplication of effort by Weaver and Gregg, though unwarranted, helped persuade Fosdick that the two programs should be cleanly separated. The memorandum of agreement that Fosdick took to the board in December 1937 (Weaver drafted it) left the natural sciences division with biochemistry, biophysics, genetics, radiation biology, general physiology, and developmental mechanics. Weaver also

[120]Weaver, "Progress report," 14 Feb. 1934, RF 915 1.7.
[121]Weaver to Tisdale, 8 Feb. 1935, p. 3, RF 915 1.2.
[122]Tisdale to Weaver, 24 Jan. 1935, RF 915 1.2.
[123]Weaver to Tisdale, 8 Feb. 1935, p. 3, RF 915 1.2.
[124]Gregg to Weaver, 10 Nov. 1936, RF 915 1.2.
[125]Flexner to Fosdick, 19 Nov. 1934, RF 915 4.41.
[126]Greene to Fosdick, 29 Mar. 1937, RF 915 1.2.

got formal approval to develop projects in physics and chemistry under the rubric of "experimental biology."[127] Thus, Weaver ended up after four years with something like his original vision of a realm of science in which biologists, physicists, and chemists could work without regard to disciplinary jurisdictions.

With experience Weaver also became more assured in his grasp of experimental biology: not just technical knowledge but an understanding of the political geography of who owned what problems and how different research schools were connected. Weaver got his education less from reading books than from traveling and talking to people. His diary records a steady growth in understanding, as he constantly asked questions about people, problems, scientific fashions, and reputations, always cross-checking the trustworthiness of his sources. He learned to decode opinions of interested parties. As he wrote to Fosdick: "[G]ood men,—even the best men,—have their own favorite irons in the fire; and the interpreting and weighing of advice is as important as its gathering." He learned that he got more disinterested advice by arranging to talk with a man "on his own back porch, as contrasted with a . . . formal interview in our offices."[128] Unlike Rose's formal and somewhat artificial classifications and maps of scientific disciplines, Weaver's knowledge was more the kind that insiders have of their own fields, a knowledge of personalities and interests as well as ideas and experimental practices. Like a tourist after a month in an unfamiliar city, Weaver began to see how main arteries and separate neighborhoods all fit together on one integral map.

Weaver felt he had turned a corner in his biological education in the spring of 1936, after composing a long description of his program for President-Elect Fosdick. He wrote to Tisdale that "the interrelationship between the various subdivisions of program takes on, with this report, its first really understandable and rational form."[129] Weaver's report is a curious document, consisting mainly of descriptions of the eight subfields in a popular science vein (with long quotations from authors like H. G. Wells), plus lists of leading workers and their locations, and technical summaries of projects.[130] Weaver obviously enjoyed showing that he could speak the specialized languages of his subfields, though Gregg thought his treatment "too full for the business-trustee and too obvious for a biologist-trustee."[131]

[127]Fosdick to Greene, 25 Mar. 1937, p. 2; Weaver to Fosdick, 29 Nov. 1937; both in RF 915 1.2. Weaver, "Program and administration," 1 Oct. 1937, pp. 21, 25–26; Fosdick to Weaver and Gregg, 10 Dec. 1937; both in RF 915 1.8.

[128]Weaver to Fosdick, 4 Aug. 1937, RF 205D 5.77.

[129]Weaver to Tisdale, 26 May 1936, RF 915 1.2.

[130]Weaver, "Progress report," 16 May 1936, RF 915 1.8.

[131]Gregg to Weaver, 10 Nov. 1936, RF 915 1.2.

Weaver's comprehension of his eight subfields was uneven, in part because some were less comprehensible than others. The most accessible were genetics, a well-defined discipline with dominant central schools, and nutrition and endocrinology.[132] General biochemistry and biophysics were more difficult for Weaver to conceptualize. Not surprisingly: biochemistry was an eclectic service discipline (Weaver followed textbook practice in organizing it by types of compounds), and "biophysics" seemed to mean anything in which methods of experimental physics were used. General physiology presented similar problems of definition: "[T]here is no other section of program which is as difficult to delimit or describe," Weaver complained.[133] The more Weaver learned, the more his eight official subfields seemed arbitrary formalities, useful "only as an administrative device." After 1936 he used them less and less.[134]

At the same time, Weaver became bolder in conceptualizing the unity of his program. To replace "vital processes" he invented or borrowed new terms like "molecular biology."[135] In his report for 1938 Weaver used this expression as an umbrella for virtually his entire program.[136] Usually, however, he used "molecular biology" to mean chemical and physical studies of biomolecular structure (not just macromolecules but any molecules of biological interest).[137] This usage most clearly reflects the etymology of the term as an analogue of "molecular physics": As molecular physics connected bulk properties of matter to atomic or molecular units, so molecular biology explained the properties of living matter in terms of biomolecular units like proteins or nucleic acid. "Molecular morphology" was a less common variant.[138]

This usage was not peculiar to Weaver: William Astbury and Charles V. Taylor, for example, used it in the same sense.[139] And Charles Slichter: Weaver was startled to discover in 1934 how similarly he and his mentor were thinking. Slichter, he noted, "speaks with customary vigor and enthusiasm concerning the probable great developments of science over the next quarter century, expressing the opinion that these will center around a new sort of molecular physics. . . . [I]t turns out that his ideas are very closely analogous to those which underlie the

[132]Weaver, "Progress report," pp. 19–25, 32–41.

[133]Ibid., pp. 8–11, 17–19. Tisdale to Weaver, 24 Jan. 1935, RF 915 1.2.

[134]"Progress report," p. 5. Weaver to Fosdick, 29 Nov. 1937, RF 915 1.2.

[135]Warren Weaver, "Molecular biology: Origin of the term," *Science* 170 (1970): 591–592. Weaver, "A quarter century in the natural sciences," RF *Ann. Rep.* (1958): 72. Weaver, *Scene of Change*, pp. 72–75.

[136]RF *Ann. Rep.* (1938): 203.

[137]Weaver, "Natural science appropriations 1933 to 1946," RF 915 2.11.

[138]Weaver to Arnett, 27 Dec. 1938, RF 915 1.3.

[139]C. V. Taylor to Weaver, 5 Oct. 1938, RF 205D 8.106.

whole program in vital processes."[140] What exactly Slichter had in mind Weaver did not say, but most likely it was some variant of his idea of "infracellular chemistry." (The context of the discussion was Slichter's scheme for a large project on the chemistry of cancer cells.) So, too, the scientific panel that assessed Weaver's program in 1938–1939 anticipated that putting biology together with chemistry and physics "may ultimately lead to a unified science of all nature." Weaver was delighted with their suggestion that he do more in enzymology, chemical embryology, and physicochemical genetics: "This is 'molecular biology,'" he rejoiced, "and already an enthusiasm of the division."[141]

Weaver avoided eye-catching labels after 1938, substituting the more ordinary "chemistry in relation to biology," etc. Perhaps he felt that unfamiliar terms might irritate the very people he wished to cultivate (he was aware, e.g., that biologists resented the term "biophysics"). By whatever name, however, the idea of "vital processes" or "molecular biology" was the key to Weaver's vision of his program and his job, which was to help chemists, physicists, and biologists to work together on the big problems of biological structure and function.

Partners in Science

Thus, by the late 1930s the elements of a new system of patronage were in place: a coherent programmatic vision (vital processes); an established managerial role between constituent and board; and vehicles of funding, like the project grant, that made possible an active partnership between grant-giver and grant-getter. Weaver's vision of "vital processes" was the keystone. It was a practical necessity for explaining to scientists why everything in their disciplines did not have an equal claim upon RF funds, and why chemistry and physics belonged in a program labeled "experimental biology." It was also essential to creating an activist role for sponsors of research. Terms like "vital processes" and "molecular biology" could not be attached to well-defined academic jurisdictions. Thus, they reduced the risk of patrons becoming captive to any particular discipline. No less important, "vital processes" was a guide to modes of laboratory practice—cooperative and transdisciplinary—that lay between individual work and formal team research. A strong programmatic vision, rooted in a conception of laboratory practice, enabled Weaver to hold his own with the trustees and their committees of review. A roomy managerial role enabled Weaver to

140 Weaver diary, 27 June 1934, RF.

141 Weaver, "Notes on report of NS committee of review," 8 June 1939; Hanson diary, July 1939; both in RF 3.26–27. "Report of the committee of review," May 1939, 915 2.12.

translate his vision of "vital processes" into a roster of individual and group projects.

Transdisciplinary programs were a great advantage not just to foundation managers but to deans and fund raisers and anyone who operated across discipline boundaries. These new versions of a "unified science of nature"—an ancient dream—were ways of conceiving the geography of science that were consonant with the emerging system of sponsored, project research: a system in which managers like Mason, Weaver, and Slichter were becoming essential actors. Indeed, ideas like "vital processes" or "molecular biology" are inseparable from their use as practical tools in managing a multidisciplinary program of projects. They were the software, so to speak, of the social machinery of sponsored research.

The system of sponsorship that Weaver helped construct was a more intimate form of the relationship that Woodward, Greene, Trowbridge, Rose, and Thorkelson had pioneered. All were activist managers, but Weaver was personally involved with individual scientists and research communities in a way that his predecessors never were. Not just in the business of designing laboratories and managing endowment funds: Weaver was involved in assessing research problems and organizing the social machinery of production—projects, schools, and transdisciplinary networks. The new generation of managers of science were active partners in the business of science in a way that had not been seen before and would seldom be seen thereafter.

The frictions that Weaver experienced with his board and official scientific advisors were symptoms of this more intimate partnership. The lines of authority in the patron-client relation were being renegotiated. Scientists were finding out how far Weaver could be trusted to respect traditional prerogatives. Weaver was seeing how far his trustees and constituents would accept an activist style of management. These conflicts, though real and important, should not be exaggerated. Certainly they were gentler than most forms of social conflict in the 1930s, and they surfaced only where traditional prerogatives were directly threatened. As, for example, when Weaver sat down with the NRC's fellowship committees to work out a shift of resources from general fellowships, controlled by the NRC, to fellowships in vital processes, controlled by Weaver. The plant physiologist William J. Robbins was especially bitter: as head of the committee that handled the biology fellowships he had most to lose.[142] In the new system of patronage, the patron-client relation was direct and unmediated, and NRC activists

[142]Weaver memos, 8 Oct. 1934, 11 Jan., 18 Apr. 1936, RF 200 170.2074 and RF 200 170.2076.

naturally felt cut out. Signs of conflicts also surfaced when scientists were invited to pronounce on RF policies, as with the 1934 panel. But these seem more a reflexive effort by official spokesmen to assert traditional values of laissez-faire and self-regulation. Dakin and Howell were worried less by what Weaver was doing than by the precedent that any kind of active patronage would set.

Thus, with Flexner and his friends nipping at their heels, Weaver and his fellow officers established working relationships with some hundreds of biologists, chemists, physicists, and even a few mathematicians and engineers. The record of these relationships reveals few signs of tension and conflict and a strong common interest in making projects. In the field it seemed less urgent to defend principles and prerogatives than it did in board and committee rooms. In the business of making projects, it was more practical things that mattered—like deciding who should get what and why.

CHAPTER ELEVEN

Biological Communities and Disciplines

WEAVER'S CONCEPTION OF "vital processes," like Rose's idea of "making the peaks higher," took on concrete meaning only in the process of making projects. Each project represents a negotiated compromise between Weaver's ideal vision of a transdisciplinary domain and the practical realities of practice within particular disciplines. Location is again the key to understanding who got what. We need to look again for regional differences in projectmaking—not geographical and political regions now but regions defined by disciplinary boundaries and subcultures, professional networks, and group commitments to research tools or practices. Historical analysis gets more complicated in the era of project grants. It is relatively easy to grasp the large-scale structure of university systems. With Weaver's project grants, we need to understand the structure, mores, and shifting research fronts of some dozen research communities. There are a lot more projects to deal with: Trowbridge and Thorkelson arranged dozens; Weaver hundreds, not including fellowships and grants-in-aid.[1]

I will focus on three groups of scientists: general physiologists, a dispersed and polyglot group of biologists who sought (in vain) disciplinary unity; "molecular biologists," chemists and physicists who migrated into physiology and biochemistry; and experimental physicists, toolmakers who saw in biochemistry and biology new uses for their technical skills. Biologists, chemists, and physicists were all essential for Weaver's vision of "vital processes," but for each group Weaver had to devise different strategies for making projects, since each one had a dif-

[1]My analysis deals only with board-approved appropriations, but they account for some 80–90 percent of expended funds. Grants-in-aid and fellowships, which did not need to be approved item by item by the board, were generally smaller than the supplementary to projects. Sometimes, however, Weaver used grants-in-aid to aid scientists or sciences peripheral to his core program.

ferent role in the practice of "vital processes." With groups that owned, so to speak, the big problems of biology, Weaver tried to develop disciplinary communities in a more or less systematic way. This was the case, for example, with experimental biologists—cell physiologists, general physiologists, geneticists. He used a similar strategy of discipline building, though less systematically, with those branches of chemistry that owned the central problems of biomolecular structure, namely, bio-organic and biochemistry.

These efforts to develop disciplines or research specialties were, on the whole, the least successful of all Weaver's ventures in sponsored research. Disciplines were too big and diverse to be fundamentally changed by even a substantial program of individual projects. Weaver lacked the resources and leverage for community development on such a scale, just as Rose had lacked the resources and leverage to fundamentally alter national university systems. Take general physiology, for example: straddling the border of physics and biology, it seemed only to lack the resources, organization, and institutional base of more successful academic neighbors. It seemed an ideal opportunity. Weaver soon learned, however, that internal factionalism and competition from established disciplines made general physiology what it was and that there was little he could do to alter these basic ecological realities. The lively pace of new projects in this field slackened markedly in the late 1930s (table 11.1). Bio-organic chemistry was a similar case, at least in the United States, as was mammalian genetics. Weaver would have liked to develop these productive new specialities, but he was frustrated by intitutionalized patterns of dispersal and territoriality.

It was different with physical chemists and physicists. They did not own problems central to "vital processes," but they did own tools that could be applied to problems of biomolecular structure and function. Here the task of patronage was not to build disciplines but to improve communication between them. The challenge for Weaver was to enable physical chemists and physicists to turn their skills to biomolecular problems. A few exceptional individuals did both physics and biology—physicist William Astbury, chemists Linus Pauling and The Svedberg, and mathematician Dorothy Wrinch, for example, the forerunners of the "molecular biologists" of the 1950s. Analysis of their projects in Chapter 12 will turn on the ways in which Weaver helped to guide and stabilize the trajectories of individual careers across different regions of science.

With most physical chemists and physicists, however, Weaver had to find ways to engage them in biological work that did not require wrenching changes in disciplinary commitments. That meant cooper-

Figure 11.1. Rockefeller Foundation projects in general physiology to 1945. Graphics courtesy of Jack Kohler.

ative research projects, in which physical scientists provided technical services, and biochemists or biologists the problems. Foundation sponsorship made possible work for which neither partner alone had the intramural resources. Cooperative projects were inherently complex, requiring communication across disciplinary language frontiers and a division of labor among strong-minded (and sometimes narrow-minded) experts. Weaver discovered through experience that laboratory instruments were a simple and effective vehicle for such projects. Extending the range of instruments usually produced results of interest to both parties. It made possible a limited partnership that did not make impossible demands on the tolerance and altruism of either partner. My analysis of physics projects in Chapter 13 will hinge on neither communities nor individual careers but on machines and how they structured the work of small multidisciplinary groups.

Thus the individual research grant became, in Weaver's hands, a flexible instrument for realizing his vision of "vital processes." Projects in general physiology and bio-organic chemistry were designed to make borderland specialities better able to survive competition with stronger academic rivals. Aid to "molecular biologists" was meant to demonstrate the viability of transdisciplinary careers. Laboratory technologies were deployed to build new channels of communication and exchange between the physical and biological sciences. Weaver used different strategies for making projects in the various subregions and subcultures of "vital processes." My account of Weaver's field practice takes a similarly pragmatic and ecological approach, focusing in turn on communities, careers, and tools.

General Physiology, European Style

"General physiology" was a term that seemed to mean whatever its diverse practitioners wanted it to mean. In 1935 Weaver tried to explain what he meant by it. As originally used by Jacques Loeb, he noted, "it referred almost exclusively to the applications of chemical and physical techniques to basic physiological problems." Gradually, however, the name was appropriated by people who were not practitioners of Loeb's style of developmental mechanics. It "came to be used for the sort of physiology a zoologist does," in contrast to the plain physiology done in medical schools. More recently still, the term had degenerated to little more than a catchall for "any kind of zoology that is not definitely morphology, taxonomy [or] genetics." The general physiology that interested Weaver, however, was the original Loeb variety "with a little tinge of the second definition."[2]

Why general physiology was so amorphous is a long story; suffice to say that its chronic identity crisis reflected the practical reality that general physiologists never succeeded in staking out their own social space between the powerful disciplines of zoology, medical physiology, and biochemistry.[3] General or cell physiologists were most numerous in the United States but also most dispersed. Most practiced their craft in departments of zoology or medical physiology, where they tended to

[2] Weaver to Tisdale, 8 Feb. 1935, RF 915 1.2.

[3] Kohler, *From Medical Chemistry to Biochemistry* (Cambridge: Cambridge University Press, 1983), ch. 11. Philip J. Pauly, *Controlling Life* (New York: Oxford University Press, 1987). Pauly, "General physiology and the discipline of physiology, 1890–1935," in Gerald Geison, ed., *Physiology in the American Context* (Bethesda, MD.: American Physiological Society, 1987). pp. 195–208.

become zoologists or medical physiologists, as Weaver observed. General physiologists were a more coherent but much smaller community in Europe, where a few eminent practitioners had managed to acquire small research enclaves on the margins of the big teaching departments—August Krogh's is one we have already met with. These showpiece laboratories produced excellent research, but without teaching service and a regular market for disciples, there were few incentives for their creators to expand beyond their own specialized lines of work.

At the time Weaver came on the scene, biochemistry was the fashion among European general physiologists. The exemplary figure was Otto Warburg, whose small but highly productive research team at Dahlem set the pace in the fast-moving field of cellular metabolism and bioenergetics. Warburg's hallmark was his uncommon ability to unite the ideas and methods of chemistry, physics, and biology. (He began his career in experimental zoology but never forgot that he was the son of the eminent physicist Emil Warburg.) He was a virtuoso performer with physical apparatus like the spectrophotometer, and was equally adept in theoretical chemical thermodynamics. In the early 1930s, he was deeply into studies of respiratory enzymes and coenzymes and the connection between respiration and the growth and division of cells.[4] Enthusiasm for Warburg's work was unbounded in the mid-1930s. Typically, Arthur Compton and Thorfin Hogness saw him leading the way to a revolution in cell biology comparable to the quantum revolution in atomic physics.[5] Weaver shared their enthusiasm.

Neither Warburg's institute nor other major centers of this mode of general physiology (e.g., Otto Meyerhof's institute at Heidelberg) received funds from Weaver's program—not surprisingly, since most were in Germany, and after 1933 the RF was loath to do business there. Weaver and Tisdale were most active and effective at the margins of European general physiology, helping to spread the new modes of practice to up-and-coming groups like David Keilin's, at Cambridge University. Keilin had recently discovered the cytochromes and was besting Warburg in a lively and public dispute over the biological significance of these respiratory enzymes.[6] However, he was still something of an outsider to general physiology. An eminent parasitologist at the time of his serendipitous discovery, his official job was to direct the

[4]Hans Krebs, *Otto Warburg* (New York: Oxford University Press, 1981). Kohler, "The background to Otto Warburg's conception of the *Atmungsferment*," *J. Hist. Biol.* 6 (1971): 171–192.

[5]Compton and Hogness, "An institute devoted to fundamental physical and chemical research on the growth and function of living cells," 13 May, 1938, RF 216D 12.177.

[6]David Keilin, *The History of Cell Respiration and Cytochrome* (Cambridge: Cambridge University Press, 1966).

parasitology research of the Molteno Institute. To work on cythochrome Keilin had to get extramural funds. Alan Gregg provided these funds initially, and Weaver inherited the project from Gregg in 1934 and saw Keilin through the crucial years of his metamorphosis from a parasitologist to a full-time enzymologist.[7]

The Molteno project remained a small one, however. Weaver would have been glad to help Keilin expand his operation, but Keilin was not particularly ambitious to run a large school, or to have a laboratory full of modern equipment, or to get credit for new quantitative techniques. (His favorite piece of equipment was a small handheld spectroscope.) He had no desire to expand into other lines of general physiology or to enter into cooperative schemes with physicists and chemists. With a few postdoctoral fellows he produced outstanding work, but his laboratory was essentially a one-man show—a continental institute in an English modality. Weaver had little to do but renew Keilin's grant and share in his distinguished achievements.

Weaver was always eager to assist in making small projects into centers of the new discipline, especially in places where general physiology was just beginning to take root and where a modest grant might make the difference between success and failure. Tisdale was always on the alert for young and talented biologists with entrepreneurial flair—like John Runnström at the Stockholm Högskola (later, University). Trained as a zoologist, Runnström had built up a small school of experimental embryology and Entwicklungsmechanik in the Loebian style. Working mainly on sea urchin eggs at the Roscoff marine station, he probed the relation between respiration and the activation of cell division—the very problem that had launched Loeb and later Warburg on their careers in general physiology. Using Warburg's microchemical methods, Runnström had found evidence that this process was more complicated than his distinguished predecessors had thought, and he was emboldened to move into the more fashionable and fast-paced fields of enzymology and bioenergetics. He dreamed of using Warburg's methods to correlate respiration with cell permeability, differentiation, regeneration, division, and growth.[8] Tisdale was impressed.

One of the first zoologists in Sweden to do such experimental work, Runnström became an embattled emissary of modernism among unreconstructed morphologists. His refusal to do even token morphological

[7]Miller diary, 23–24 Oct. 1934; Keilin to O'Brien, 25 Oct. 1934; both in RF 401D 42.546. O'Brien diary, 11 June 1932, RF 401A 20.264.

[8]Runnström, "Memorandum III," 29 Sept. 1931, RF 800D 6.66. Runnström to Tisdale, 29 Dec. 1938, RF 800D 6.68.

research, while it won high praise from Warburg, may have cost him at least one academic chair and slowed his advancement at Stockholm.[9] For Tisdale and Weaver, Runnström's international connections, solid record of achievement, entrepreneurial ambition, and expanding vision of his discipline added up to an ideal opportunity to create a center of experimental biology for all of Sweden. By 1936 no fewer than twenty-five researchers were engaged in some aspect of Runnström's ever-expanding program. In 1938 the RF made one of its rare capital grants for a new laboratory. Its support was crucial in realizing Runnström's dream of a national center in the new biology.[10]

In France such a center already existed: the Institute for Physico-Chemical Biology of the Rothschild Foundation, which brought together under one roof physicists, chemists, and biologists. The four biologists were remarkable for their relative youth (thirty-one to forty-eight years) and their devotion to the experimental style of biological research. It was rare in France for younger workers in new fields to be in charge of their own departments, especially in biology, which was dominated by traditional morphologists. (They thrived on the flourishing market for teachers of traditional biology in the French colonies.) The man who created these four positions, the physiologist Andre Mayer, thought it quite an achievement for France; Tisdale, who knew all about Parisian science politics, thought so too.[11]

All four members of Mayer's group were already familiar to the Paris office. René Wurmser had been one of the first IEB fellows; the others had had fellowships from the RF. Emmanuel Fauré-Fremiet had an RF grant-in-aid, and Wurmser and Boris Ephrussi were both applying for grants. Wurmser was planning to study chemical thermodynamics at Berkeley; Louis Rapkine wanted to spend time with Keilin. Ephrussi had close connections with the Morgan school at Caltech, where his work on embryology and genetics was warmly admired—he had been there in 1934 as an RF fellow. Tisdale had only to suggest that the several grants and applications be combined into an expanded and inte-

[9]Tisdale to Jones, 29 Oct. 1931; Warburg to Jones, 14 Sept. 1931; both in RF 800D 6.66. Runnström to Tisdale, 29 Dec. 1938, RF 800D 6.68. Pasquale Pasquini, "To the memory of John Runnström," *Acta Embryol. Exp.* (1971): 120–136, esp. p. 122. Vera Runnström-Reig, personal communication, 1990. Runnström was made professor of experimental biology in 1932, when he was forty-four.

[10]Runnström to Tisdale, 24 Sept. 1935, 23 Dec. 1936, 29 Dec. 1938; Miller diary, 10–15 Aug. 1936; Weaver to Miller, 24 Sept. 1936; Sven Tunberg to Tisdale, 8 Dec. 1936; Weaver diary, 22 Apr. 1938; all in RF 800D 6.66–68.

[11]Tisdale diary, 17 Sept. 1935; Miller diary, 18–19 June 1935; both in RF 500D 12.127.

grated group grant, which would enable them to develop as an ensemble what they had already begun as individuals.[12]

The researches of the Rothschild group, though diverse, all took a molecular and biochemical approach to cell function. Wurmser, the most chemically oriented, worked with the biologist Rapkine on the thermodynamics of respiratory enzymes. Embryologist Fauré-Fremiet was reworking his previous histological problems in terms of protein structure and behavior. Boris Ephrussi, who had been trained in classical Entwicklungsmechanik, was working with the geneticist George Beadle applying techniques of tissue culture and transplantation to the developmental genetics of *Drosophila*. It was just the sort of imaginative, cross-disciplinary work that Weaver and Tisdale hoped would give shape to a new discipline of general physiology.[13]

Runnström's group had more need of active nurturing by Tisdale and Weaver. Runnström's interests, despite his good intentions to expand, tended to remain narrowly embryological. He was not at home in chemistry and was inexperienced in managing a large and diverse group. His colleagues lacked the biochemical sophistication of the Rothschild group. His program needed to be diversified, and interdisciplinary connections had to be made. The stimulus for these changes came in part from Tisdale and Harry Miller in the Paris office. They and Weaver provided a good deal of timely guidance and encouragement, and occasionally discouragement: for example, when Runnström requested funds to take his group on their customary summer pilgrimage to Roscoff for the sea urchin season. Tisdale and Weaver declined to underwrite the excursion, partly on principle but partly, too, because it would have sustained Runnström's attachment to marine zoology.[14] By paying for newer enzymological lines but not traditional ones, Weaver and Tisdale hastened Runnström along his changing career trajectory.

Gradually, Runnström delegated the purely embryological work to a colleague, Sven Hörstadius, and by 1938 he was deeply engaged in the study of respiratory enzymes and the bioenergetics of biosynthesis and growth. He gave up sea urchins for yeast, a far more convenient organism for biochemical work and one that was not seasonal—essential for competing in a fast-moving research front. He never converted com-

[12]Tisdale to Weaver, 29 July 1935, RF 500D 12.127.

[13]A. Mayer to Tisdale, 2 Mar. 1936; Wurmser reports, 13 Oct. 1936, 18 Nov. 1937, 5 July 1939; Fauré-Fremiet report, 26 Oct. 1938; Ephrussi report, 25 Apr. 1939; all in RF 500D 12.127. Jan Sapp, *Beyond the Gene* (New York: Oxford University Press, 1987), ch. 5.

[14]Tisdale diary, 1 Nov. 1934; Tisdale to Weaver, 16 Nov. 1934; Weaver to Tisdale, 21 Jan. 1935; all in RF 800D 6.66–67.

pletely to the Warburg style: indeed, he seemed to revert to his former embryological interests after 1940, when direct contact with Tisdale and Weaver ceased. But in the 1930s, thanks in part to RF funds and Tisdale's coaching, his school was probably the most important center of general physiology in Scandinavia, with its own distinctive blend of biochemistry and biology.[15]

Tisdale and Weaver also helped to arrange collaborations between Runnström and physical scientists like Erik W. Hulthen (experimental physics), Oscar Klein (theoretical physics), and Ragnar Nilsson (organic chemistry). Team efforts were motivated partly by the practical need for experts in unfamiliar physical techniques, like spectroscopy and conductivity measurement.[16] However, Runnström also believed that multidisciplinary teamwork was the modern thing to do and would pay off in unexpected ways. He had been struck during a visit to Woods Hole by the readiness of American zoologists to engage in cooperative research, and was inspired to import the American style of practice to Sweden. It was easier said than done. Cooperation between scientists in different disciplines was not routine in any European university, and Runnström had just about given up hope when Hulthen knocked on his door, having been similarly inspired by a sojourn in Niels Bohr's institute, where an RF-sponsored cooperative research program was underway. Neither one knew quite where to start, but Tisdale, who happened to be in Stockholm just then, urged them to pick a problem that one of them at least knew something about. That, plus the prospect of support if they could devise a plan, did the trick. When Tisdale returned a week later, the two had agreed on a joint project on the photochemistry of respiratory enzymes. Within months Hulthen was building a calorimeter, and Runnström was teaching himself thermodynamics.[17] By 1936 they had diversified into the electrophysiology of nerve, metabolism and membrane potentials, photochemical activation, and the effect of radiation on growing yeast cells.[18]

Tisdale was also instrumental in arranging for collaboration between Runnström and Ragnar Nilsson. The problem here was Hans von Euler, the autocratic head of the organic chemistry institute and reigning prince of enzymology. Euler's efforts to control Runnström and Nilsson's project and budget—"up to his old tricks," Tisdale noted—was just the sort of territorial behavior that made cooperative

[15]Runnström reports, 1936–1943, RF 800D 6.66–.68.

[16]Miller diary, 28, 30 Jan. 1935, RF 800D 6.67.

[17]Tisdale diary, 1 Nov. 1934; Miller diary, 28, 30 Jan. 1935; all in RF 800D 6.66–67.

[18]Runnström to Tisdale, 14 June, 24 Sept. 1935; 15 May, 23 Dec. 1936; 29 Dec. 1938. Hogness Report, 16 Aug. 1937, RF 800D 6.67–68.

research a rarity in Europe. Runnström began to doubt that collaboration with a chemist was worth the trouble, but Tisdale was on to von Euler and had no trouble preventing him from grabbing control. Soon Runnström was reporting that Nilsson's chemical expertise had moved his research on cell physiology "on to a new plane and . . . in a larger air and with wings."[19]

The exceptional talent and resources that Tisdale found at Stockholm were rarely found elsewhere in Europe. At Copenhagen, Weaver and Tisdale hoped that August Krogh would play a central role in the cooperative project with Bohr's physicists, but he remained a somewhat marginal actor, sticking with his specialized line of research on membrane permeability. Tisdale also scouted the Netherlands for opportunities, and a project was undertaken that involved cooperation between physicists and experimental biologists at Utrecht and Delft. It was not a notable success, however. Dutch biologists were divided and contentious—as were, apparently, Dutch academics in general, owing to the close proximity of universities in the major cities and their bitter fights for students and limited state resources. It would have been better, Tisdale thought, if funds were concentrated in a single center, but obviously there was nothing he could do about that.[20] A long history of provincial rivalries in that corner of Europe had created institutional structures and interests that impeded cooperative projects. Tisdale had better luck at Berne, where a small project was arranged with physiologist Alexander von Muralt, on the application of optical methods to biomolecular structure and function of muscle and nerve.[21]

With his unequaled knowledge of European scientists and his experience in brokering multidisciplinary projects, Tisdale could take an active role in developing general physiology in Europe. He was constrained, however, by the way the field had developed institutionally. There were few general physiologists in university institutes of zoology and physiology, which guarded the gates through which students passed and resources flowed. This was the result of the European strategy of cultivating new disciplines in a few showpiece institutes. While these marginal centers avoided confrontations with entrenched academic interests and produced research of admirable quality, they were not widely imitated. Nor did they make mainline institutes of physiology and zoology more open to new things. Opportunities for systematic discipline building thus were constrained. It was different in the

[19]Miller diary, 28, 30 Jan. 1935; Weaver diary, 26 May 1935; Runnström to Tisdale, 14 June, 24 Sept. 1935, 15 May, 23 Dec. 1936; all in RF 800D 6.67.

[20]Tisdale diary, 4 Oct. 1934, RF 1.2 650D unprocessed file.

[21]On Muralt's project, RF 803D 4.44–45.

United States, where general physiologists tended to be already assimilated (or encysted) in traditional departments of zoology.

General Physiology in the United States

With all its limitations, the strategy of elite enclaves sufficed for a time to keep Europe ahead of the United States in general physiology and a few other small but highly productive specialities (theoretical physics, bio-organic chemistry, and enzymology, e.g.). American scientists had been trying hard to close these gaps, hoping to exploit America's overall advantage in size and wealth, and its unequaled investment in mass higher education. As Thorfin Hogness observed in 1938, not one of the recent Nobel prizes in general physiology had gone to an American. He thought it likely, however, that the political crisis in Europe would shift the center of this field to the United States, just as it had atomic physics.[22]

It may seem odd that Americans were not competitive in general physiology, since they had outstripped Europe in most other branches of experimental zoology. The reason has to do with the way in which new specialties developed, inside traditional departments of zoology. When experimental ideals became widely fashionable in the early 1900s, practices that had developed in embryology and physiology were extended to all the traditional zoological fields: heredity metamorphosed into genetics, cytology into cytochemistry, biogeography into ecology, natural history into animal behavior. Existing fields of zoology were thus updated, without the interjection of new specialties to threaten the morphological mainstream. While physiological and biochemical methods were assimilated, physiology and biochemistry were not. Zoologists resisted efforts to appoint physiologists and biochemists, for fear that they would bring with them the alien culture of medical schools and eventually crowd out traditional zoologists. Thus, general physiologists found their best career chances in medical school departments of physiology and biochemistry. Those that opted for careers in zoology departments were obliged to do so as practitioners of one or another of the older zoological specialties.

The result was that American general physiologists outnumbered their European colleagues but were also more diverse and contentious, and less productive of work at the cutting edge. Their primary loyalties were to zoological specialties, not to general physiology per se. There were no dominant figures like Warburg and Meyerhof to define a new

[22] A. Compton and Hogness, "An institute . . . ," 13 May 1938, p. 5, RF 216D 12.177.

mainstream at the margin, because the American habit of assimilation made it difficult. Weaver's American clientele were somewhat older than Tisdale's and more likely to be known for achievement as teachers or authors of textbooks. Their research interests were more morphological than biochemical, and closer to traditional experimental histology or zoology. Weaver was no less constrained than Tisdale but in different ways.

Joseph Bodine's project at the University of Iowa exemplifies one way in which general physiology developed within a department of zoology. Bodine was a Middle-American version of John Runnström: an ardent advocate of experimental methodology in a traditional research line, namely, entomology. Bodine specialized in the early development of the grasshopper egg. Called to head the Department of Zoology at Iowa in 1929, Bodine formulated a grand program of research in cytology, embryology, and cell physiology, using the grasshopper egg as a standard organism, much as Runnström and others used the egg of the sea urchin. It was an appropriate substitution, given Bodine's location in the middle of a sea of grass and corn.

Bodine's ambitions were encouraged by the university's administrators, who hoped to make zoology a showcase of modern practice and an inspiration to other science departments. Outside intervention was the key to improvement: Bodine's zoological colleagues at Iowa, as in most midwestern state universities, were mainly older men, teachers little interested in research. The idea was to dilute the Old Guard with young biologists pursuing modern lines of research. In 1927 the University had created a research professorship for Emil Witschi, a young Swiss embryologist from the school of Hans Spemann, who was in the United States as a Rockefeller fellow. Bodine was then brought in to reorganize the department (Witschi was too individualistic and temperamental to do it), hire new faculty, and inaugurate a program of research.[23] It was a pattern that Rose and Thorkelson had often seen and encouraged with departmental grants. Weaver, too, saw it as a potential showpiece of modern practice in general physiology.

By the time Hanson and Weaver first visited Bodine in 1933–1935, almost everyone in the department had been drawn into work on grasshopper eggs (except Witschi, who had very different ideas of cell development and resented the new man). The agenda of Bodine's school was to systematically gather quantitative data on the morphology, physiology, and genetics of the "normal cell," thus establishing a baseline for research on cells of any sort. The grasshopper seemed an ideal standard organism for cell physiology: its large

[23]Hanson diary, 11 Sept. 1933, 23 Nov. 1934, RF 218D 2.15–16.

cells were easy to manipulate, and embryos of pure lines could be produced in large quantities. By 1939, Bodine was operating a grasshopper factory capable of producing eggs and embryos in any stage of development at any time of year. Bodine was just the person, Weaver thought, to give general physiology a more coherent social and intellectual shape. Bodine's program encouraged joint research by teams of experts, and it trained large numbers of young biologists in quantitative methods. A standard organism promised to give a common focus to an inchoate field, as *Drosophila* had to genetics.[24]

Bodine was a typical, even stereotypical, American academic entrepreneur, with more zeal than polish. Weaver did not know quite what to make of him at first: "He shows, at times, lack of poise and judgment in his enthusiasms. He labors under the unfortunate impression that it is up to him to 'sell' his department and his work."[25] Weaver could see, however, that Bodine was an able organizer and had a gift for picking talented young researchers and instilling in them his own enthusiasm for quantitative methods. Everyone said so. Not everyone approved of Bodine's "factory method" or expected his group to make great discoveries; but even his detractors admired his ability to produce quantities of useful data and well-trained students. Francis Schmitt, for example, wrote that he "runs one of the best shows in general physiology in this country."[26]

The Iowa school might have taken the lead in importing Warburg's style of general physiology. It was a major component of Bodine's diverse program, which included the enzymology of differentiation and the bioenergetics of cell growth and division, as well as permeability and osmotic regulation, experimental embryology, and biophysics. In practice, however, Bodine preferred morphological work. Most of his new faculty, like he himself, were cytologists not physiologists, and his one biophysicist, Gordon Marsh, lacked Bodine's driving ambition.[27] Bodine was just a few years too old to be swept up in the new fashion of biochemistry and biophysics. He had one foot in the old and one in the new.

Weaver and Hanson actively encouraged Bodine to move more quickly in the direction of biochemistry and biophysics, as Tisdale did Runnström. They declined to renew a grant for Bodine's older lines of

[24]Bodine to Weaver, 18 Mar. 1933; Weaver diary, 2 Nov. 1933; both in RF 218D 2.15. Bodine to Hanson, 11 Feb. 1935, RF 218D 2.17. F. O. Schmitt to Hanson, 14 Dec. 1939, RF 218D 2.19.

[25]Weaver diary, 2 Nov. 1933, RF 218D 2.15.

[26]Schmitt to Hanson, 14 Dec. 1939; G. H. Bishop to Hanson, 5 Dec. 1939; C. E. McClung to Hanson, 22 Nov. 1939; all in RF 218D 3.19.

[27]Bodine to Weaver, 18 Mar. 1933; Weaver diary, 2 Nov. 1933; both in RF 218D 2.15. Bodine to Hanson, 11 Feb. 1935, RF 218D 2.17.

embryological research, arguing that it was the university's responsibility to support established research lines. At the same time, they encouraged Bodine to apply for a renewal for new and more modern studies of the electrophysiology of the developing embryo, respiratory enzymes, and enzyme activation in differentiation. Already begun in a small way, these lines did become more central to Bodine's program after 1939, in good part because of Weaver and Hanson's gentle but firm application of the method of the carrot and the stick.[28] However, Bodine did not join Runnström and Warburg in the avante garde of general physiology.

Bodine's enthusiasm for new practices was genuine, but cytology and embryology were more familiar and accessible. His training and his role as head of a diverse department of zoology set limits on his desire to join the European research front. So too, perhaps, did his investment in the grasshopper as a standard organism. Just as the choice of the sea urchin egg kept general physiologists connected to the morphological traditions of marine zoology, so too the grasshopper egg kept Bodine's group tied to traditional entomology. Entomology was a large, prestigious, and paying enterprise in farm-belt universities, thanks to a vigorous market for entomologists in state and federal agricultural agencies. In this context, Warburg's style of general physiology did not confer a competititive advantage in making careers, as it did where it was valued as academic high culture. Weaver's strategy of importing elite modes of European practice was unsuited to the system in which Bodine operated.

If Bodine exemplifies the role of general physiology within zoology, Francis Schmitt's program at Washington University was Weaver's chief hope in the context of medical physiology. Formally in the Department of Zoology, Schmitt was in fact a card-carrying member (Ph.D. 1927) of Joseph Erlanger's powerful school of medical neurophysiology. Only thirty-one years old in 1934, Schmitt was equally strong in physiology and physical science. As an NRC fellow, he had studied chemical thermodynamics with G. N. Lewis, biophysics with A. V. Hill, and biochemistry with Warburg and Meyerhof. He was one of the two Americans who, everyone agreed, fit the ideal type of a biophysicist as one who was a physiologist among physiologists and a physicist among physicists. (The other was Detlev Bronk.) To Weaver and Hanson, Schmitt seemed just the person to build an American school of general physiology.[29]

And Washington University seemed just the place. Virtually the creation of Rockefeller philanthropy, it had been consciously developed by

[28]Bodine to Hanson, 2 Nov. 1939; Hanson to Bodine, 9 Oct. 1939; Hanson diary, 24 Oct. 1939; all in RF 218D 3.19.

[29]Hanson diary, 22 Mar. 1934, 22 Nov. 1937, RF 228D 5.54 and 5.56.

Flexner, Rose, and Thorkelson as a showpiece of modern medicine and science in the American Southwest. Money from the RF's fluid research fund had helped Schmitt inaugurate his research program (it was the impending termination of this fund, in fact, that prompted Schmitt to apply to Weaver for his own grant). The proximity of Erlanger's group in the medical school more than compensated, in Weaver's mind, for the absence of a strong tradition of experimental zoology at Washington University.

Schmitt worked mainly with the giant squid axon, the neurophysiologist's equivalent of the standard fruit fly, sea urchin, or grasshopper, Large and easy to manipulate, it was ideal for physical and chemical research. Unlike most neurophysiologists, who concentrated on the mechanics of the nerve impulse, Schmitt had a very broad vision of what could be done with the squid axon. Visiting in 1934, Weaver and Hanson were astonished by the breadth of his ambition—to correlate the electrical, metabolic, and respiratory activities of nerve cells, no less. He was as adept in Warburg's spectroscopic and enzymological methods as he was in the methods of electrophysiology. The biochemistry of nerve pigments and intermediary metabolism were also grist for Schmitt's restless and omnivorous mind. On the biophysical side he was obtaining, with a novel amplifier and oscillograph, data that challenged some fundamental views of A. V. Hill's school. A new line of work in the mid-1930s was the study of the macromolecular structures of the nerve axon and nerve sheath, using X-ray diffraction, flow birefringence, ultracentrifugation, and other new techniques for determining the shape of protein bundles and monolayers. Equally the theoretician and experimentalist, model builder and biologist, Schmitt seized upon every instrument he could for his multisided development of general physiology.[30]

The organization of Schmitt's group reflected his eclectic research style. His brother Otto was the chief instrument maker. Then a graduate student in physics, Otto was invariably described as a "genius," referring to his knack, indeed obsession, for building instruments.[31] Someone like that was vital to a program so dependent on new instrumentation. On the chemical side, Richard Bear was a mirror image of Schmitt: trained in theoretical and physical chemistry, he surprised Schmitt by becoming a convert to physiology.[32] Schmitt also expanded

[30]Schmitt to C. Graves, 6 Feb. 1934, 4 Mar. 1936; Schmitt, "General outline of a three year program of research on the nature of the irritable mechanism," 4 Mar. 1935; all in RF 228D 5.54–55. Schmitt, "Proposed tentative program of research 1938–1943," 17 Jan. 1938, RF 228D 5.59.

[31]Hanson diary, 19 Nov. 1934, 14–17 July 1937, RF 228D 5.54 and 5.56.

[32]Schmitt to Graves, 4 Mar. 1935, RF 228D 5.55.

into experimental embryology, in a joint project with Victor Hamburger who, like Witschi, was a scion of Hans Spemann's school at Freiburg.[33] Schmitt's group resembled the biologists of the Rothschild Institute in Paris, with Schmitt himself, a biophysical Alec Guiness, playing all the leading roles.

For Weaver, Schmitt's project was close to ideal, with a brilliant, polymathic entrepreneur, a committed team of physicists, chemists, and biologists, and dependable institutional support. There was little Weaver and Hanson had to do beyond inviting Schmitt to ask for more money.

Weaver was an active participant, however, in the creation of MIT's new department of "biological engineering," to which Schmitt was called as head in 1941. Few projects aroused Weaver's enthusiasm as this one did. None came closer to realizing his dream of giving coherent institutional shape to the idea of "vital processes."

The idea of a new discipline of "biological engineering" was the brainchild of MIT's president, Karl Compton, and of John Bunker, a bacteriologist and biochemist in MIT's department of public health. It was the biological component of Compton's plan to bring MIT's science departments back into the mainstream of fundamental research.[34] Of MIT's once eminent programs in applied biology and public health, only food technology and bacteriology remained strong, and the MIT biologists had failed entirely to keep up in biophysics and biochemistry. Compton's idea was to build programs in these fields on the foundation of the burgeoning market for chemists and physicists in the laboratories of hospitals and public health agencies.[35] Compton also meant to capitalize on MIT's strength in biomedical instrumentation and electrical engineering, people like physicist George Harrison, who designed spectroscopes for clinical-chemical analysis, and engineer John G. Trump, who built high voltage X-ray generators for hospital radiology units.[36]

Compton assumed that physicists and chemists would take the lead in creating the new profession of "biological engineering." MIT biologists were unlikely to change their ways, he thought, unless forced to by

[33]V. Hamburger, "Proposal for a program of joint research," 18 Jan. 1938, RF 228D 5.59.

[34]Compton to RF, 11 Feb. 1937, KTC 57.40. Servos, "The industrial relations of science."

[35]Weaver diary, 16 Oct. 1937; Gregg diary, 20 May 1941; both in RF 224D 1.13. Gregg to Compton, 26 Nov. 1941, RF 224D 1.17. Compton to RF, 11 Feb. 1937, and exhibit A, KTC 57.40.

[36]Compton to RF, 11 Feb. 1937, pp. 1–2, 36–37, exhibit A, p. 1; Compton, "Biological engineering," 1 Mar. 1938, exhibit D; Compton to Prescott et al., 4 Mar. 1936; all in KTC 57.40–41.

strong leaders imported from outside. The new degree program was in fact a general course in physics and chemistry, with special emphasis on biological and medical applications and on training in physical instrumentation. Compton's plans for new appointments included a physicist and an electrical engineer, organic and colloid chemists, and biologists with strong interests in physics and chemistry. "Biological engineering" was a mechanism for recruiting and training physical scientists for careers in the growing health industry. The new department would be a place where chemists, physicists, engineers, and biologists could work together on basic problems of biomedical science.[37]

It is easy to understand why Weaver was so excited by Compton and Bunker's scheme when he first was told about it in late 1937. Compton's plan, he realized, offered a way of getting around the problem that frustrated all his efforts to develop general physiology and biophysics: namely, the absence of recognized, stable careers for those with an intellectual interest in such interdisciplinary work. As he wrote Fosdick:

> Physics departments quite understandably prefer to hire and promote men who will enhance the reputation of the department among physicists . . . [and] are, therefore, very loath to take on men interested in applying physics to biological problems. Somewhat less (but still recognizable) prejudice exists . . . with the placing and advancement of biophysicists in . . . departments of physiology.[38]

The idea was to develop "a new combination of facilities—one could almost say . . . a new discipline and a new profession." It was, indeed, an institutional embodiment of Weaver's idea of "vital processes." Weaver had long hoped that some bold visionary would invent practical ways of encouraging transdisciplinary practice. Compton's scheme was "the first really important opportunity of this sort we have had."[39]

Weaver took an unusually active role in shaping the new program in 1938–1941. Compton was too bent, he thought, on keeping MIT's old-fashioned biologists out and on building up physics. Weaver was more concerned to keep physicists and electrical engineers from co-opting biology to their own parochial ends. In the matter of new appointments, Weaver challenged Compton's emphasis on physical scientists. He exhorted Compton to appoint prominent biologists "to minimize the danger that the biological fraternity [might] treat . . . 'biological engineering' in a technological institute as something possibly a little

[37]Compton to RF, 11 Feb. 1937, pp. 1–2, KTC 57.40.

[38]Weaver to Fosdick, 23 Feb. 1940, RF 224D 1.16.

[39]Ibid. Weaver to A. N. Richards, 29 Mar. 1940; Weaver to Compton, 2 Feb. 1940; both in RF 224D 1.16.

queer, a little strangely outside of their own historical framework of ideas and organizations." Compton resisted: the whole point was to go beyond biologists' traditional view of their field. Weaver thought that physicists and electrical engineers simply could be borrowed as needed from their own departments; Compton argued for permanent appointments for them within biology.[40] In the end the two agreed that the core of the department would be biophysics and general physiology rather than biology or physics.[41] It was an equal blend of Compton's local agenda and Weaver's vision of "vital processes."

Weaver also took an unusually active role in selecting a biophysicist and biochemist to head the new department, and his unique knowledge of the general physiology network proved most useful. He knew whose star was rising and whose was falling, who had entrepreneurial knack and who did not, who had a broad vision of biology and who were narrow specialists. He leaked the news to Compton that Detlev Bronk was movable, and told Bunker how Albert Szent-Györgyi might be wooed. He discouraged Compton from considering Princeton's E. Newton Harvey (no entrepreneurial flair) and Hudson Hoagland (too risky—he had gotten on the wrong side of Herbert Gasser in a physiological fight). And so on.[42] Weaver was surprised when Francis Schmitt showed an interest in the job and was delighted when an offer was made and accepted. He was careful to take no part in the actual negotiation, however, to avoid any appearance of helping one RF project raid another.

Schmitt quickly put his personal stamp on the new department. Public health was eliminated soon after his arrival, and sanitary engineering and food technology a few years later. The name "biological engineering" was dropped for plain "biology," and new appointments were made in biophysics, enzymology, and chemical microbiology. General physiology remained the core of the MIT department, and physical instrumentation the hallmark of Schmitt's style. (MIT's concentration of talent in instrument making was mainly what persuaded Schmitt to come.)[43] Weaver was of course eager to help and made a

[40]Compton, "Notes on conference" 15 Oct. 1937, KTC 57.34. Weaver to Compton, 2 Feb., 11 Apr. 1940; Compton to Weaver, 5 Feb., 14 Apr. 1940; all in RF 224D 1.16.

[41]Compton, "Biological engineering," 23 Feb. 1939, exhibit A, KTC 57.41. Compton to Weaver, 5 Feb. 1940, RF 224D 1.16. Three existing groups would be transferred to the new department (bioelectrical engineering, microbiology, and nutritional biochemistry); two new groups would be recruited (in electrophysiology and radioisotopes).

[42]Weaver to Compton, 15 Feb. 1940; Bunker memo, 8 Jan., 10 Feb. 1940; Weaver to Richards, 21 Mar. 1940; Compton to Weaver, 5 Feb., 12 Mar. 1940; Weaver diary, 9, 27 Feb. 1940; Hanson diary, 31 Oct. 1940; all in RF 224D 1.16. Bunker memo, 10 Feb. 1940, KTC 57.36.

[43]Compton, "Biological engineering," 23 Feb. 1939, exhibits B and D, KTC 57.41.

large grant to develop electron microscopy for research on the microstructure of nerve, muscle, and genes.[44] Weaver was delighted with what he saw on his first postwar visit to Schmitt's department: "The work in molecular biology [n.b.] involves the closest interlocking of a variety of experimental techniques."[45] Schmitt's updated form of general physiology came closest to realizing Weaver's ambition to give institutional shape and stability to this core field of "vital processes."

Unsuccess Stories: Cytologists and Others

Schmitt was the exception among Weaver's American projects in general physiology, however. Most practitioners were devoted to a narrowly specialized technique or problem and lacked entrepreneurial know-how. Institutional support was hard to arrange for a field that was not a regular academic department. However, Weaver's failures reveal better than his successes the limits of discipline building in a field like general physiology.

Take the case of Kenneth Cole, a talented experimental physicist and virtuoso in the precise measurement of impedances in nerve fibers. Ambitious to make his mark in medical biophysics and well placed in the prestigious department of medicine at Columbia, Cole easily got Weaver's notice and support. Yet the project was a bust. Cole never expanded beyond his highly specialized line, and his blunt refusal to teach or even learn physiology alienated him from his medical sponsors. Instead, he badgered Weaver to set him up in a "rather pretentious development in biophysics" in Columbia's Department of Physics. Weaver pointed out to Cole that success in biophysics depended on combining the physical and physiological sides equally, but to no avail. Cole's research suffered from his self-imposed isolation. He accumulated exquisitely accurate data but without any real sense of their biological meaning. Such behavior was precisely what kept biophysics from becoming a real discipline, Weaver thought.[46]

Cytology and anatomy were other fields in which Weaver saw opportunities for general physiology but which failed to live up to expecta-

---

Bunker, report, 17 May 1941, RF 224D 4.35. Weaver diary, 27 Feb. 1940; Compton to Weaver, 12 Mar. 1940; Miller diary, 31 Oct. 1940; all in RF 224D 1.16. Bunker to Compton, 15 Apr 1939; Compton to Bunker, 18 Apr. 1939; both in KTC 57.35.

[44]Schmitt to Hanson, 22 May 1941; Schmitt, "The role of the electron microscope," 17 May 1941; Bunker to RF, 17 May 1941; Hanson diary, 3 Feb. 1941; all in RF 224D 4.35.

[45]Weaver diary, 14 Nov. 1946, RF 224D 2.19.

[46]Weaver diary, 24 Nov. 1937; "Appraisal," Dec. 1938; Bronk to Hanson, 4 Oct. 1935, 20 Oct. 1941; all in RF 200D 133.1650.

tions. These were large and powerful disciplines in the United States, both in departments of biology and in medical schools. There were strong movements in both to assimilate new experimental and quantitative methods, led by people eager to shed the stodgy image and routine service roles of classical anatomists. Experimental ideals, however, did not mean that cytologists became cell physiologists; more usually it led them to focus on the study of cellular fine structure. Improving microanatomical methods was the fashionable research front in the 1920s and 1930s.

Judging from the people Weaver dealt with, experimental anatomy and cytology attracted enthusiasts who promoted one method or instrument with a single-minded, messianic zeal. Edmund V. Cowdry, Gordon H. Scott, and Robert Chambers were anatomists with Ph.D. degrees who found employment in medical schools as professors of cytology or histology. All were one-idea men who claimed to have the key to cell biology. Cowdry and Scott were advocates of microincineration, a method of obtaining local concentrations of inorganic elements by measuring the intensity of absorption spectra from tiny areas of incinerated tissue slices.[47] Chambers was the inventor and promotor of the micromanipulator with which it was possible to dissect out organelles of individual cells. All were more interested in perfecting and popularizing their techniques than in solving biological problems.

Cowdry and Chambers were almost stereotypical inventors: suspicious and combative, demanding and complaining, yet almost devoid of practical entreprenurial know-how. They gave Weaver no end of trouble, which he suffered with almost endless patience. Chambers worked at City College in New York, only blocks away from the foundation's offices, and seemed always to be knocking on Weaver's door with arguments why Weaver should set him up in an "Institute of Cellular Physiology." Often he required Weaver's aid in managing his complicated relationships with deans and potential or wavering patrons.[48] He never seemed to know what exactly he wanted or how to get it, even when Weaver coached him in framing a proposal and using the RF's pledge to lever matching funds from other foundations. In the end, Weaver simply stepped in and did all the fence mending and arm twisting himself, a practice he usually tried hard to avoid.[49] But he had a soft spot for unworldly enthusiasts like Chambers, and for such people he was willing to risk becoming too personally involved.

[47]Scott proposal, 21 Mar. 1936, RF 228D 4.53.

[48]Weaver diary, 16 Apr., 27 Sept. 1935, 25 May 1937; Hanson diary, 15 May, 16 Nov. 1937; Chambers to Weaver, 7, 11 Dec. 1937; all in RF 100A 104.1266–1267.

[49]Weaver diary, 24 Feb., 17 Mar. 1938; Weaver to Harry Chase, 18 Mar. 1938; all in RF 200A 104.1267–1268.

Chambers was too much the individualist and one-idea man ever to lead an institute of cell physiology; Weaver never took his pipe dream seriously. But Weaver shared Chambers's enthusiasm for physical instrumentation. He, too, believed that novel techniques were likely to produce new and unexpectedly fruitful knowledge. Even if Chambers himself was unable to identify significant biological problems with his machine, others might. Weaver helped Chambers because he wanted people trained to use the micromanipulator, and that Chambers could do, with a little help. So, too, with Cowdry and Scott, who also dreamed of founding a comprehensive program in cell physiology. Gordon Scott seemed to have a broader vision than Chambers. With the help of local physicists he had acquired a spectroscope and an ultracentrifuge and was hoping to build an electron microscope. It was, according to one visitor, a "well-coordinated attack on cell chemistry."[50] However, there was no comparable expansion on the physiological side of Scott and Cowdry's group, and when it turned out that the physicists were not as actively engaged as Scott had pretended, Weaver and Hanson lost interest in expanding Cowdry's project. Cowdry complained bitterly to Weaver that he got no grants while others did (Schmitt, he probably meant); but his grant was not renewed.[51]

Limits of Discipline Building

Of all the biological sciences, general or cell physiologists seemed the closest to Weaver's ideal practitioners of "vital processes." They had the strongest claim to key problems of biological structure and function, and seemed the most open to cooperation with physical scientists. Weaver had high hopes in 1934–1935, but these had largely faded by the end of the decade. Late projects lack the programmatic thrust of the first group. They were eclectic: some little more than individual grants-in-aid. With one exception (MIT), they were occasional grants to older practitioners, to recognize past contributions or to keep alive specialized technical skills.[52]

General physiologists did not become the biological core of a program in "vital process" because they were not a well-defined community. The lack of territorial integrity greatly complicated the process of making projects. Weaver and his staff found it almost impos-

[50]B. M. Patten diary, 17–20 Apr. 1935, RF 228D 4.53.

[51]Scott to Weaver, 8 Feb. 1936; Cowdry to W. M. Marriott, 21 Apr. 1936; Cowdry to Weaver, 22 Mar. 1935; Hanson diary, 15 May 1936; all in RF 228D 4.53.

[52]On Chambers, e.g., see "A unique laboratory in the city," Feb. 1938, RF 200A 104.1268.

sible to get consistent opinions about the quality of individuals' research, or even about what or who belonged in the RF's program. Almost no other group gave Weaver such conflicting advice. More than once he had tried to get "useful cross ratings of the general physiologists" but had been baffled: "[P]ractically every person on the list is spoken of in very high terms by some and in strongly adverse terms by others."[53] Dissension made it hard for Weaver to identify likely opportunities and sell them to the trustees, who liked unambiguous testimonials from undisputed authorities. The problem was compounded by the presence on the RF board of physiologists who were themselves partisans, like Herbert Gasser. (Gasser was the kind of physiologist who believed that biochemists dealt with "dead stuff" and that biophysical methods like X-ray diffraction would never be of any use to biologists.)[54] Discord among the experts lengthened the odds that a project would produce results that could be generally agreed to be fundamental—a necessity for Weaver's own credibility with Fosdick and the trustees.

Evidence of fragmentation and infighting is everywhere. Joseph Bodine exhorted Hanson that Chambers "is not a physiologist, never was a physiologist, and never will be a physiologist."[55] The field encouraged self-styled iconoclasts and gadflies like Lewis V. Heilbrunn, from whose critical lash only Francis Schmitt and Detlev Bronk escaped unscarred.[56] Heilbrunn in turn was widely disparaged for his antiquated notions of physical chemistry, and for having made his career not in research but in writing textbooks and nurturing his students—who were numerous and adoring.[57] Merkel Jacobs was another member of the older generation respected for his teaching but considered out of date scientifically.[58] Bodine ran afoul of many people, but especially Frank Lillie, who deeply resented Bodine's outspoken criticism of his work on hormones and development. (Bodine dismissed endocrines as a passing fad and argued that grasshopper eggs were better for research than mammalian embryos because they had no endocrines to "complicate" the process of development.)[59]

These feuds were also carried on among the experts—Lillie among

[53]Weaver to J. W. M. Bunker, 18 Apr. 1940, RF 224D 1.16.

[54]Weaver diary, 25 Sept. 1935, RF.

[55]Hanson diary, 27 Mar. 1939, RF 200A 104.1268.

[56]Hanson diary, 22 Mar. 1934, RF 228D 5.54.

[57]Hanson diary, 19 Mar. 1943; W. C. Allee to Hanson, 27 Mar. 1943; S. C. Brooks to Hanson, 2 Apr. 1943; C. W. Metz to Hanson, 21 Mar, 1943; N. Harvey to Hanson, 25 Mar. 1943; all in RF 1.2 241 1.1.

[58]Schmitt to Hanson, 1 May 1944, RF 241D 3.33.

[59]Weaver diary, 2 Nov. 1933, RF 218D 2.15.

them—who reviewed Weaver's program in 1938. In the general chorus of praise for Weaver's program, general physiology was one discordant note. Bodine's and Chambers's projects were two of the three projects singled out for public criticism, and Bodine's was the only one the panel pronounced unworthy. Lillie was Bodine's strongest critic, but the rest agreed that he was not a general physiologist of the first rank.[60] This was troublesome, because Bodine's grant was up for renewal, and Weaver was disinclined to drop one of his few plausible projects in biology. Aware of the personal and factional jealousies among general physiologists, Weaver and Hanson took pains to gather opinions about Bodine (many of them favorable), then decided to renew. It is likely, however, that the contentiousness of the biologists on the review panel was one reason why Weaver took so few new initiatives in general physiology after that date.

Weaver and Hanson were a good deal more broad-minded and generous than the general physiologists themselves. They knew and talked regularly with everyone and may have understood the complex web of personalities and factions better than any partisan. They could spot personal bias and were more tolerant of one-idea men. They valued the older generation of teachers for the quality of their disciples, even if their research was not a la mode. Both Lewis Heilbrunn and Merkel Jacobs received grants on the basis of their books, teaching, and professional service, despite lukewarm reviews of their research from the experts.[61]

Weaver could hardly have picked a more difficult field for an experiment in discipline building, but his problems were inherent in the strategy itself. Taking a whole discipline as his objective, Weaver was dependent on insiders for definitions of quality and significance. When insiders could not agree, Weaver could get no clear sense of priorities. There was little he could do to push strategic lines of research or assist individuals whom their colleagues might regard as outsiders or incompetents. The strategy of community development left Weaver at the mercy of disciplinary politics. In fact, his best results in general physiology tended to come out of projects in which biologists were not the principles. The chemist Thorfin Hogness, for example, began by exploring the biological uses of spectroscopy and ended up becoming a leading American practitioner of Warburgian

[60]Hanson diary, 3 Nov. 1939, 6–14 Nov. 1939; Hanson to Taliaferro, 25 Oct. 1939; all in RF 218D 3.19. "Report of the committee of review, appraisal and advice," 1938, pp. 29–30, RF 915 2.12.

[61]Hanson diary, 19 Mar. 1943, and correspondence (Heilbrunn), RF 1.2 241 1.1. Correspondence (Jacobs), RF 241D 3.32–33.

general physiology (more on him later). The odds were more in Weaver's favor when he did not have to pick future trend setters for a whole discipline.

The frustrations that Weaver experienced with general physiology were not peculiar to that field. It was the same story wherever he tried to systematically develop a discipline or specialty. No matter what the subject matter, disciplines seemed to be either too big and established, like biochemistry, or too underdeveloped and diffuse, like mammalian genetics. Weaver would have liked to develop mammalian genetics: halfway between basic biology and "psychobiology," it was neither too removed from practical concerns nor too close to risky areas like eugenics and human genetics.[62] It was relevant to Gregg's program but did not compete with him, and it seemed as if modest investments would have large effects. Weaver soon discovered, however, that the field was "surprisingly neglected," with "shockingly few" centers of training and practice. With Harvard's William E. Castle near retirement, there was no center devoted to research on mammals. There were exceptional individuals, like Sewall Wright (Chicago), Leslie C. Dunn (Columbia), and a few others in biology departments, but no centers like Rollins A. Emerson's group in maize genetics at Cornell or Morgan's *Drosophila* school and its many clones. It was, Weaver noted, "a miniature duplicate" of the problem that Gregg faced with psychiatry: before projects could begin, there had to be substantial investment in institutions, training, and infrastructure.[63] But large-scale institution building had been ruled out in the financial crisis of 1933–1935, and the system of individual projects was unequal to such a task.

Bio-organic, or natural products chemistry was a similar case, and a more serious one for Weaver than mammalian genetics, because in Weaver's view it belonged with general physiology at the core of the program of "vital processes." Bio-organic chemists had jurisdiction over the big problems of molecular structure and function in the same way that general physiologists had over the big problems of cell structure and function. In Weaver's unorthodox geography of science they were an essential part, with general physiology, of a transdisciplinary science of vital processes. Bio-organic chemistry was also, like general physiology, a German specialty, thanks to the special symbiosis between universities and synthetic chemical industries. In the German system natural products chemistry was the acme of organic chemistry—prestigious, well funded, and attractive to the most talented and ambitious structural organic chemists. Not so in the United States, where the

[62]"Agenda for special meeting, 11 April 1933, pp. 79–80, RF 900 22.168.

[63]Weaver, "Progress report," 16 May 1936, pp. 23–25, RF 915 1.8.

organic chemical industry languished in the shadow of the German patent monopoly until after World War I. Weaver thought he could help bring bio-organic chemistry in the United States up to par with Europe, but that proved an elusive goal. It was not just the shortage of talent, Weaver found, but the lack of a social system for recruiting talents to the field, and internal rivalries among chemists.[64]

In the United States, organic chemists were decidedly second fiddle to physical chemists. The brightest young recruits became physical chemists, attracted by the large market for teachers of general chemistry and the prestige of a theory-driven science.[65] From the top of the chemical pecking order, physical chemists looked down upon structural organic chemists as mere craftsmen and fact gatherers. One leading organic chemist, Frank C. Whitmore, told Weaver that "the physical chemists have had control so long that they have almost convinced everyone that there is no such thing as fundamental research in organic chemistry."[66] Weaver had ample opportunity to observe how the caste system was maintained. So few organic chemists received NRC fellowships, he complained, that many young talents thought it not worth their while even to apply. He suspected it was bias, and the fellowship board's secretary, the physicist Floyd K. Richtmyer, unwittingly confirmed his suspicion when he told Weaver that organic chemistry was a "card index" science that produced vast numbers of facts by "turning a crank . . . —how could it attract good minds?"[67] At Princeton, Hugh S. Taylor told him, organic chemists almost never got graduate fellowships, so the best students naturally opted for physical chemistry.[68] Again and again Weaver was told of major chairs in organic chemistry remaining vacant or being filled by people whom organic chemists complained were not members of their tribe.[69]

Further complicating Weaver's task was the fact that leading American organic chemists took little interest in the biological side of the field.

[64]Weaver, "Progress report," 14 Feb. 1934, pp. 9–10; "Progress report, 16 May 1936, pp. 10–11; both in RF 915 1.7–8.

[65]Servos, *Physical Chemistry from Ostwald to Pauling: The Making of a Science in America* (Princeton: Princeton University Press, 1990).

[66]Whitmore to Weaver, 17 May 1934, RF 200 170.2074. Hanson diary, 15 Jan. 1936, RF 200D 158.1940.

[67]Richtmyer to Weaver, 8 Feb. 1935; Weaver to Richtmyer, 22 Jan. 1935; both in RF 200 170.2075.

[68]Taylor to Weaver, 5 Dec. 1936, 16 Jan. 1937, 4 May 1938; Weaver diary, 22 Oct. 1936; all in RF 200D 154.1892–1893.

[69]Weaver diary, 4 Nov. 1937, RF 205D 5.78. Weaver to Richtmyer, 22 Jan. 1935, RF 200 170.2075. Weaver diary 8 Dec. 1933. RF. The British chemist Robert Robinson thought Americans underrated their up-and-coming talents. Weaver diary, 15 June 1936, RF 401D 38.498.

Those who did had strong connections with the pharmaceutical and food industries and worked on substances that had potential commercial value, like vitamins or hormones.[70] The chemistry of proteins, nucleic acids, hemoglobin, or chlorophyll—the problems most relevant to Weaver's program in "vital processes"—had no commercial value and were left to the Europeans. There was nothing in the United States that resembled the Anglo-German elite of bio-organic chemists, like Adolf Butenandt, Leopold Ruzicka, Max Bergmann, Robert Robinson, Ian Heilbron, and Alexander Todd.

American organic chemists saw Weaver's program as a chance finally to get resources long denied them, but they were not interested in working with biologists. Self-interest, sharpened by a sense of injustice, made them aggressive but not very canny or diplomatic grant-seekers. The defensiveness and partisanship of synthetic organic chemists like Roger Adams and Frank Whitmore made them, from Weaver's view, untrustworthy advisors. Thorfin Hogness warned Weaver that their advice would be "less useful than otherwise" and suggested he rely on Europeans or people in other disciplines, who had a less parochial outlook.[71] He was right. Seeking advice from Frank Whitmore on possible candidates for a new post in bio-organic chemistry at Caltech, Weaver was lectured on the importance of appointing someone "definitely identified with their group," and he came away determined to seek advice elsewhere.[72]

Lack of statesman-like advice from community leaders was even more damaging in the case of organic chemistry than with general physiology, because the RF trustees took more persuading that chemistry was a legitimate field for Weaver's division. They did not share Weaver's belief that bio-organic chemistry was an integral part of "vital processes," and were uneasy when Weaver talked about systematic development of a physical science in a program of "experimental biology." They willingly gave money to the occasional brilliant chemist like Linus Pauling, but regarded such grants definitely as "out of program."[73]

Most of Weaver's bio-organic grants in the 1930s were to European stars. In the United States, grants were made to Princeton, to refurbish Lauder Jones's laboratory, and to Caltech, where Noyes's dream of appointing an organic chemist was finally realized. Not coincidentally, both projects were engineered by distinguished physical chemists,

[70]Dean Stanley Tarbell and Ann Tarbell, *Essays on the History of Organic Chemistry in the United States, 1875–1955* (Nashville, Tenn.: Folio, 1986), chs. 15–16.

[71]Weaver diary 14 August 1937, RF 205D 5.78.

[72]Weaver diary, 4 Nov. 1937, RF 205D 5.78.

[73]Fosdick to Greene, 25 Mar. 1937; RF 915 1.2. Weaver, "Notes on report of NS committee of review," 8 June 1939, RF 915 3.26.

Hugh Taylor and Linus Pauling, who had a strong personal interest in organic chemistry. Weaver helped push the Caltech project in the bio-organic direction. He took a strong stand against a synthetic chemist of the Roger Adams type (too narrow, too commercial) and put Pauling in touch with the young protein chemist Carl G. Niemann, a protégé of Max Bergmann.[74] Weaver hoped that these two projects would become centers of research and training in bio-organic chemistry, as the MIT project was for general physiology. But they were hardly the large development that Weaver had envisioned for the core of his program in "vital processes."

General physiology and bio-organic chemistry were the two disciplines that Weaver tried hardest to develop in a systematic way. He succeeded here and there—at Stockholm, Paris, Iowa, Caltech, MIT—but, on the whole, Weaver's experience suggests that the strategy of discipline building was not well suited to the problems of late-blooming disciplines that had to grow up in the interstices of entrenched academic departments. Weaver had neither the resources nor the leverage to alter this established system of allocating responsibilities and resources, anymore than Rose and Trowbridge did in the 1920s. Nor was the individual project grant the appropriate instrument for making disciplinary peaks higher. Disciplines, with their particularistic interests and rivalries, were just not amenable to change through small grants and gentle nudging. Weaver's style of patronage worked better at the margins of disciplines than at their centers; it was better at creating new networks of communication than redesigning systems of recruitment and reward.

All in all, the 1930s system of patronage was better suited to developing individual careers. Even where he was trying to build disciplines Weaver had his greatest effect on the lives of a few individuals like Runnström, Ephrussi, Bodine, Schmitt, or Niemann. This was even more the case with the little group of physical chemists and physicists who found themselves, with Weaver's help, on the ground floor of "molecular biology."

[74]Pauling to Weaver, 1, 21 June 1937, 23 Feb. 1938; Weaver to Pauling, 16 June 1937; all in RF 205D 5.77 and 6.79.

CHAPTER TWELVE

Weaver and the Biomolecular Set

NO ONE CIRCA 1940 could have predicted the double helix, genetic code, and other revelations of "molecular biology." But looking back from that vantage point, Weaver felt that he had been present at the beginning of a great movement in science, helping to make it possible. The events of 1953 seemed to vindicate his vision of "vital processes."[1] How could he have felt otherwise? The odd sounding labels that Weaver espoused became names to conjur with in the era of molecular biology. Transdisciplinary careers became an ideal to which the brightest aspired (though they remained uncommon). Something like Weaver's idea of a multi-disciplinary field of "vital processes" was for some a reality. Weaver's judgment of people also seemed vindicated by events. The people he had gambled on in the late 1930s became recognized as major and minor prophets of molecular biology—William Astbury, the Svedberg, Linus Pauling, George Beadle, Dorothy Winch. In the late 1940s, the roster of Weaver's projects included most of the important people doing X-ray crystallography of macromolecules, Dorothy Crowfoot Hodgin and Max Perutz most notably, and many of the early workers in biochemical and bacterial genetics. Weaver was there at the beginning, and he was everywhere.

But why was he there, and what was he doing? What indeed were the protomolecular biologists doing there? We need to beware of hindsight. As Pnina Abir-Am has observed, the so-called structural school of research on macromolecules was less a "school" in the 1930s than a collection of people who worked on the structure of macromolecules for a variety of reasons in different disciplinary contexts. So, too, with the early practitioners of what came to be known as molecular genetics.

[1]Weaver, "A Quarter century in the natural sciences," RF *Ann. Rep.* (1958): 72. Weaver, *Scene of Change*, pp. 72–75. Weaver, "Molecular biology: Origin of the term," *Science* 170 (1970): 591–592.

It is all too easy to impose an order and historical continuity of which participants could hardly have been aware. The greater the success story, the greater the temptation to co-opt the past into invented myths of creation.[2]

It is no less a mistake to assess Weaver's record as patron as if it were simply a chapter in the constructed or the deconstructed histories of "molecular biology." What that term came to mean after 1953 was not what it meant to Weaver in 1938. We should not buy uncritically into a success story and laud Weaver for uncanny foresight; by the same token we should not, if we think molecular biology a story gone awry, impute to Weaver a bias for the physical sciences and a "reductionist" view of their relation to biology.[3] Weaver did not see the relation between physical and biological scientists as a one-way street. His vision of "vital processes" was more a holistic than a hierarchical view of science. Biophilosophical debates had nothing to do with Weaver and little to do, I think, with what was happening along the border between physical chemistry and biology in the 1930s. The question is, What brought so many "biological-molecular" people like Astbury, Wrinch, and Pauling to the problem of macromolecules circa 1930?[4] And how did Weaver get so intimately involved in their odd careers? Forget for a moment where it all led to: What did the principles think they were doing at the time?

I think we should see Weaver and the biomolecular set (if I may so call them) as players in a passing phase in the history of protein chemistry. My argument in brief is this: the research front of protein structure was changing around 1930 in such a way as to invite outsiders to think that they had the tools and vision to make fundamental contributions. For a few brief years, incentives to transdisciplinary poaching were unusually strong, and poachers were attracted. What attracted Weaver, however, was not the scientific game so much as the poachers themselves, whose bold disregard of discipline boundaries he recognized as exemplary of the kind of scientific careers and practices that he hoped to promote.

Until the late 1920s, the problem of the structure and behavior of

[2]Robert Olby, *The Path to the Double Helix* (Seattle: University of Washington Press, 1974), ch. 16. Abir-Am, "Themes, genres and orders of legitimation in the consolidation of new scientific disciplines: Deconstructing the historiography of molecular biology," *Hist. Sci.* 23 (1985): 73–117. Eric Hobsbawm and Terence Ranger, eds., *The Invention of Tradition* (Cambridge: Cambridge University Press, 1983).

[3]Abir-Am, "The discourse of physical power and biological knowledge in the 1930s."

[4]Astbury's adjectival "biological-molecular" seems more apt since the people to whom it applies were not biologists but chemists and physicists. Astbury to Tisdale, 13 Dec. 1937, RF 401D 46.596.

proteins had been the domain of organic and biochemists. Proteins (and nucleic acids) were believed to be aggregates of small molecules. At the same time, it was recognized that proteins, so ubiquitous in physiologically active tissues, must have special features of structure or shape that gave them the varied and remarkable properties possessed by enzymes, membranes, genes (nucleic acids were thought to be inert support for active proteins), muscle, and nerve. Organic and colloid chemists were encouraged to think that they might solve the problem of proteins by straightforward determination of their molecular structure or state of aggregation. In the late 1920s, however, it became apparent that proteins were macromolecules of definite but varied structure. That realization weakened the claim of organic and colloid chemists to ownership of the protein problem; it seemed unlikely that their methods would bear fruit anytime soon. At the same time physical chemists were tempted into the field. They possessed the physical tools for measuring the overall size and shape of macromolecules—the cause, it was presumed, of the remarkable physiological properties of proteins.

Encouraged by this presumption (and by ignorance of the difficulties), these outsiders sought shortcuts to big questions: not to the biochemists' goal of complete structural determination but to understanding the relation between macromolecular and physiological function. For a few years ownership of the protein problem was up for grabs, and new people perceived the problem in new ways. But only for a few years: by the early 1940s it was apparent that there were no shortcuts to the physiological behavior of proteins and nucleic acids. The problem reverted to its original owners, the protein biochemists, joined by X-ray crystallographers who now aspired to determine not just the molecular shape but the primary molecular structure of the protein chain.[5] Some of the poachers retired from the field in the 1940s, disabused of their hope that chemical physicists might possess the key to the relation between molecular shape and physiological behavior—the secrets of life. Others took part in the revival of that peculiar mode of practice in the 1950s following the discovery of the double helix and the mechanics of genetic replication.

It was not the protein problem per se that first attracted Weaver to the biomolecular group. He only gradually became aware of proteins as a unifying theme.[6] Weaver did not plan the biomolecular group of pro-

[5]Olby, *Double Helix*. Joseph Fruton, *Molecules and Life* (New York: Wiley, 1973). Fruton, "Early theories of protein structure," in P. R. Srinivasan, J. S. Fruton, and J. T. Edsall, eds., *The Origins of Modern Biochemistry: A Retrospect on Proteins*, vol. 325 (New York: Annals N. Y. Acad. Sci., (1979), pp. 1–20. Edsall, "The development of the physical chemistry of proteins, 1898–1940," in Srinivasan et al., *Origins*, pp. 53–76.

[6]RF *Ann. Rept.* (1973): 207–213; (1940): 198–209.

jects. He picked up Svedberg's grant from Gregg when Gregg had to concentrate on psychiatry. Pauling's project was left over from Mason's reconnoitering in chemical physics. Astbury and Wrinch were recruited not because they were working on proteins but because Weaver and Tisdale were eager to have a physicist and a mathematician on their roster. What caught Weaver's attention was not the biomolecular problem but the biomolecular *people.* He was attracted by the prospect of recruiting to the program in "vital processes" physical scientists who had an imaginative interest in biology. Only gradually did he perceive that they were working essentially on one problem.

Weaver and Tisdale identified with their biomolecular clients in an unusually personal way, which is not surprising given the similarity of their career experiences. They were all poachers and boundary crossers, Weaver and Tisdale no less than Astbury and Wrinch. Wanderers from mathematics and chemical physics joined paths for a time with a mathematician who had gone even further afield to become a patron of wanderers. The process of making projects was also more personal in such cases. Weaver had to ease problems of people who were making risky transitions from one career to another and inventing odd transdisciplinary roles. Weaver offered timely encouragement, access to new professional networks, and financial protection from disciplinary pressures that brought most mavericks back to the mainstream. He encouraged Pauling to give his imagination a looser rein, while doing his best to restrain Wrinch's unbridled enthusiasms. Astbury, Svedberg, Pauling, and Wrinch exemplified Weaver's ideal of "vital processes." Their vivid personalities caught his interest in the human side of scientific practice.

The common thread of the biomolecular projects is the similarity of careers: each of the four was expert in a physical (or mathematical) technique and, just about the time Weaver appeared on the scene, was becoming interested in extending their expertise to novel biochemical problems. It is a pattern characteristic of young careers. For Astbury (b. 1898), X-ray diffraction of wool fibers was his first independent research after an apprenticeship as a physicist in the school of William H. Bragg. Pauling (b. 1901) was about the same age but more precocious, a full professor (at age twenty-eight) with a remarkable record of achievement in the theory of chemical bonding and the precise measurement by X-ray diffraction of bond lengths and bond angles. Wrinch (b. 1894) left behind a respectable body of research on the theory of potentials and boundaries when she took up mathematical biology in 1933. Svedberg (b. 1884) was the oldest and most accomplished of the four, but his work with the ultracentrifuge from the mid-1920s was really a second career, quite unlike the traditional colloid chemistry of his first. Weaver did not induce any one of the four to take up unconventional research; but his support made it possible for all to pursue new careers with greater confidance and purpose.

Astbury: Fibers and Molecules

The seed of Astbury's interest in protein fibers was planted in 1926, when Bragg asked him to prepare some X-ray diffraction pictures for a popular lecture on "the imperfect crystallization of common things." It was a little trick Bragg used to get his young disciples interested in problems outside the mainstream of X-ray physics; it ensured that the major fields of application of X-ray diffraction were colonized with alumni of his school. Bragg pushed Astbury to accept the new lectureship in textile physics at the University of Leeds in 1928. And he put Tisdale on Astbury's trail when Tisdale came to inquire about physicists who might be recruited to the RF's new program in vital processes.[7]

Astbury had already worked out his basic techniques and ideas by the time Tisdale first came to call in May 1934 (the crucial papers appeared between 1930 and 1933). He had modified Bragg's ionization spectrometer for biophysical work, making it possible to take many measurements quickly and conveniently, though with less precision than physicists were used to. (Bragg's instrument, which was very precise but very laborious to use, was fine for physicists because they worked only on a few very simple inorganic substances.) Astbury had also hit upon the big idea that would inform all his subsequent work: namely, that protein molecules were flat, chainlike structures with open hexagonal folds resembling the closed hexagonal rings of polysaccharides. He arrived at this structure not by chemical means but by analogy with the reversible stretching of hair. Stretched and relaxed wool fibers, he had discovered, gave quite different diffraction patterns—the so-called alpha-beta transition of keratin. His picture of the molecular architecture of protein chains mirrored the extensibility of the wool fiber.[8]

Naturally Astbury was eager to see if his theory applied to other proteins, especially those with interesting behaviors—but that meant entering the alien realm of physiology. Just a week before Tisdale first visited him, Astbury and his friend John Desmond Bernal had completed a study of an enzyme (pepsin), and he was wondering how he might find out what made chromosomes extend and contract in

[7]Tisdale diary, 9–11 Mar. 1936, RF 401D 46.596. Olby, *Double Helix,* pp. 255–258.

[8]J. D. Bernal, "William Thomas Astbury 1898–1961," *Biog. Mem. Fell. Roy. Soc.* 9 (1963): 1–35, see pp. 3, 6–10, Olby, *Double Helix,* chs. 3–5.

mitosis. What really captured Astbury's imagination, however, was muscle—the most remarkable extensible fiber of them all. He had tried some years before to get X-ray diffraction patterns from intact muscle, but the long exposure to X-rays destroyed the sensitive tissues. Other researchers, too, had failed, but Astbury hoped that the alpha-beta theory would be a master key for deciphering the imperfect diffraction data. If muscle fibers contracted as hair did, then he had only to look for the telltale signs of the alpha-beta transition. Characteristically, Astbury was sure he was on the brink of an extraordinary discovery in physiology.[9]

Tisdale could hardly have appeared at a more opportune moment, especially since Astbury's newfound enthusiasm for physiology was outrunning his institutional support in the Textile School. Astbury's lectureship and laboratory existed only because of a substantial grant from the Worshipful Company of Clothworkers, who had been sold on the idea that basic research was essential to the long-term health of the woolen cloth industry. Astbury was paid to work on wool, and it would have been awkward for the university if he now switched to materials like muscle, nerve, or genes that were of interest only to academic biologists. It was a problem Tisdale knew how to solve. He proposed that the RF provide the means for Astbury to undertake a second line of work in fundamental biology, while the research on wool would continue with industrial and university support. Accustomed to acting as an intermediary, Tisdale arranged things with the university's vice-chancellor and with officials of the clothworkers company. Both expressed enthusiasm for his plan provided they did not have to pay for the new lines of research. In July, a modest grant from the RF provided funds for an assistant trained in physiology and for the development of X-ray instruments especially designed for work with intact muscle or chromosomes.[10]

Tisdale expected astonishing things, and he was not disappointed. Toward the end of 1934 Astbury and Sylvia Dickinson began to work with films of purified myosin, the chief protein of muscle, and to their delight discovered that these films, when stretched and relaxed, displayed the same diffraction patterns as stretched and unstretched hair. Steamed, the films contracted spontaneously, almost like muscle. Was it possible, Astbury wondered, that hair, with its sulfur cross-linkages, was a "vulcanized" muscle protein? Later he elaborated this idea into a

[9]Tisdale diary, 14 May 1934, RF 401D 46.594.

[10]Tisdale diary, 14 May, 29 June 1934; Astbury to Tisdale, 14 July 1934; Astbury, "Memorandum concerning proposed x-ray research on the molecular structure of biological tissues and related bodies," 18 July 1934; Tisdale to Weaver, 31 May, 18 July 1934; Tisdale docket, 24 July 1935; all in RF 401D 46.595. Olby, *Double Helix,* pp. 41–46.

general theory that the most primitive organisms began with only one protein, which during evolution became many, each adapted to a different specialized use.[11] For the next ten years Astbury surveyed proteins from various sources, with increasing support from the Rockefeller Foundation.

Weaver and Tisdale were amazed by Astbury's boundless intellectual ambition and his surprising lack of interest in institutional aggrandizement. He liked to work at the bench and never aspired to became the director of a large institute, as did so many of Weaver's successful clients. For Weaver that was ideal: "A. is most careful and conservative on the financial side," he noted, "but seems to have an imaginative and philosophical capacity relative to the scientific problems. He is clearly a man who can be trusted."[12] Weaver and Tisdale liked Astbury's unpretentious and spontaneous manner and the vivid metaphors he used to explain his ideas to nonexperts. They admired his combination of bold speculation and exact measurement. Though well aware that his ideas and popular language were controversial, they trusted Astbury's transparent enthusiasm and lack of guile. Meeting him for the first time in June 1936, Weaver noted in his diary: "Not all of A's ideas will prove to be correct, but he is a great pioneer and will open up leads of great importance."[13] The same might be said of Weaver's own idea of "vital processes," of which Astbury was an exemplary practitioner.

Svedberg: The Ultracentrifuge

Svedberg's conversion to protein research had occurred almost a decade before Weaver came upon the scene, in the early years of his work on the ultracentrifuge. Svedberg had begun to develop high-speed centrifuges in the hope that they might prove a convenient and exact method for measuring what was then believed to be the key to the properties of colloidal solutions: namely, the distribution of particle size. When it turned out, to his astonishment, that most proteins came in only one (very large) size, Svedberg realized that he had an instrument that was uniquely suited for the exact measurement of molecular weights, a central problem for protein chemists. The ultracentrifuge changed Svedberg's life. Colloid chemistry was left behind as he be-

[11] Tisdale diary, 1 Dec. 1934; Astbury to Tisdale, 9, 14 July 1934, 23 Jan., 4 Feb. 1935; Astbury, "Memorandum," 18 July 1934; Astbury, "X-ray researches into the molecular structure of biological tissues and related substances," to Tisdale, 10 June 1935; all in RF 401D 46.595. Bernal, "Astbury," pp. 10–11.

[12] Weaver diary, 9 May 1935, RF 401D 46.595.

[13] Weaver diary, 17 June 1936; Weaver to Tisdale, 29 July 1935; A. Muralt to Tisdale, 14 June 1936; all in RF 401D 46.596.

came fascinated, even obsessed with the design and construction of ultracentrifuges. Pressing for ever-higher speeds, Svedberg perfected a machine that took up a whole room with its complex optical system for continuously measuring sedimenting material. An oil-turbine drive was tested in 1926 and was in routine use by 1927.

Measuring protein size and shape proved no less captivating. By 1928 Svedberg was systematically combing the animal and plant kingdoms for proteins to put in his new machines. Not surprisingly, this biomolecular survey got him thinking about the possibility of a molecular measure of phylogenetic relationships. Respiratory proteins were especially well suited to his approach because they were easy to get in pure form and were considered highly species specific. He was surprised and delighted when most proteins appeared to be made up of different multiples of a unit protein with a molecular weight of 17,500—the so-called Svedberg unit. By 1929, Svedberg was convinced that he was on the track of a master key to understanding the molecular architecture of life.[14]

Weaver first met Svedberg during his tour with Lauder Jones in 1933 and noted with satisfaction that he had "now gone over entirely to biological problems."[15] The orbit of Svedberg's career was in fact at its biological apogee. He had published his last colloidal paper in 1926, and some two dozen papers on proteins had appeared since then, eleven in 1931 alone. His work on molecular evolution was leading him deeper into biology. Svedberg told Miller of plans to produce interspecies hybrids of marine invertebrates, to see if proteins "Mendelized."[16] He hoped to identify "mutations" caused by X-rays on chromosomal proteins by changes in the association of their molecular units. Biochemists had not yet begun to doubt the reality of the "Svedberg unit" (they would soon). Svedberg's enthusiasm for biology would never be greater. That, and his virtual monopoly in ultracentrifuge technology, made his project almost ideal for Weaver: "an unusual combination of mechanical genius, of patience and experience, and of imagination." Weaver also responded, as did everyone, to Svedberg's personal warmth and charm: "a most pleasing personality," he noted in his log.[17]

[14]Claesson and Pedersen, "Svedberg." Pedersen, "The Svedberg and Arne Tiselius: The early development of modern protein chemistry at Uppsala," in Giorgio Semenza, ed., *Selected Topics in the History of Biochemistry* (Amsterdam: Elsevier, 1983), vol. 1, pp. 233–281. Boelie Elzen, "Two ultracentrifuges: A comparative study of the social construction of artefacts," *Soc. Stud. Sci.* 16 (1986): 621–662. Olby, *Double Helix*, ch. 2.

[15]Weaver diary, 8 May 1933, RF 800D 7.75.

[16]Miller diary, 29 Jan. 1935, RF 800D 7.76.

[17]Weaver diary, 8 May 1933, RF 800D 7.75.

Weaver was also impressed by Svedberg's new laboratory, completed in 1931 with the aid of the IEB's 1928 gift. It was "a scientist's dream," Weaver marveled, lovingly designed, with "every detail anticipated and provided for [and] . . . done with a real artistic flourish," down to tasteful interior decoration. Weaver admired the four medium-speed ultracentrifuges that were used for precise measurement of molecular weights. (Four were needed because each measurement took up to three weeks of continuous operation.) And there were the two impressively large high-speed machines, each in its own well-shielded room (rotors were exploding with some regularity), and capable of producing forces up to one million times gravity. The experimental machines were used mainly to perfect design and engineering, but very high speeds were also useful for detecting small amounts of heterogeneous material. Weaver thought Svedberg's laboratory was just about the most interesting one he had ever been in.[18]

Unlike Astbury, Svedberg did not depend on the RF for support of the biological side of his work. There was no conflict with his official academic duties: the Nobel prize, a personal professorship, and support from local foundations left him entirely free to do whatever research took his fancy. There was little for Weaver and Tisdale to do but think up ways for Svedberg to spend more money and expand his operations. They hoped, for example, that he might cooperate with the local physicists but did not press the point when Svedberg genially intimated that the physicists at Uppsala were quite impossible to work with. The RF's support, regularly renewed, was designated equally for development of the ultracentrifuge and for work on the structure of proteins.[19]

Mathematical Biology: Dorothy Wrinch

Dorothy Wrinch also hit upon the protein problem serendipitously, in about 1933, while trying to apply potential theory to the boundary changes of chromosomes in cell division. Accustomed to vigorous self-help (she was a militant feminist), Wrinch began strenuous efforts to give herself an education in the life sciences: in embryology with the Paris biologists, experimental cytology with J. B. S. Haldane, physical and colloidal chemistry at University College London, and protein chemistry at Cambridge. Back at Oxford in 1935, she was hard at work

[18]Weaver diary, 8 May 1933; Svedberg, "Summary of recent work," 22 Jan. 1937; Jones memo, 18 Mar. 1931; O'Brien to Gregg, 27 Oct. 1931; all in RF 800D 7.75. Claesson and Pederson, "Svedberg," pp. 606–607.

[19]Weaver to Jones, 15 Feb. 1934; Tisdale diary, 14 Apr. 1934; Miller diary, 29 Jan. 1935; all in RF 800D 7.76.

on plant morphology when Miller came to recruit her to the program in vital processes. Weaver was especially keen to sign on mathematicians, and Wrinch was one of the very few who were in earnest about biology. Miller invited her to send him a proposal.[20]

Weaver was delighted. Though he did not know Wrinch personally, he had followed her work on potential theory since the mid-1920s. Like Miller, he was impressed by her "almost fanatical" efforts to learn biology and by her wide acquaintance with people in other disciplines.[21] The only impediment, it seemed, was that Wrinch was having to earn a living by routine tutorial work—nothing money could not remedy. Weaver proposed a grant of $2,550 a year for five years for research, travel, and networking. That caused some raised eyebrows among his fellow officers: a commitment of five years was almost unheard of at the time. However, Weaver gave his personal assurance that Wrinch was a good risk, emboldened, no doubt, by his knowledge of mathematics and his ardent wish that she would show everyone what mathematicians could do for biology. The similarity of their careers made Weaver identify in a very personal way with Wrinch's project—he told her so himself.[22]

Some of Wrinch's ideas were quite plausible: for example, the idea that cell division was driven by changes in the electrostatic potential of dipolar proteins in chromosomes and cell membranes—a neat application of F. G. Donnan's ionic theory. Potential theory was something Wrinch knew about.[23] The trouble began when Wrinch plunged headlong into fields in which she had no experience at all, like protein chemistry. It began almost immediately.

Early in 1936 Tisdale received an excited letter from Wrinch hinting of a great discovery, the soon-to-be notorious "cyclol" theory of protein structure. It had not taken Wrinch long to realize that the morphology of cell division was far less easy to describe in mathematical terms than she had expected. She was struck, however, by the ability of chromosomes to change from a highly extended to a highly compact form. Having recently learned of Astbury's work, she saw an analogy between

[20]Wrinch to Miller, 27 Feb. 1935; Miller to Hanson, 17 Jan. 1935; Miller diary, 28 Jan. 1935; all in RF 401D 38.497. Abir-Am, "Synergy or clash: Disciplinary and marital strategies in the career of mathematical biologist Dorothy Wrinch," in Pnina Abir-Am and Dorinda Outram, eds. *Uneasy Careers and Intimate Lives: Women in Science, 1789–1979* (New Brunswick: Rutgers University Press, 1987), pp. 239–354.

[21]Wrinch to Miller, 27 Feb. 1935; Tisdale diary, 15 May 1935; Weaver to R. G. Harris, 24 Dec. 1935; staff conference, 20 Sept. 1935; all in RF 401D 38.497.

[22]Tisdale diary, 15 May 1935; staff conference, 20 Sept. 1935. Weaver to Wrinch, 7 Oct. 1935, RF 401D 38.497.

[23]Wrinch to Miller, 27 Feb. 1935; Wrinch to Weaver, 3 June 1935; both in RF 401D 38.497.

chromosomes and Astbury's "supercontracted" keratin fibers. Searching for the most compact geometrical form that a protein chain could assume, she hit upon a symmetrical sheet of hexagonal rings formed (on paper at least) by the formation of covalent bonds between the amino acid units. Wrinch cast membranes and chromosomes to the winds and threw herself into making the "cyclol" structure into a general theory of proteins. "Marvelous to relate," she wrote Weaver, "it turns out to be a problem in the mathematical theory of groups!" She had already begun hounding Tisdale for a chemical laboratory and assistants to synthesize a cyclol under her direction, which Tisdale assured her was out of the question. Then Weaver got the treatment: "She continually raises the question of direct [chemical] assistance under her supervision—'Oh, Dr. Weaver! If I only had a handsome young man to twiddle the test-tubes for me' "[24] Wrinch abandoned mathematical biology for good only a few months into her five-year project.

Weaver and Tisdale responded cautiously, not wishing to encourage her incessant pleas to be set up as an organic chemist. Weaver did not like the way her ideas "seemed to be sending out sprouts," and the skepticism shown by chemists and physiologists was not reassuring.[25] But Wrinch just brushed aside Tisdale's suggestion that she stick to what she knew something about. Weaver too got nowhere when he tried to engage her in professional conversation about her ideas of a "new" topology. Meeting her for the first time in June 1936, Weaver saw a person not at all like the one he had imagined: "W. is a queer fish," he noted, "with a kaleidoscopic pattern of ideas, ever shifting and somewhat dizzying. She works . . . in the older English way, with heavy dependence on "models" and intuitive ideas. On the basis of her former work in mathematics and physics WW had anticipated a more systematic analytical procedure; and finds it very hard to judge W."[26]

Wrinch was never persuaded to return to mathematical biology. She retreated behind her mathematical expertise when pressed by critics, but in fact she was trying to do organic chemistry with little more than an eye for apparent structural analogies. Mathematics played no role in the cyclol theory. Wrinch was the only one of the biomolecular group who did not combine theory with exact measurement. For a few years

[24]Weaver diary, 15 June 1936; Wrinch to Tisdale, 10 Jan. 1936; Tisdale to Wrinch, 13 Jan. 1936; Tisdale diary, 9–11 Mar. 1936; Weaver to Tisdale, 24 Jan. 1936; Tisdale to Weaver, 11 Feb. 1936; all in RF 401D 38.498. Wrinch to Weaver, 18 Mar. 1935, RF 401D 38.497. Wrinch to Tisdale, 26 Feb. 1936, RF 401D 37.471.

[25]Tisdale diary, 9–11 Mar. 1936; Weaver to Wrinch, 27 Mar. 1936; both in RF 401D 38.498. Weaver to Tisdale, 24 Mar. 1936, RF 401D 37.471.

[26]Weaver diary, 15 June 1936, RF 401D 38.498.

she troubled the waters of protein chemistry, while Weaver and Tisdale puzzled over the wildly different assessments they were getting of her work and wondered how they were going to arrange a graceful exit.

Linus Pauling: Structural Chemistry

Linus Pauling was the last of the biomolecular group to take up proteins, almost four years after the RF first began to support his work on chemical bonding in 1932.[27] His distinctive approach to macromolecular structure came directly out of his earlier work on the chemical bonding of small molecules. The hallmark of Pauling's chemical physics was a symbiosis of theory and exact measurement. On the theoretical side he discovered ways of solving the Schrödinger wave equation for hybrid orbitals and bonds that had a mixed ionic and covalent character. On the experimental side, he deployed X-ray diffraction to measure bond lengths and bond angles in complex inorganic salts of simple organic compounds, and he developed the method of electron diffraction to measure the architecture of organic compounds in the vapor state. Exact data on interatomic distances aided the solution of complex wave equations, and theory in turn gave clues and shortcuts to interpreting complex diffraction data.[28]

Pauling's unusual combination of physics and chemistry, theory and exact measurement, enabled him to enter (to create, really) a new region of science. As he put it to Weaver:

> [T]here is a large and most important field in which men with the point of view of chemistry rather than physics will make a . . . sensible use of physical methods. Spending two years upon a detailed x-ray study of crystal structure of a single chemical substance . . . [is not] the proper activity of a chemist; he should be willing to work on a more descriptive level, make wider use of empirical relationships, and expand his study to a much wider range of substances.[29]

[27]Weaver, "Appraisal," Feb. 1939, RF 205D 5.81. L. Pauling, "A program of research in structural chemistry," Apr. 1932, RF 205D 5.70.

[28]L. Pauling, "Fifty years of progress in structural chemistry and molecular biology," *Daedalus* 99 (1970): 988–1014. Pauling, "Fifty years of physical chemistry at the California Institute of Technology," *Ann. Rev. Phys. Chem.* 16 (1965): 1–14. Gustav Albrecht, "Scientific publications of Linus Pauling," in Alexander Rich and N. Davidson, eds., *Structural Chemistry and Molecular Biology* (San Francisco: Freeman, 1968), pp. 887–907. R.J. Paradowski, *The Structural Chemistry of Linus Pauling* (Madison: University of Wisconsin Press, 1972).

[29]Weaver diary, 5 Apr. 1935; Pauling, "A program," Apr. 1932, p. 2, RF 205D 5.70.

Until his own students came of age he had almost no competition—hence Pauling's astonishing output of fundamental work in the late 1920s and 1930s.

To keep up that pace of production required organization and resources, and, like his mentor A. A. Noyes, Pauling soon revealed a marked taste for entrepreneurship. Visiting Pasadena in 1933, Weaver observed that Pauling's group was "essentially an institute, in the European sense, of theoretical chemistry." That is, it was an interdisciplinary research school embedded in an academic chemistry department, with a strongly programmatic agenda and a polyglot array of bench workers. There were full-time research associates in mathematics, X-ray and electron diffraction, and theoretical physics, and Pauling hoped to get an organic chemist. He had a special research budget of $4,500 from Caltech and nine postdoctoral fellows, including virtually every one of the new generation of theoretical chemists.[30] Although it was not "vital processes," Pauling's work was quite close in spirit to the trans-disciplinary ideals of Weaver's program.

At the time of Weaver's visit, Pauling was getting set to launch out from simple organic structures to complex molecules of biological interest—or at least he talked about it. The simpler problems of inorganic crystals and ionic bonding had become routine assignments for Pauling's numerous graduate students. The real challenge, he told Weaver, was large physiologically active natural products; in his application in 1932 he had even hinted at the possibility of tackling hemoglobin and proteins. No doubt Pauling was trying to catch his patron's eye with a biological lure; but there is also evidence that his interest in biology was genuine. In 1930, for example, he gave a seminar in T. H. Morgan's group on mathematical modeling of chromosome crossing-over.

In fact, Pauling's distinctive practice of chemical physics made the leap from simple to very complex organic molecules shorter than it might seem to traditional organic chemists or biochemists. Pauling's strategy was to first measure and calculate the structure of the simple molecules from which large structures were made up, then to infer from that knowledge the conformation and electronic properties of the complex molecules of interest to biologists. In this way he hoped to avoid the technical problems that made orthodox chemical approaches so slow and laborious. Proteins, for example, were too complex to determine directly by X-ray diffraction, but their structure might be

[30]Weaver diary, 23–25 Oct. 1933; Pauling, "A program of research," Apr. 1932; Pauling, "Brief account of research," 24 Oct. 1933; all in RF 205D 5.71. Servos, "The knowledge corporation."

deduced from knowledge of the geometry and electronic structure of amino acids. Likewise, the structure of the heme molecules of hemoglobin and chlorophyll might be deduced from knowledge of the resonance structures of pyrrole, four of which were combined to make the porphyrin ring. It is no accident that Pauling chose pyrrole for his earliest study of resonance hybridization.[31]

This leapfrog approach to bio-organic structure meant, of course, that most of Pauling's time and the RF's money were spent on small organic molecules of no immediate interest to biologists. Nor was there any guarantee that the work would ever be anything but pure structural chemistry. Weaver's other projects in basic chemical physics never developed beyond the simplest organic molecules: for example, George Kistiakowsky's project in chemical thermodynamics and Robert Mulliken's in molecular spectroscopy. Disciplinary rewards were powerful incentives for chemists not to drift toward biochemistry. It would have been reasonable to expect Pauling's project to be limited in the same way as the others in chemical physics. Yet it did not. In 1934 Pauling reported to Weaver that he was spending almost all his own time on porphyrins. The renewal of his grant in 1934 and 1935 specified that the money was for work on hemoglobin and other molecules of interest to biochemists, a significant change from the initial grant of 1932.[32]

Many things impelled Pauling's career on a biomolecular trajectory: Pauling's romantic fascination with the unsolved mysteries of physiological function; his desire to prove the power of chemical-physics methods in the hardest possible cases; the fluid, multidisciplinary subculture of Caltech—all were important incentives. So too, it appears, was Pauling's relation with Weaver and the practical necessities of funding a large and diverse research operation.

The fact is that Pauling was becoming dependent on Weaver to support his large research group, as local support dried up in the Depression. Millikan watched helplessly as $120,000 disappeared from Caltech's income when the Fleming Trust collapsed in 1932. After that Pauling saw very little of the $4,500 he had been getting from Millikan for his research. It was a desperate situation. Pauling tried to maneuver Weaver into making up the deficiency, but Weaver saw through that

[31]Pauling, "A program of research," Apr. 1932; "Brief account of research in chemistry," 24 Oct. 1933; both in RF 205D 5.70–71. Pauling, "Fifty years of progress," p. 1002. Judith R. Goodstein, "Atoms, molecules, and Linus Pauling," *Soc. Res.* 51 (1984): 691–708.

[32]Pauling to Weaver, 25 Sept. 1934; "A description of researches," in Pauling to Weaver, 26 Nov. 1934; both in RF 205D 5.73.

ploy and instead showed Pauling how the RF's pledge of $10,000 could be used to pry $5,000 out of Millikan.[33] This cost-sharing arrangement had far-reaching effects on Pauling's research agenda, because it entailed a functional division of labor. Weaver made it clear that the RF's support was specifically for the new, biomolecular part of Pauling's work—the work that he did himself. The pure chemistry, being now routine and not in the RF's program, was Caltech's responsibility. The foundation's sponsorship was a counterweight to the disciplinary forces keeping Pauling in pure chemical physics. It made the biomolecular side of Pauling's work an explicit obligation. Weaver never directly pressed Pauling to be more biochemical. He did not need to. The pressure was built into the patronage relation; it was inherent in being part of a program in "vital processes." With all the members of the biomolecular group, the RF's patronage gave added impetus to careers already moving in a more physiological orbit.

An unexpected discovery early in 1936 also drew Pauling deeper into protein work. As part of his study of the literature on hemoglobin, he repeated some well-known experiments on its reversible reaction with oxygen. To his surprise, he observed a dramatic change in the character of the bonding between the porphyrin ring and its central iron atom. Just as Astbury observed the contraction of myosin films and thought of muscle, so it struck Pauling that respiration and oxygen transport might be understood in terms of changes in the conformation of the protein or globin part of the hemoglobin molecule.[34] Pauling's enthusiasm for this line of thought was further aided and abetted by the protein chemist Alfred Mirsky, who was visiting Caltech and who also liked to connect molecular and physiological behavior: for example, protein denaturation in vision, muscle fatigue, fertilization, and cell division. A whole new program of research came into view.[35]

Early in 1937 Pauling set an assistant, Gustav Albrecht, to measuring the bond lengths and bond angles of amino acids. Characteristically, Pauling also tried to anticipate experiment by calculating the structure of polypeptide chains from what he knew theoretically about resonance in amide bonds. He was disappointed: the calculated structures failed to fit Astbury's X-ray data. Setting theory aside, he continued the slow

[33]Pauling to Weaver, 25 Sept., 26 Nov. 1934; Hanson diary, 14 Nov. 1934; both in RF 205D 5.73. Noyes to Weaver, 12 Apr. 1936, RF 205D 5.75.

[34]Weaver diary, 5 Apr. 1935, 6 Mar. 1936; Pauling, "Report," 10 Nov. 1936; Pauling to Weaver, 29 Jan. 1937; all in RF 205D 5.74–76. Pauling, "Fifty years of progress," p. 1002; and "Fifty years of physical chemistry," pp. 12–13.

[35]Miller diary, 25–27 Sept. 1935; Weaver diary, 6 Mar. 1936; Pauling report, 10 Nov. 1936; all in RF 205D 5.74–75.

process of measuring the conformation of amino acids and simple peptides.[36] Weaver urged Pauling to publish his ideas on protein structure anyway: "[I]t would certainly seem to me," he wrote, "that the very substantial amount of rigorous and experimental publication which you have to your credit would . . . bear the burden of a little speculation."[37] But Pauling thought not. Weaver was right, as it turned out: Astbury's interpretation of his diffraction data was at fault, not Pauling's theory.

Weaver among the Molecular Biologists

The biomolecular approach to protein structure reached its apogee in the late 1930s. Everything seemed to point to the existence of a few fundamental principles: the "Svedberg unit," Astbury's alpha-beta transition, Wrinch's cyclol. Pauling was convinced that proteins had a regular coiled shape, which could be deduced from a knowledge of hydrogen bonding. Biomolecular ideas invaded even the inner sanctum of protein chemistry when biochemist Max Bergmann proposed that numerical regularities in amino acid composition reflected periodicities in amino acid sequences.[38] Communication among the chief protein groups intensified as their labors seemed to converge. Rivalries also intensified as the stakes went up and evidence accumulated for and against the variant pictures of the mastermolecule.

Communication was especially vital to widely dispersed workers whose professional orbits did not normally cross. Weaver and Tisdale were well-placed to serve as communicators, and they were essential players in the biomolecular network. Tisdale got Svedberg to help organize an informal conference on proteins.[39] He persuaded Astbury to make a tour of protein laboratories in the United States, persisting gently when Astbury seemed more inclined to stay at his bench, and suggesting that Woods Hole would be a good place to go since all the biologists would be there. He put Astbury in touch with Ross Harrison, whom he knew was eager to apply Astbury's ideas to his own research in embryology. Weaver drew up Astbury's itinerary and supplied funds.[40] The RF paid for the 1938 conference on proteins at Cold Spring Har-

[36]Pauling to Weaver, 29 Jan. 1937, 29 Jan. 1938, "Report," 10 Nov. 1936, RF 205D 5.75. Pauling, "Fifty years of progress," pp. 1002–1004. Olby, *Double Helix,* pp. 272–278.

[37]Weaver to Pauling, 2 Mar. 1938, RF 205D 6.79.

[38]Fruton, *Molecules and Life,* pp. 163–165.

[39]Tisdale diary, 2 Nov. 1934, RF 800D 7.76.

[40]Tisdale diary, 8 Dec. 1936; Tisdale to Weaver, 18 Dec. 1936; Astbury to Tisdale, 29 Dec. 1936; all in RF 401D 46.596.

bor and for three European conferences designed to bring together physical scientists and biologists.[41]

From time to time Weaver entertained the idea of systematically developing an international "invisible college" of protein workers. J. D. Bernal raised the idea with Tisdale in 1936 or 1937, and Tisdale talked with René Wurmser about getting scattered groups to concentrate on one "standard" protein and a common agenda of key problems. (Wurmser liked the idea, but nothing was ever done about it.) Svedberg favored a big cooperative project to bring to bear every possible kind of physical instrumentation on the protein problem.[42] (It was a recapitulation of his grand attack on colloids in 1920.) Weaver was already supporting several international exchanges, two with Astbury's group and one with Wrinch, and thought it would be possible to develop an invisible college simply by being more systematic in grants for travel and cooperation. He took fright, however, when Jacque Errera, a member of Bernal's circle, suggested that Dorothy Wrinch might be assigned the role of traveling catalyst to keep the network together. That was too much for Weaver: "She would certainly travel, and . . . might be a catalyst," he mused, "but I am afraid that her influence might . . . be explosive rather than cohesive." He decided that it would be better if he and Tisdale took "a receptive and interested role rather than any very active part in the planning of any organized group."[43]

Tisdale and Weaver did far more than just make site visits, read reports, and arrange the details of grants and renewals. They were also confidants, sounding boards for new ideas, providers of encouragement and moral support. Tisdale had a particularly warm relationship with Astbury, sharing the ups and downs of his enthusiasms. In 1935 Astbury wrote elatedly of his new theory of protein denaturation, which led him to hope that the fibrous and globular proteins had a common structural basis. A year later he shared with Tisdale his despondency when Pauling and Mirsky published their own theory of denaturation and depicted his work as merely confirmatory.[44] Astbury was no less open about the ideas that went awry—like the "universal protein" that turned out to be an artifact (a mixture of cholesterol and proteins), or his theory that myosin would be found in the electrical organ of tor-

[41]C. H. Waddington, "Some European contributions to the prehistory of molecular biology," *Nature* 221 (1969): 318–321.

[42]Tisdale diary, 23 Dec. 1937, 21 Nov. 1938, RF 500D 12.127.

[43]Weaver diary, 2 Oct. 1937; Weaver to Tisdale, 14 Oct. 1937; Tisdale diary, 6 Oct. 1937; all in RF 401D 38.499.

[44]Astbury to Tisdale, 15 July 1935; Tisdale diary, 6 Dec. 1936; both in RF 401D 46.596.

pedo fish, not to contract, as in muscle, but to store electrical charge (it was not and it did not).[45] On the occasion of his election to the Royal Society in 1940, Astbury wrote to his friend and patron:

> I am not writing you this note for any bombastic reason—I just want to share it with you—that's all—for I doubt whether it would have happened without your good offices in securing the support of the Foundation. I'm eternally grateful for this, whether you like it or not! They call you "Father Christmas" on the Continent, and you were a real Father Xmas to me. So let me share this pleasure with you.[46]

Tisdale too felt that he had been an essential participant in Astbury's adventures.[47]

Weaver was somewhat more formal than Tisdale in his outward manner; personal gestures from him were more liable to be mistaken as signs of favoritism, which he had constantly to guard against. But with the biomolecular group he relaxed his guard: for example, when Pauling offended friends of Noyes, then ill and dying, and precipitated a confrontation with Millikan by too aggressively pressing chemistry's claim to a larger share of Caltech's resources. Weaver happened to be in Pasadena just as the crisis was coming to a head, and Pauling asked him for his candid personal opinion. Assured that Pauling really wanted advice, Weaver put most of the blame on Pauling's tactlessness and lack of consideration of Noyes's frailties. Pauling immediately agreed that Weaver was right and took his advice in making amends with Millikan. Weaver's intercession may well have headed off a serious rift among the Caltech chemists. Pauling was deeply appreciative of Weaver's blunt but constructive counsel.[48]

With the biomolecular group the social machinery of sponsorship became more informal and personal. As trespassers outside their own disciplinary communities, people like Astbury and Pauling came to rely on sympathetic outsiders like Weaver who shared their belief that trespassing was a good thing. Weaver and Tisdale too were trespassers among biochemists and physiologists—romantics who did not respect conventional jurisdictions. They shared with their clients an interest in opening new channels of communication; they, too, spoke a language that was designed less for fellow experts than to inform and engage out-

45Astbury to Tisdale, 24 Feb. 1937, 8 Dec. 1936, RF 401D 46.596.

46Astbury to Tisdale, 17 Mar. 1940, RF 401D 46.598.

47Tisdale to Astbury, 3 Apr. 1940; Astbury to Weaver, 22 Apr. 1940, and to Miller, 19 Apr. 1940; all in RF 401D 46.598.

48Weaver diary, 5 May 1936, 30, 31, Jan. 1937; Pauling to Weaver, 16 Apr. 1937; all in RF 205D 5.76.

siders. Weaver borrowed Astbury's colorful and popular idiom. Impressed by the "freshness, interest and sense of adventure" of one of Astbury's long letters to Tisdale, Weaver proposed to circulate it among the trustees as a "confidental monthly report."[49] He, too, used popular and metaphorical language to catch the imagination of the trustees and to unite the diverse experts who cohabited, not without friction, the domain of "vital processes."

Weaver and Tisdale were not detached observers of the rising controversies over "vulcanized muscle," cyclols, and primordial proteins. Partisans of no one group, they were certainly partisans of the biomolecular enterprise. As its sponsors, they too had a good deal to lose should it be discredited. They tended to discount criticism as expressions of personal or factional jealousies. When biochemist Ross Gortner told Hanson that the Svedberg unit was "a mystical number which does not exist in nature," Hanson noted that Gortner seemed to feel that the RF had helped European colloid chemists build magnificent laboratories while leaving Americans (i.e., Gortner) to make do.[50] Miller discounted criticism of Wrinch's cyclols, aware that there was "some little prejudice against DW as a woman."[51] He expressed disbelief when Bernal told him that all Svedberg's data on molecular weights was "absolutely meaningless." (Bernal hastened to add that he was not disparaging Svedberg's experiments, only his theory.)[52] Listening to Swiss organic chemists damn Wrinch up and down, Tisdale thought how natural it was for chemists to repel poachers from a field they claimed as their own.[53] Weaver and his staff always took expert opinion with a grain of salt, but with the biomolecular group they went beyond common caution. They were advocates. They liked the bold and imaginative style of the biomolecular practitioners: it was congruent with their own mentality as outsiders, and it embodied the organizational ideal of "vital processes." They tried, consciously or not, to protect this style of doing science from attack by specialists.

On rare occasions Weaver did get personally involved in biomolecular controversies. In 1938, for example, he persuaded Pauling to make a critical test of Wrinch's cyclol model. Evidence was mounting against it: in 1937 Maurice Huggins presented strong evidence for a hydrogen-bonded structure (Wrinch assumed covalent bonding), and Weaver knew that most organic chemists agreed. Yet other people whose judgment he trusted, like Thorfin Hogness, and biochemists Edwin Cohn

49 Weaver to Tisdale, 11 Feb. 1937, RF 401D 46.596.
50 Hanson diary, 17–18 Dec. 1935, RF 800D 7.76.
51 Miller diary, 16 Nov. 1936, RF 401D 38.498.
52 Miller to Weaver, 10 Nov. 1939, RF 800D 7.77.
53 Tisdale diary, 2–3 June 1937, RF 401D 38.499.

and Rudolf Peters, still seemed to think that Wrinch might be essentially right if wrong on details. So Weaver asked Pauling to make a thorough examination of the arguments for and against the cyclol theory. Pauling, he noted, was "one of the few persons who will not be in the slightest awed by W[rinch]'s facility in mathematics and mathematical physics." Pauling agreed to talk to Wrinch and write a report.[54] The meeting took place in early 1938, and the report was devastating, not just for the cyclol idea but for Wrinch's methods. Pauling discovered, as Weaver had, that Wrinch was less persuasive in person than in print. He noted that her fluency with chemical terminology hid a lack of real knowledge of chemical principles. She did not understand how structures were inferred from laboratory evidence. When pressed on the evidence for and against a hydrogen-bonded structure, she asserted that she was not interested in any particular structure but only in showing that structures could be deduced mathematically from the cyclol postulate.[55] Pauling's critique, published in 1939, exploded the cyclol and cut short Wrinch's brief, brilliant foray into the protein field.

Weaver was not surprised by Pauling's report. However, he and Tisdale remained surprisingly sympathetic to Wrinch. Even after the stormy protein conference at Cold Spring Harbor in 1938, when only Irving Langmuir came to her rescue, Weaver remained open-minded. He had to admit, however, that if her scientific stock continued to go down it would be impossible to ask the trustees for a renewal.[56] Doubtless Weaver's personal role in Wrinch's ill-fated odyssey from mathematics to biology to bio-organic chemistry made him reluctant to abandon her. Wrong or not, Wrinch represented an ideal of practice that was dear to him.

Weaver and Tisdale were well aware that practicing physiology or biochemistry without a license was a risky business, especially for people like Astbury and Wrinch who entered as would-be settlers upon other people's territory. (Pauling and Svedberg remained within the boundaries of physical chemistry and experienced little territorial conflict.) Tisdale worried that Astbury, aided and abetted by the RF, was fast making himself ineligible for chairs of physics, and urged Weaver to consider permanent support. (Astbury seemed less concerned than his anxious patrons: he did not want a chair in pure physics.)[57] A more

[54]Weaver to Wrinch, 17 Sept. 1937; Wrinch to Weaver, 24 Sept. 1937; Weaver diary, 1 Nov. 1937; Pauling to Weaver, 6 Mar. 1938; all in RF 401D 38.499–500.

[55]Pauling to Weaver, 6 Apr. 1938, and "Report on the work of Dorothy Wrinch," 31 Mar. 1938, RF 401D 38.500. Abir-Am, "Synergy or clash," pp. 263–269.

[56]Tisdale to Weaver, 25 Apr. 1938; Weaver to Tisdale, 16 Sept. 1938; Weaver diary, 14 Mar. 1938; all in RF 401D 38.500.

[57]Tisdale to Weaver, 7 Jan. 1937; Miller diary, 20 Oct. 1937; both in RF 401D 46.596.

serious worry was the violent reaction against Astbury by the British physiological establishment. Not surprisingly they did not take kindly to Astbury's analogy between the contraction of muscle and the stretching of wool, or to experiments which blasted living tissue with steam and X-rays. Astbury's naive enthusiasm rubbed them the wrong way, and they resented his posing as a physics missionary bringing the true religion to unenlightened physiologists. As he wrote to Tisdale, he had been "struggling against the 'status quo' for six years now" in the hope that "the medical sciences will in the end accept physics as an invaluable and loyal partner."[58] Tisdale worried that it would be years before Astbury had enough solid work to overcome physiologists' ingrained hostility.

These were important practical matters for Weaver and Tisdale. Tactless alienation of biologists would discredit the transdisciplinary idea of "vital processes" and make it harder for Weaver to justify grants to physical scientists. Also, territorial conflicts impeded cooperative projects. Astbury had a hard time finding biological coworkers. He ascribed it to "the usual biological prejudices," and in the case of J. B. Bateman, a student of Joseph Barcroft, he was probably right. Bateman seemed eager at first but then abruptly begged off, denigrating experiments with "dried or steamed-heated or hammered muscle" and demanding to have complete control of the physiological side of the work. Clearly someone had warned him off, probably Barcroft. It is also apparent, however, that Astbury did not make it easy for biologists to work with him; Sylvia Dickinson was the exception, and she proved hard to replace.[59]

Astbury's relations with the physiologists did improve. A. V. Hill, the one physiologist who had been sympathetic and encouraging from the start, thought Astbury had become more sophisticated. Tisdale thought it equally likely that it was "a crack in the biologists' armor, said armor being a hypothetical shell made up of the belief, too frequently expressed, that scientists without fundamental training in biology have nothing to contribute to biology."[60]

In Wrinch's case it was the chemists who were outraged, and their relations, unlike Astbury's with the biologists, went from bad to worse. Protein chemists like A. C. Chibnall were unbendingly hostile. Synthetic organic chemists were more open-minded, hedging their bets just in

[58]Astbury to Tisdale, 14, 9 July 1934; Tisdale to Weaver, 7 Feb. 1935; all in RF 401D 46.594–595.

[59]J. B. Bateman to Astbury, 7 July 1934; Astbury to Tisdale, 9, 14 July 1934, 29 Dec. 1936, 18 Jan., 4 Feb., 13 Dec. 1937; all in RF 401D 46.594 and 46.596.

[60]Tisdale diary, 9–11 Mar. 1936, RF 401D 46.596. Astbury to Tisdale, 14 July 1934, RF 401D 46.594. But see Bernal, "Astbury," pp. 7–8.

case the cyclol structure did turn out to be correct. Robert Robinson confessed to Tisdale that Wrinch had revealed a potentially important field for synthetic organic chemists that they, "to the[ir] shame," had overlooked. (He rushed an application to Tisdale for a large project in the synthesis of artificial polypeptides.)[61] The time of good feeling, however, was short. Wrinch was constantly in and out of Robinson's laboratory, trying to cajole his assistants into working on her ideas and sorely trying Robinson's patience. Weaver became aware just how bad things were when a member of Robinson's group, to whom he mentioned Wrinch's name, "blushed furiously and had to draw on the deepest reserves of his English character to keep from being profane . . . concerning what the Oxford chemistry crowd thinks of her."[62] When Robinson finally did test the cyclol hypothesis, with negative results, Wrinch just went to his Swiss rival, Leopold Ruzicka, claiming that Robinson had refused to help. That was the last straw.[63] Within a few years her relentless lobbying and self-deceptions had alienated even chemists like Max Bergmann, who had started out on her side.[64]

Weaver tried to make Wrinch see that she was harming her cause by propagandizing it so aggressively, but succeeded only in increasing her agitation.[65] He kept trying to persuade her to publish less, travel less, stop preaching, and concentrate on getting solid evidence for her ideas, but all in vain. Finally giving up on chemists (they were "too complicated"), Wrinch launched a campaign to convert physicists and physical chemists to her cause. But even well-wishers like Bernal were finally put off by her willful misuse of physical evidence. He feared that she would discredit the whole field of X-ray crystallography in the eyes of biochemists and physiologists: "[K]nowing her chemistry is wrong, they will conclude that X-ray analyses are somewhat crazy too."[66]

[61]Robinson to Tisdale, 3 Mar. 1936; Tisdale diary, 13–14 Mar. 1936; both in RF 401D 37.471. Weaver diary, 15 June 1936; Wrinch to Tisdale, 10 Jan. 1936; Tisdale diary, 9–11 Mar. 1936; Wrinch memo n.d. [Jan 1936]; all in RF 401D 38.498. Weaver diary, 14 Mar. 1938, 401D 38.500. Tisdale wondered if Robinson really was interested in natural proteins, but Weaver was willing to take the world's most distinguished bio-organic chemist on his own terms.

[62]Weaver to Tisdale, 14 Oct. 1937, RF 401D 38.499.

[63]Weaver diary, 30 Oct. 1936, 1 Feb. 1938; Robinson, "Report," 6 Dec. 1938; Tisdale diary, 3 May 1938; Robinson to Miller, 16 May 1939; all in RF 401D 37.471–473.

[64]Tisdale diary, 9–11 Mar. 1936, 2–3 June 1937; Weaver diary, 14 Mar. 1938; Miller diary, 16 Nov. 1936, 25 June 1937; all in RF 401D 38.498–500. Weaver diary, 1 Feb. 1938, 30 Oct. 1936; Tisdale diary, 3 May 1938; all in RF 401D 37.471–472. But for positive views of Wrinch, see Weaver memo, 23 Apr. 1937; Tisdale diary, 30 June 1937; Hanson diary, 6, 11–12 May 1937; all in 401D 38.499.

[65]Weaver diary, 30 Oct., 30 Nov., 14 Dec. 1936, RF 401D 38.498.

[66]Tisdale diary, 21 Nov. 1938, 6 Oct. 1937; Miller diary, 19 July 1937; Weaver to Wrinch, 17 Sept. 1937; all in RF 401D 38.499.

Wrinch just could not help herself: she believed in the power of publicity—a belief, Abir-Am suggests, that was born of her experience in the feminist movement. Her knack for vivid presentation always got the attention of the press but only enraged would-be colleagues.

Conflicts of the sort that Wrinch and Astbury experienced were constitutive of the biomolecular style of science, only aggravated by quirks of personality. Wrinch set off alarms only when she began to act like a chemist, not a mathematician. Astbury did the same when he began to work like a physiologist with intact muscle rather than muscle protein. Another of Weaver's biomathematical favorites, Nicolas Rashevsky, triggered the same violent reaction from the Chicago physiologists.[67] To chemists and physiologists it looked like these interlopers were trying to poach on their disciplines. People whose disciplinary turf was not threatened were far more tolerant of Wrinch's abrasive personality: physicists like William Bragg or Niels Bohr, for example. (Tisdale reported that the Copenhagen group was "much excited" by Wrinch's ideas, and Bohr had his mechanics build oversized cyclol models.) Wrinch was also well received by physical chemists, and Irving Langmuir made quite a nuisance of himself on her behalf.[68]

Such conflicts naturally worried Weaver and Tisdale, but also strengthened their resolve to support the unorthodox souls who dared to make transdisciplinary careers. Hence Weaver's extraordinary patience with Wrinch, long after he knew full well that she could only bring further embarrassment. Tisdale, being closer to Oxford, lived in fear of her visitations, but Weaver seemed gladdened when Wrinch let fly "a flash of her old time vigor."[69]

Biomolecular approaches to the protein question went out of fashion about 1940. No doubt the dramatic collapse of Wrinch's cyclol and Bergmann's periodic hypothesis helped discredit all such theories. Also, years of experimental and theoretical work seemed not to be bearing out the grandiose hopes for dramatic revelations. Svedberg was discouraged by the growing disbelief in his idea of a unit protein and by the failure of some experiments on protein "mutations." He turned to measuring the molecular weights of other kinds of macromolecules, like polysaccharides and synthetic industrial polymers.[70] In the long

[67]Weaver to Taliferro, 1 July 1936; Hutchins to Fosdick, 20 Dec. 1937; Talliaferro report, 18 Dec. 1937; all in RF 216D 8.108. Hanson diary, July 1939; Weaver diary, 5 Sept. 1939; both in RF 216D 8.110. Rashevsky had a grant-in-aid from the RF.

[68]Weaver diary, 30 Nov. 1936; Hogness report, 16 Aug. 1937; Langumir to Weaver, 24 Oct. 1938; Tisdale diary, 21, 28 Nov. 1938; all in RF 401D 38.498–500.

[69]Weaver diary, 14 Mar. 1938; Weaver to Tisdale, 16 Sept. 1938; both in RF 401D 38.500.

[70]Tisdale diary, 2 May 1938; Svedberg to Tisdale, 17 Oct. 1938; Svedberg to Weaver, 9 Oct. 1939, and report, 23 Nov. 1939; all in RF 800D 7.77.

run, Svedberg was more attached to his ultracentrifuge than to any of the problems for which it could be used. In 1937 Thorfin Hogness described him as "very much of a mechanical engineer" and noted that most of the protein work was done by assistants.[71] Astbury kept on with proteins, trying vainly to defend his original variant of the folded chain against a growing array of skeptics and competitors. But he seemed more interested in finding new and harder things on which to try his skills in X-ray diffraction: synthetic polymers, for example, and very complex biological materials like cell walls and chromosomes.[72] The war only accelerated a trend in Astbury's work back toward problems relevant to agriculture and industry.

Pauling, too, dropped the general problem of protein structure and, inspired by Karl Landsteiner, began to use serological reactions to measure the physiological size and shape of protein molecules. Although Pauling speculated about the mechanism of immunity, his work in immunochemistry is best seen as yet another technique for precisely measuring molecular dimensions–a biomolecular variant of Pauling's previous work with X-ray and electron diffraction.[73] The slow production of X-ray diffraction data on amino acids was continued by Robert B. Corey, but Pauling's invention of the alpha helix and pleated sheet conformations of proteins was a whole decade away. Of some hundred publications bearing Pauling's name between 1934 and 1940, only about eight had to do with proteins and hemoglobin, the rest were structural chemistry.

In sum, practitioners of the biomolecular approach to the protein problem had fallback positions in their experimental practices, and as the early promise faded those who could fell back. Wrinch persisted in promoting her cyclol, but the world was no longer listening. The RF continued to support these projects through the 1940s—even Wrinch's, with a terminal fellowship. It is a reminder that Weaver's deepest concern was not to solve particular problems but to help individuals who exemplified a cross-disciplinary style of scientific practice.

Meanwhile, protein chemists and X-ray crystallographers reclaimed

[71]Hogness report, 16 Aug. 1937, RF 800D 7.77. Elzen, "Two ultracentrifuges." See also RF 252D 1.10–14.

[72]Astbury to Miller, 12 Oct. 1939; Astbury to Weaver, 22 Apr. 1940; A. Melland to Weaver, 30 May 1940; all in RF 401D 46.597–598. Bernal, "Astbury," pp. 17–20. Astbury hoped, by comparing intact chromosomes with their component nucleic acid and nucleoprotein, to discover a relationship analogous to that between muscle fiber and myosin.

[73]Pauling to Weaver, 12 Jan. 1939, RF 205D 6.81. On the immunochemical project, see RF 205D 7.91–99. Lily E. Kay, "Cooperative individualism and the growth of molecular biology at the California Institute of Technology, 1928–1953" (Ph.D. diss., Johns Hopkins University, 1986).

the ground as the biomolecular set moved on to other things. As they saw it, the biomolecular episode of the late 1930s was an unfortunate aberration—"confusion," one participant later called it, that was quickly forgotten as chemists got on with the real work of chemical analysis.[74] This view assumes, of course, that the two groups were doing the same thing with protein structure, but that was not the case. The biomolecular set were not looking for shortcuts to chemical structures so much as for relations between molecular shape and physiological activity. As Astbury put it, molecular biology was "the search for the comparatively few essentials that must be involved in the building-up of living things."[75] The biomolecular set took the protein problem into a new domain of practice, in which chemical physicists did not need to be licensed by biochemists to work on biochemical problems, and where they did not need to see problems as biochemists did. Svedberg talked about a "new scientist": physical chemists or biologists who knew the other's field so well they could work in it instead of just borrowing from it.[76] That was also Weaver's ideal of "vital processes." In the mid-1930s the biomolecular approach to the protein problem was the most appealing and handy vehicle for realizing that ideal.

Conclusion

What then can we say about Weaver's role in the history of molecular biology? Less whiggishly, what was Weaver doing in the early stages of a movement that a decade later became "molecular biology"? He was, I have argued, promoting the careers of individuals who exemplified a style of transdisciplinary science: chemical physicists, whose investment in new kinds of instrumentation and research strategies took them unexpectedly across discipline boundaries. What drew Weaver to individuals like Astbury, Svedberg, Pauling, and Wrinch was the congruence of their careers with his own ambition to foster transdisciplinary research on "vital processes." In so doing, Weaver became a participant in a peculiar episode in the history of protein chemistry, in which the problem of protein structure was perceived in a new way, as a problem of biomolecular function. For a few years, a traditional field of biochemistry was open to other kinds of practitioners, outsiders with quite different conceptions of the problem and with novel experimental practices. Weaver's career as a manger of science happened to coincide with this episode, and his support for biomolecular poachers enabled them to

[74]Fruton, "Early theories of protein structure," pp. 13–14.
[75]Astbury, "Report," 27 Oct. 1938, p. 7, RF 401D 46.597.
[76]New York *Times,* 29 Oct. 1937, RF 800D 7.77.

pursue their transdisciplinary careers further than would otherwise have been possible. Sponsorship by the Rockefeller Foundation did not inspire but did intensify a biomolecular episode that has to be understood in context, rather than as a precursor of the more far-reaching events of 1953 and beyond.

A contextual analysis also illuminates Weaver's role in the prehistory of the genetics side of molecular biology: especially, his support of geneticist George W. Beadle, who with Boris Ephrussi and Edward L. Tatum invented key experimental practices of biochemical genetics. (The first, in 1934, involved the transplantation of genetically different tissues in *Drosophila;* the second, in 1941, was the genetic and biochemical analysis of mutants in the bread mold, *Neurospora.*)[77] Weaver was again at a strategic point and a crucial moment, not because he foresaw a molecular genetics but because his career and Beadle's were steered into similar orbits by an impulse to reconnect genetics with embryology, physiology, and evolutionary biology. Beadle was one of many geneticists in the 1930s who were trying to invent experimental ways to "put the gene back in the whole organism." Weaver, looking for projects in the field of genetics that exemplified the ideals of "vital processes," found what he was looking for in the work of people influenced by this holistic movement in biology.[78]

The connection between Weaver's program and Beadle's work in developmental genetics comes into focus if we look at Weaver's other projects in genetics. They seem at first sight a mixed lot, but on closer inspection reveal a pattern of investment in work that connected genetics with other biological disciplines. Göttingen zoologists Alfred Kühn and Ernst Caspari (Beadle and Ephrussi's chief competitors) received grants-in-aid for their work on transplantation in the meal moth, *Ephestia.*[79] Geneticists C. Leonard Huskins and Sheldon Reed got funds to apply the transplantation method to the development of coat color in mice (a far more convenient organism for developmental genetics than flies, Huskins tried to persuade Weaver).[80] Other, seemingly unrelated projects reveal a concern with genetics and the whole

[77]G. W. Beadle, "Recollections," *Ann. Rev. Biochem.* 43 (1974): 1–13. Beadle, "Genes and chemical reactions in Neurospora," *Science* 129 (1959): 1715–1719. Olby, *Double Helix,* ch. 8 Lily E. Kay, "Selling pure science in wartime: The biochemical genetics of G. W. Beadle," *J. Hist. Biol.* 22 (1989): 73–101. Kohler, "Systems of production: *Drosophila, Neurospora,* and biochemical genetics, *Hist. Stud. Phys. Sci.* 21 (1991), in press.

[78]Kohler, "Systems of production."

[79]Tisdale to Weaver, 18 July, 10, 28 Aug. 1934; Kühn to RF, 2 July 1935; Kühn to Tisdale, 8 Jan., 6 Apr., 14 Sept. 1936, 19 Oct. 1937; all in RF 717D 13.123.

[80]C. L. Huskins memo, 4 Apr. 1936, in Morgan to Weaver, 28 Apr. 1936; Huskins to Hanson, 17 Dec. 1937, 30 Mar. 1939; all in RF 427D 13.120. Board minutes 3629, 40072, RF 427D 13.118. Hanson to H. S. Jennings, 4 Jan. 1939, RF 427D 13.121.

organism: the genetics of disease resistance and physiological vitality and reproduction, genetic variability and geographical distribution, interspecies breeding and hybridization, and so on.[81] There is a similar pattern in Weaver's nutrition projects, many of which had to do with the effects of diet on the vitality and life cycle of animals.[82]

We are not accustomed to think of Weaver as an advocate of holistic biology, and in the philosophical sense he was not. He was a holist, however, when it came to scientific practice. He was drawn to projects that involved different branches of biology because they exemplified the transdisciplinary ideal of "vital processes." So many of Weaver's projects in genetics and nutrition had to do with the whole organism because such work reunited biological disciplines that had grown too specialized. That holistic impulse is what brought Weaver into Beadle's career just at the time that he was inventing experimental methods for doing developmental and biochemical genetics. Beadle wanted to put the gene back in the whole organism; Weaver wanted to reconnect the biological and biochemical disciplines.

Weaver and Hanson were quick to appreciate the significance of the door that Beadle had unexpectedly opened into the "ultra-modern field lying between genetics and biochemistry." Weaver hastily pushed through a grant for the *Neurospora* project in 1941. It was a special case, he wrote Fosdick, because Beadle's discovery had "resulted, in a major part, from circumstances which we helped to create" (through grants-in-aid to Ephrussi and Tatum, and group grants to the Caltech and Stanford departments).[83] Weaver's support for Beadle's new line of work came at a critical moment. The phenomenal productivity of the *Neurospora* method, and its increasingly biochemical emphasis, was rapidly outgrowing Beadle's local resources, which came largely from the biology department. New funds earmarked for the *Neurospora* work enabled him to expand his group and hire a full-time biochemist. A

[81]Tisdale diary, 4 May 1939; E. W. Lindstrom to Hanson, 1–7 July 1939, 24 Apr. 1940; all in RF 219D 2.7. T. Dobzhansky to Hanson, 17 Jan. 1934, 1 Nov. 1935, RF 205D 7.86–87. Hanson diary, 1 Jan. 1936; R. Cleland to Hanson, 8 Dec. 1938, 28 Nov. 1929; all in RF 200D 142.1759. M. R. Irwin, "Genetics project," 8 Oct. 1935; Weaver to Miller, 1 Oct. 1935; Weaver diary, 7 July 1934; Irwin to Weaver, 11, 14 Feb. 1938, and proposal; all in RF 200D 161.1997–1998.

[82]Maynard report, 30 Nov. 1935; C. M. McCay report, 17 Apr. 1940; Hanson diary, 17 Dec. 1935; all in RF 200D 137.1687. H. C. Sherman report, 30 Mar. 1937; Hanson diary, 16 Dec. 1935; both in RF 200D 133.1643.

[83]Weaver to Fosdick, 24 Dec. 1941; Beadle to Weaver, 28 Nov. 1941; Hanson diary, 15–18 Dec. 1941; Beadle memo, 18 Dec. 1941; all in RF 205D 10.141. Weaver diary, 17 Feb., 14 Mar. 1941; Hanson diary, 25 Dec. 1941, 26 Feb. 1942; all in RF 205D 10.135–136.

"mutant hunting room" was soon in operation in which novel biochemical mutants were being mass produced.[84]

With Beadle as with Astbury, Pauling, Svedberg, and Wrinch, Weaver's timely intervention sustained a career in transit into new kinds of transdisciplinary practice. What attracted Weaver was not the problem of genes, proteins, or metabolic pathways, but the prospect of breaking open disciplinary constraints on careers and practices. Weaver was not trying to reduce biological ideas to chemistry and physics, or to limit biologists' practices to what could be done with physical methods. His vision of scientific patronage and practice was holistic and integrating. Whether or not the molecular biology of the 1950s was liberating or restrictive (probably, it was both), it evolved out of movements that were holistic and ecumenical.

Weaver's idea of "vital processes" remains the key to understanding his role in the prehistory of molecular biology. Weaver envisioned a realm of scientific practice in which chemists, biologists, physicists, and mathematicians could pursue problems of "vital processes" beyond the limits of what any one discipline would normally permit. It was a personal vision, but it derived from a tradition of transdisciplinary, programmatic research that flourished in the land-grant universities of the upper Middle West, with regional variations at Caltech, MIT, and wherever there was sponsored research. It was, indeed, another of the varied manifestations of the holistic, integrating trend that was a hallmark of science between the wars. Weaver did not know developmental genetics or macromolecular chemistry well enough to foresee where they might lead. But he did see, in the activities of people like Beadle, Astbury, and others, a kind of career and a mode of practice that he was sure would pay dividends. Weaver did not foresee what "molecular biology" would become—how could anyone?—but he was a midwife at what was later seen to be its birth.

[84]Beadle to Hanson, 24 Feb. 1942, 3 Apr. 1943; Beadle, "Progress report," 24 Sept. 1942; all in RF 205D 10.142–143. Beadle, "Genetics and metabolism in Neurospora," *Physiol. Rev.* 25 (1945):643–663. Kay, "Selling pure science."

Chapter Thirteen

Instruments of Science

Most chemical physicists, of course, had neither the capacity nor the desire to do biochemistry or physiology themselves. Pauling and Astbury were the rare exceptions, and had Weaver dealt only with people willing to make major changes in their careers, he would not have had many in his program. In fact, the roster of projects involving physical chemists and physicists was a very long one (fig. 13.1). Most involved collaboration with biochemists or physiologists, and most strikingly, almost all were organized around a laboratory technology, often of a novel sort. The record is remarkable: for about a decade there was hardly any major new instrument that Weaver did not have a hand in developing. In addition to the ultracentrifuge and X-ray and electron diffraction, the list includes electrophoresis, spectroscopy, electron microscopy, heavy and radioactive isotopes, and particle accelerators.[1] These were probably the most systematic and productive group of projects in Weaver's program, even though they involved the greatest proportion of people who were not biologists. Did Weaver have a hidden agenda? Was his early training as a mechanical engineer breaking out? Why instruments?

The answer lies in the process of grant-making: Weaver's chief aim was to get physical and biological scientists to work together on vital processes, and it turned out that instruments were a highly practicable and fruitful vehicle for doing that. Instruments defined a relatively straightforward division of labor and did not require that either party be deeply versed in the other's arcane skills. Physicists did not need to go back to school in physiology to apply their know-how to physiologi-

[1]Kohler, "Rudolf Schoenheimer, isotopic tracers, and biochemistry in the 1930's," *Hist. Stud. Phys. Sci.* 8 (1977): 257–298. Lily E. Kay, "The Tiselius electrophoresis apparatus and the life sciences, 1930–1045," *Hist. Phil. Life Sci.* 10 (1988): 51–72.

Figure 13.1. Rockefeller Foundation projects in biophysics to 1945. Graphics courtesy of Jack Kohler.

cal problems. Biologists did not need to be experts in electronics and machine design to get access to new laboratory technologies. Instruments made it easier to bring together people who did not want to venture far or for long outside the boundaries of their own disciplines. Good organization and communication were essential for a project to work, as we shall see, but drastic changes in career strategies were not. Instruments made possible a limited partnership based on an exchange of services, in which many people were able to participate. Weaver did not set out to systematically develop new instruments. It was just that instrumentation projects were eminently doable.

To understand how these projects worked we need to look at how new instruments are invented and introduced into practice. Instruments, like scientific careers and problems, have life cycles, and the social organization required to make effective use of new instruments changes as arcane and unfamiliar gadgets become routine fixtures of laboratories. Often new instruments are created to solve some small, specialized problem, or simply in the hope that important uses will turn up. The successful few are eventually produced commercially for routine use in a wide variety of fields. In between there is a period of gradual improvement and spread through informal learning of craft skills, as people learn by experience what new instruments are and are not good for.[2] It is in this middle stage that transdisciplinary collaboration becomes essential, and that instruments themselves become an incentive for physical and biological scientists to venture outside their disciplinary subcultures.

Just as the protein problem, in one stage in its life cycle, drew chemical physicists out of their accustomed orbits into a novel kind of biomolecular work, so for a few years physical instruments drew experimental physicists into cooperative projects with biomedical scientists.

This same middle period in the development of instruments was also a window of opportunity for Weaver. Although he believed that physical instruments were an inherently good thing for biologists because they fostered quantitative habits, that by itself was not a sufficient reason for him to invest in them. The timing was crucial. It was specifically the middle phase in the life cycle of new laboratory technologies that offered a role for Weaver because it was then that the collaboration of

[2]Yakov M. Rabkin, "Technological innovation in science: The adoption of infrared spectroscopy by chemists," *Isis* 78 (1987): 31–54. Peter Galison, "Bubble chambers and the experimental workplace," in Peter Achinstein and Owen Hannaway, eds., *Observation, Experiment, and Hypothesis in Modern Physical Science* (Cambridge, Mass.: MIT Press, 1985), pp. 309–373. Timothy Lenoir, "Models and instruments in the development of electrophysiology, 1845–1912," *Hist. Stud. Phys. Sci.* 17 (1987): 1–54.

physical and biological scientists was most essential. Weaver was hesitant to get involved in developing an instrument too early, before its payoff for biologists had been proved. And he lost interest fast when instruments became commercialized and routine. Only for a few years were instrument projects exemplary of the ideals of "vital processes." And only in those years were they congruent with the social process of making grants. If Weaver invested too early in the life cycle of an instrument, he ran the risk that physicists would use RF money to do physics. If he invested too late in the cycle, he ran the risk of helping biochemists and biologists do routine production with standard methods. With the trustees looking over his shoulder, Weaver could not afford even the appearance of investing systematically in pure physics or in instruments for their own sake. The optimal time to invest was when applications to biology were certain but not yet realized, when doing research could be combined with developing new instrumentation. The character of the work was crucial, not the instruments per se. Weaver's faith in instruments and quantitative methods was abiding, but the moments when it was practicable to organize projects around instruments were fleeting.

It is an interesting question (to answer it would take another book) why so many new instruments appeared in 1925–1935 that were adaptable to biomolecular problems. Probably this clustering was a consequence of heightened activity on the border between physics and chemistry; it is no coincidence that the new discipline of "chemical physics" was also taking shape in the early 1930s. Most of the new instruments were devised for research not on biological materials but on the solid state: colloidal particles (ultracentrifuge), crystals and metals (X-ray diffraction), solid surfaces (electron microscopy), solid-liquid interfaces (radioisotopes). Physical chemists were moving beyond classical work on solutions toward problems closer to quantum physics (low-temperature research, e.g., or quantum chemistry). So too in physics, interest in the interaction of radiation and matter was extending out from atomic structure to the structure of solid materials and intermolecular forces. New research fields along the boundary between chemistry and physics were being rapidly occupied in the late 1920s, and in the process many new kinds of instrumentation were invented or adapted to molecular phenomena. It was a short step from there to biomolecular problems.

For some physical scientists, adapting new instruments to the life sciences was nothing new. Colloid chemists, for example, had always seen the biomedical sciences as an important source of problems and allies. Their strategy of discipline building in the early 1920s was to make themselves and their methods indispensable to as many established fields as possible, especially medicine. For mainstream chemists and physicists, changing job markets created new incentives to look beyond their disciplinary boundaries. The boom and bust in the industrial market for chemists after World War I inspired a deliberate campaign by the American Chemical Society and the Chemical Foundation to develop the market for chemists in the medical sciences. A decade later the American Physical Society and the new American Institute of Physics followed the chemists' example when growing industrial markets for Ph.D. physicists stalled in the Depression. Chemists' and physicists' growing interest in biology in the 1930s was one of the results of a decade of organizing and consciousness raising.[3]

These movements primarily involved experimentalists and analysts. The brilliant intellectual style of quantum theoreticians like Erwin Schrödinger has overshadowed this broader and perhaps more significant movement of experimentalists from chemical physics into biology and medicine. Experts in optics, electronics, instrument design, and precision measurement looked to the biomedical sciences for practical, not cultural, reasons—for jobs and research funds, not philosophy and quantum conundrums. Weaver's own perceptions about physics and biology were undoubtedly also shaped by the propaganda campaigns of the 1920s. As an applied mathematician, he had himself been a part of the movement to make physics less exclusive and more open to transdisciplinary connections. In the 1930s, his program in vital processes took advantage of and sustained the flight from industrial to biomedical uses of physical methods. Technical and social trends thus synchronized the life cycles of a generation of laboratory technologies and the careers of a generation of people who invented, used, and sponsored them. Weaver appeared on the scene in the role of patron just when many new laboratory technologies were emerging out of pure physics into the realm of biomedical application, and just a few years before they went into commercial production in the 1940s.[4] In this interval, biologists still needed physicists to build instruments and keep them in running order, modify them for biological materials, and make sure they were used safely. No less did physicists need biomedical problems to keep students and graduates employed and to provide grist for the process of improving designs for mass production. For a few years incentives for collaboration were at a maximum, and collab-

[3]Spencer Weart, "The physics business in America, 1919–1940: A statistical reconnaissance," in Nathan Reingold, ed., *The Sciences in the American Context: New Perspectives* (Washington: Smithsonian Institution Press, 1979), pp. 295–358. David Rhees, "The chemists' crusade: The rise of an industrial science in modern America, 1907–1922" (Ph.D. diss., University of Pennsylvania, 1987).

[4]Rabkin, "Technological innovation."

oration made instruments ideal vehicles for Weaver's promotion of transdisciplinary scientific practices.

Weaver and other private patrons were indispensable players in this stage in the development of laboratory technologies. As instruments outgrew the disciplines in which they were originally developed, instrument builders outran their intramural funding. Departments of physics or chemistry were not so altruistic as to subsidize their biological brethren. It was an ideal opportunity for an outside patron whose raison d'être was to expedite what disciplinary territoriality impeded. Extraterritorial funds enabled physicists to do work not directly relevant to (and fundable by) their own departments. For those inexperienced in cooperative research, Weaver's broad experience in organizing such work was no less important a resource.

Weaver's transdisciplinary resources and organizational know-how became less crucial as the new instruments were commercialized. The whole point of commercialization is to make it easy for anyone to use instruments without having to find willing physicists or to organize complicated and expensive multidisciplinary teams. It is no wonder that Weaver lost interest when that stage was reached. Commercial instruments were no less productive of results, but they were no longer vehicles for new modes of transdisciplinary practice, and that to Weaver was what mattered.

Two groups of projects, in spectroscopy and in radioisotopes, reveal most clearly how Weaver's interest evolved with the life cycle of laboratory technologies. The spectroscopy projects were Weaver's first experiment in an activist style of patronage. Confidant in his professional knowledge, he recruited spectroscopists with little heed to the personal and organizational problems of cooperative research. With radioisotopes, in contrast, Weaver was surprisingly skeptical at first of their potential in biomedical research and suspicious of accelerator physicists and their professed interest in biology. Why would they prefer a service role in biology to being in the thick of the new physics? Why indeed, but why did Weaver think that spectroscopists were more sincere in their desire to serve? Their projects as it turned out, were among the least successful, while the radioisotope projects were among the most productive and innovative.

The answers to these questions become clear if we keep in mind how the uses and meaning of instruments change as the technologies evolve. The spectroscope was not central to a fashionable research front in pure physics when Weaver came on the scene. Accelerators and radioisotopes, in contrast, were essential for work in the new and highly productive field of nuclear physics. Accelerators were ambiguous machines: biomedical applications had to compete with research on nuclear reactions. Spectroscopes were unambiguously machines for working on other people's problems—hence Weaver's suspicion of the former and his uncritical adoption of the latter. His initial perceptions changed, however. As accelerators became larger and more powerful, there was a period of a few years in which making isotopes for biologists was as appealing to accelerator designers as problems in pure physics. Accelerators then became an apt vehicle for Weaver's program. It turned out, meanwhile, that the most productive uses of the spectroscope were not in complex biological problems but in simple chemical ones. Thus, laboratory technologies moved in and out of conjunction with Weaver's program as they moved through their life cycles, just as biomolecular careers criss-crossed Weaver's as they evolved.

Best Laid Plans: Spectroscopy

Of all the instruments that Weaver helped to promote, the spectroscope was the one in which he took the most personal interest. Spectroscopic biology was his first venture as a manger of science, and he pursued it with all the zeal of an enthusiastic beginner. He did not wait for promoters to come to him but systematically recruited spectroscopists, combing lists of NRC fellows for likely converts. He circulated a form letter to physicists, explaining that the Rockefeller Foundation was making a survey of optical methods in biology and medicine, and inviting proposals, statements of plans, and information that might be passed along to like-minded enthusiasts. Weaver and Tisdale went so far as to suggest particular problems that the RF would be interested in supporting, such as analysis of body fluids, vitamins, hormones, enzymes, and clinical problems.[5] None of Weaver's predecessors in the RF had been so bold.

Weaver also took steps to organize a loose network of researchers into a coherent field. He commissioned Alexander Hollaender, a bright young NRC fellow (later director of biology at the Oak Ridge National Laboratory), to compile a bibliography of spectroscopic techniques and biomedical applications. At first Weaver thought of Hollaender's bibliography only as a guide for his own personal use, but he soon realized that it would also be useful to researchers.[6] Like the bibliographies compiled by the NRC's division of physics in the 1920's, Hollaender's bibliography was meant to entice talent away from familiar lines into new productive ones by easing access to the latest literature. A few years

[5]Weaver to J. S. Foster, 31 Jan. 1934, RF 427 14.126.

[6]Weaver to G. Harrison, 27 Nov. 1934, RF 200D 142.1749. On Hollaender's bibliography, see RF 200D 156.1918–1924.

later Weaver engaged Thorfin Hogness to survey European centers of research in spectroscopy.[7]

It seems that Weaver meant the spectroscopy program to be a test case of the viability of his role as an activist manager of science. In presenting the first spectroscopic project to the board in 1934, he invited the trustees to approve the general principle that officers could take the initiative in defining programs and making projects.[8] He was taking the bit in his teeth, breaking out of the limits imposed by his own inexperience and the trustees' suspicions of officers who displayed a too independent spirit. He wanted the spectroscopic projects to set a precedent. He would in time learn to get what he wanted to more subtle ways.

It was not hard to guess why Weaver chose spectroscopy to be his test case. It seemed an ideal technology for bringing physicists and biologists together: familiar, reliable, and available in every physics department. Also, spectroscopists were already accustomed to a service relationship with theoretical physicists and chemists. Like analytical chemists they were used to making careers as analysts in industry and government laboratories. (The National Bureau of Standards and the oil refining industry were centers of applied spectroscopy.)[9] Above all, there was no prestigious research front to compete with extradisciplinary service as there was in other branches of physics. Arthur Compton told Tisdale that spectroscopy was unlikely to yield more fundamental insights and that it had become mainly a matter of filling in blanks in tables of data. For this reason, he thought, spectroscopists would do well to concentrate on biological and medical applications.[10] Thus, Weaver could be confident that spectroscopists would not use RF funds to bootleg pure physics.

Most of Weaver's recruits were well-known spectroscopists and had prior experience in biomedical work. John Stuart Foster, for example, had worked on lead poisoning with clinicians at McGill University.[11] L. S. Ornstein, the leading man in the quantitative measurement of line intensities, had been providing occasional services for clinicians before Tisdale sought him out at the University of Utrecht. Tisdale assured him that support would be forthcoming if he could find a suitable biologist to work with on some fundamental problem.[12] At the University of

[7]Weaver diary, 1 Aug. 1936; Weaver to Hanson, 2 Aug. 1936; all in RF 216D 12.168.

[8]Docket 34063, 16 Feb. 1934, RF 216D 12.166.

[9]Rabkin, "Technological innovation."

[10]Tisdale diary, 30 Nov. 1934, RF 216D 8.106.

[11]Hanson to Foster, 21 Dec. 1933; Weaver to Foster, 31 Jan. 1934; Foster to Weaver, 2 Feb. 1934; all in RF 427 14.126.

[12]Tisdale to Weaver, 12, 22 Oct. 1934; L. S. Ornstein to Tisdale, 2 Nov. 1934; Weaver to Tisdale, 14 Sept., 8 Nov. 1934; all in RF 1.2 650D unprocessed file.

Michigan, spectroscopist Harrison M. Randall was already working with clinicians to develop quantitative spectroscopic methods for measuring trace metals in blood. Weaver picked up their project with alacrity when Gregg ruled it insufficiently medical for his program.[13]

MIT's George R. Harrison was an especially important addition to Weaver's program. His laboratory was organized to provide routine analytical services to researchers in Boston's hospitals and medical schools. Harrison specialized in designing new instrumentation and procedures for spectroscopic analysis and had unequaled shop facilities. An ambitious entrepreneur, he wanted to make MIT a world center of applied spectroscopy. Annual summer conferences and short courses in spectroscopic technique attracted scientists from diverse fields and gave them practical skills in spectroscopic analysis. Weaver was an active participant in these conferences, learning the science and making contacts with leading practitioners. So when clinician Kenneth Blackfan asked Harrison to do the routine analysis for a study of iron in anemia, Harrison knew just where to turn for funds.[14] A project was quickly consumated.

In some cases Weaver took the initiative in putting together the elements of a project—for example, at the University of Rochester, where Lee DuBridge had just arrived as the new head of physics. DuBridge had built an ultrasensitive detector for a critical test of mitogenetic radiation, and Hanson had observed at AAAS meetings that DuBridge was the one physicist who always attended the biologists' informal meetings. Strong medical departments and a special school of optical engineering (financed by the Eastman Kodak Company) made Rochester an ideal site for a project, and Weaver urged DuBridge to send him a proposal. DuBridge never got around to it, though he assured Weaver that biophysics would be a central part of his program for rejuvenating his department.[15]

At Chicago, Weaver moved even more aggressively to get a project started. From a list of terminating NRC fellows, Weaver picked two promising young plant physiologists, Elmer S. Miller and Frederick Zscheile, who were experienced spectroscopists but only marginally employed. Through Frank Lillie, Weaver identified senior researchers, Thorfin Hogness and the biochemist Fred Koch, who were willing to supervise research on quantitative analysis of vitamins, enzymes, and

[13]Gregg to L. H. Newburgh, 19 May 1933: H. M. Randall to Weaver, 7 Feb. 1934; both in RF 200D 159.1958.

[14]Harrison to Weaver, 5 Aug., 19 Nov. 1934; Weaver to D. L. Edsall, 7 Aug. 1934; K. T Compton to Hanson, 3 June 1936, 17 July 1937; all in RF 200D 142.1749.

[15]Weaver to DuBridge, 27 Apr. 1934; DuBridge to Weaver, 2 May 1934; Hanson diary, 28–29 Feb. 1934; 28–30 Dec. 1937; all in RF 200D 161.1977.

steroid hormones. Miller and Zscheile were easily persuaded to drop their own research on photosynthesis to join the new project.[16] Tisdale also tried to get Arthur Compton involved, but Compton was too absorbed in his cosmic ray work to rise to any lesser bait.[17]

Spectroscopy: Learning from Experience

Most of the spectroscopy projects did not live up to Weaver's high hopes, usually because of a lack of real cooperation and leadership. At Michigan, Harrison Randall and Ora Duffendack, the senior physicists, paid too little attention to the newly fledged Ph.D.s to whom they assigned the project. The quantitative measurement of metallic elements in body fluids, though technically simple, produced only routine data. The more demanding part of the project, infrared analysis of peptide structure, was too hard for inexperienced apprentices and got nowhere. Weaver saw what was happening, and Miller urged Duffendack to make sure that the project did not "degenerate into a sort of lazy relief job for two or three physicists who would otherwise be out of work." But there was little they could do. Weaver later admitted it was "somewhat visionary" to expect success without the active personal participation of seasoned researchers.[18] Robert Woodward would have sympathized.

The McGill project was even more embarrassing. After a year's work, John Foster discovered by accident that the high concentrations of lead he had been finding came from lead solder in the needles used to draw samples. Foster blamed the clinicians and withdrew from the project, but in fact the fault was partly his. Foster had no knowledge of biology and no desire to acquire any. He was interested only in the theory of quantitative spectroscopic analysis and in perfecting an instrument for routine medical analysis.[19] Since he did not care what samples he analyzed or for what purpose, it is not surprising that he took so long to detect an obvious blunder. A similar pattern obtained in other projects, though without such dire results. Ornstein happily carried out routine analyses without any idea of what biological problems were being investigated. Luckily he had an able and experienced partner in the microbi-

[16]Mason to Weaver, 12 Jan. 1934; Weaver diary, 8–10 Jan. 1934; E. S. Miller to Weaver, 20 Jan. 1934; Weaver to Lillie, 22 Jan. 1934; Lillie to Weaver, 25 Jan. 1934; Hogness et al, "Proposed project," 24 Jan. 1934; all in RF 216D 12.166.

[17]Tisdale diary, 30 Nov. 1934, RF 216D 8.108.

[18]Weaver to Miller, 1 Oct. 1935; Hanson diary, 10 Oct. 1935; Weaver assessment, n.d. [c. 1938]; all in RF 200D 159.1959.

[19]Tisdale diary, 6 Aug. 1935; Miller diary, 28 Oct. 1935; Hanson diary, 28 Feb. 1937; all in RF 427 14.127.

ologist Albert J. Kluyver, who made good use of Ornstein's skill to do some important work on chemiluminescence and the quantum theory of photosynthesis.[20] Neither project, however, advanced the art of organizing collaborative research in physics and biology. Foster and Ornstein could just as well have been commercial analysts.

George Harrison had no personal interest in biology either, but his entrepreneurial ambitions made him alert to the need for real communication. Some months into the project he told Weaver that their "principle discovery so far has been that of the great need for closer touch between the medical and spectroscopic laboratories." He assigned one of his physicists to spend a few hours each week assisting the clinicians and criticizing their methods of taking samples; in turn, a clinician came to MIT to observe how spectroscopic measurements were done.[21] Harrison later wrote that his most important discoveries were not scientific but organizational—how an atmosphere could be created that encouraged collaboration between physicists and biologists.[22]

In fact the collaboration was never as close as Harrison claimed: less close, Karl Compton admitted, than other joint projects with local clinicians. The applied spectroscopy laboratory was superbly equipped and organized to provide routine medical analysis, but it was more like a commercial service than a biophysical research laboratory. The scientific results were humdrum and did not develop along fundamental lines, as Weaver had hoped. The Harvard clinicians were eager to expand on the clinical side but did not rise to Weaver's invitation to propose more fundamental biochemical or biophysical lines.[23] Harrison's group did their best work on what most interested them as physicists, namely, instrumentation. In the late 1930s, they invented a monochrometer for the infrared and ultraviolet regions and designed a continuously recording spectrophotometer. Weaver was delighted, especially since it was he who had gotten Harrison interested in inventing such a machine.[24] But communication between physicists and clinicians remained a chronic problem of almost every spectroscopic project.

What went wrong? Weaver's inexperience was partly to blame: in his

[20]Ornstein and Kluyver reports, 12 July 1935, 28 Apr. 1937; Tisdale diary, 4 Nov. 1935; all in RF 1.2 650D unprocessed file.

[21]Harrison to Weaver, 19 Nov. 1934, RF 200D 142.1749.

[22]Harrison to Weaver, 23 Apr. 1936, RF 200D 142.1750.

[23]Harrison to Compton, 3 July 1935; Harrison to K. D. Blackfan, 29 Apr. 1936; Harrison to Weaver, 23 Apr. 1936; all in RF 200D 142.1749–1750. Weaver to C. P. Rhoads, 19 Oct. 1937, RF 200D 155.1913.

[24]Harrison to Weaver, 22 Sept. 1938, 20 Mar. 1939; Weaver to Harrison, 23 Sept. 1938; Compton to Hanson, 3 June 1936, 17 July 1937; all in RF 200D 142.1750.

eagerness to sign up physicists, Weaver paid too little attention to what makes collaborations work, namely, personal interest and commitment. Spectroscopists were naturally more interested in instrumentation than in biologists' problems. And a service relation, while it had an advantageously low threshold for participation, had the disadvantage of being so undemanding that neither party was forced to understand what the other was really doing. In hindsight, spectroscopy was not the most suitable technology for an experiment in transdisciplinary practice. Ultraviolet and infrared spectroscopy proved more useful for fingerprinting simple organic compounds, like the fractions of petroleum distillates, than for analyzing the structure of complex biological molecules.[25] And the technology was too simple, too routinized to force physicists and biomedical scientists into a closer rapprochement.

The Chicago project was the exception to the rule: it moved from routine analysis to fundamental biochemical problems, and induced Thorfin Hogness to become a convert to general physiology. The Chicago group displayed none of the problems of communication that bedeviled projects run in tandem by physicists and biologists, in part, perhaps, because the principles were all chemists or biochemists and spoke a common language. At first Hogness simply analyzed the diverse materials provided to him by Fred Koch, a nutritional biochemist, who hoped to hit upon something as spectacular as provitamin D, which had been discovered by spectroscopic analysis some years before. Hogness soon tired of Koch's single-minded search for new steroid hormones. He shared with physicists a keen (and somewhat arrogant) sense of mission in bringing biologists up the standards of physical chemistry. Not content to simply provide analytical services, he dreamed of replacing traditional animal assays with quick, cheap, and precise spectroscopic methods. As his understanding of biological problems improved, he began to talk about spectroscopic biology as a "virgin" field, "entirely unexplored," pregnant with undreamed of discoveries.[26] Reviving an earlier interest in photosynthesis, he began casting about for "significant cooperative work with biologists and organic chemists."[27]

Weaver entered the picture at this critical point, suggesting that Hogness might like to take a six-month tour of European laboratories at the foundation's expense to survey what was being done in spectroscopic biology at centers like Otto Warburg's institute. This was just

[25]Rabkin, "Technological innovation."

[26]T. Hogness, E. Z. Kraus, F. C. Koch, "Proposed project," 24 Jan. 1934, RF 216D 12.166.

[27]Hogness to Weaver, 28 Jan., 28 May 1936, RF 216D 12.168.

what Hogness most wanted at that moment, and he returned a convert to Warburg's style of enzymological research. With a new grant from Weaver, Hogness signed up some enzymologists, and within a few years his laboratory had become the leading center of Warburg's style of general physiology in the United States.[28] Prophesying a revolution in biology comparable to quantum physics, Hogness and Arthur Compton dreamed of a great institute for cell biology, endowed by the RF, to which even Otto Warburg himself might be tempted to transfer his allegiance.[29]

Apart from Hogness, however, the spectroscopy projects were an embarrassment to Weaver. Intended to demonstrate what activist patronage could do, they only revealed how many things could go wrong with planned, programmatic research. If the experiment suggested anything, it was that Simon Flexner and Herbert Gasser were right in thinking that mathematicians had no business messing about in biomedical research. The panel of experts who reviewed Weaver's program in 1938 found nothing to praise in the spectroscopic grants.[30] Individual scientists were more openly critical, as Weaver well knew from an experience in 1933 at a dinner party of eminent medical scientists:

> [B]eing still rather green at my job . . . and very full of beans about the application of physical techniques to medical problems, and perhaps not as discrete in my relations with the medical profession as I later found it desirable to be, I launched forth into an enthusiastic and vigorous discussion of the possible usefulness of spectroscopic methods in medicine. . . The doctors present at once told me the regulation story that physicists simply did not understand . . . what an inconceivably complex mixture of substances one had to deal with inside the human body, so that my remarks were more amusing than practical. They challenged me, in fact, to tell them what I would really do with a spectroscope in medical research.[31]

Weaver proceeded to describe how normal and diseased materials could be systematically separated into fractions, analyzed spectrophotometrically, and the results correlated with specific diseases. The doctors, he felt, considered his speech "more the effect of the liquid which had been imbibed in the meantime than the effect of any serious

[28]Weaver to L. Carmichael, 16 Sept. 1937; Hogness to Weaver, 13 May 1938; both in RF 216D 12.169–170.

[29]Hogness, A. Compton, "An institute . . . ," 13 May 1938, RF 216D 12.177.

[30]"Report of the committee of review," 1938, p. 26, RF 915 2.12.

[31]Weaver, oral history, pp. 695–700.

mental process, and certainly of any competence to discuss that kind of problem."[32]

This episode stuck in Weaver's mind, a reminder of his difficult apprenticeship as a manager of science. He broke off the spectroscopy experiment after a few years, but remained very sensitive about its failure and took every opportunity to insist that the idea was sound and his critics shortsighted. In 1937, for example, C. P. ("Dusty") Rhoads called from the Rockefeller Institute in a state of some excitement about a scheme for using spectroscopy to find out if products of normal steroid metabolism might be carcinogenic. Weaver reminded his friend how, at that earlier occasion at Scarsdale, he had been one of the most energetic in trying to deflate his enthusiasm for spectroscopy. Rhoads had forgotten the episode, but Weaver recorded in his log his belief that it may have been the seed of Rhoads's bright idea.[33] The experience of the spectroscopy projects did not dampen Weaver's taste for activist sponsorship. But it did teach him to be discrete and not to push particular instruments or problems too hard or to venture too far beyond what scientists considered practicable. Also he learned about the importance of good communications and organization in cooperative research. That knowledge he applied to good effect in projects involving other kinds of laboratory technology, like radioisotopes.

Accelerators and Isotopes

It was the cyclotron that most captured the imagination of biomedical scientists in the mid-1930s, and the flood of new isotopes they poured forth. "Unlimited" was the usual word for the expected uses of radioisotope tracers in transport, metabolism, membrane and nerve physiology, and so on. The euphoric mood was captured by A. V. Hill's vivid analogy between the revolution wrought in biology by the microscope: with the microscope, biologists had for the first time seen individual cells; with isotopes, they could "see" individual atoms. Picked up by Ernest Lawrence and others, the image caught on like a promotional slogan.[34]

Weaver, however, was in no hurry to jump on the cyclotron bandwagon. He wondered if radioisotopes really would produce the

[32]Ibid.

[33]Weaver diary, 28 Sept., 15 Nov. 1937, RF 200D 155.1913. Weaver, oral history. Weaver continued to fund the occasional project in spectroscopic biology, including Rhoads's project.

[34]Weaver diary, 25 Jan. 1937; Lawrence to Weaver, 10 Nov. 1937; Weaver to Fosdick, 23 Nov. 1937; all in RF 205D 12.178. Schmitt to Shaffer, 14 Apr. 1939, RF 1.2 228D 1.7.

biological wonders that everyone expected, and turned down at least six applications for accelerator projects in his first few years. The field was so new and scientific interest in it so keen, he later explained, "that we could hardly afford to expose ourselves to general attack for this purpose."[35] If the projects failed, in other words, it would be a far more public and damaging failure than with, say, spectroscopy. What worried him, it seems, was that grants for accelerators and isotopes would be perceived as grants for pure physics. Unlike spectroscopy, nuclear physics was a highly competitive research front. Every university department wanted an accelerator, and there was a lively competition to be the first to make new isotopes. It was not unreasonable for Weaver to think that physicists who professed a greater interest in biological applications than in basic nuclear physics were either fooling themselves or trying to fool him. In fact, the vast majority of new isotopes were heavy metals of no biological significance. (It was some years before radioisotopes of carbon, nitrogen, and oxygen were invented.) Physicists' predictions of biological uses tended to be vague on particulars, and many early experiments were repetitious and inconsequential from a biological point of view. Should the biological uses of radioisotopes prove not to be as "unlimited" as people expected, the RF was certain to be criticized for building machines for nuclear physicists under the guise of a program in biology.

Gradually, however, Weaver was drawn in. Between 1935 and 1945 he supported nine accelerator projects, or about one in six of all such machines built in that period. They form three distinct clusters. Two early cyclotron projects broke the ice: Niels Bohr's at Copenhagen, and Frédéric Joliot's at the Collége de France. In these projects pure physics and biological applications were about evenly balanced. The European projects were followed in the late 1930s by four American projects: Washington University, and the universities of Minnesota, Rochester, and California. These were predominantly for biomedical applications, as was the last of this group, at Stockholm University in 1945. The final cluster included Ernest Lawrence's great 184-inch machine in 1940 and a Van de Graaf machine at MIT. In these final two projects, perfecting the engineering design was the aim, and they were sold to the trustees as once-only grants in pure physics.

The timing and shifting purpose of these projects reflected the rapidly changing technology of accelerators. As the size and power of these machines increased, they intersected with different research fronts in physics and the life sciences. The first relatively small machines were of great interest to physicists but very little to biomedical

[35]Weaver memo, 9 Nov. 1936, RF 226D 2.20.

scientists. They were too small to produce usable amounts of radiation and radioisotopes for biomedical research. They were perfectly suited, however, to discovering nuclear reactions and filling in the inventory of isotopes in the periodic table—one of the liveliest games in physics in the early 1930s but not one in which Weaver was able to play. (Weaver did regret, however, that Ernest Lawrence had not asked the RF to support his early work on the invention of the cyclotron.)[36]

When accelerators became extremely large and powerful in the mid-1940s, they crossed the threshhold of a new research front into the world of "mesotrons" and other subnuclear particles previously seen only in small numbers in cosmic rays. Though the large cyclotrons were not actually used to produce such particles until after World War II, physicists realized their potential well before that. (Cosmic ray physicist Arthur Compton worried, e.g., that large cyclotrons would put him out of business.) After 1940 the use of accelerators for basic biochemical and biophysical research was overshadowed by high-energy physics, with its new military connection. The technology outgrew Weaver's program.

It was in the middle phase of scaling-up that accelerators were best suited to basic biochemical and biophysical research, and thus to Weaver's program. In the late 1930s, the inventory of isotopes was largely complete and the inventory of subnuclear particles not yet begun. For these few years, cooperative research with biomedical scientists could compete for the attention of accelerator physicists with basic research in their own discipline. The waning of public enthusiasm for physical and industrial science and the rising prestige of medical research in the Depression reflected and accentuated this change in the use and meaning of accelerator technology.

These mid-sized "medical" or "production" cyclotrons were not unambiguously suited to Weaver's program, however, because they were also becoming useful to clinicians and radiologists for radiation therapy. Routine clinical application of isotopes and radiation was of no more interest to Weaver than was nuclear physics, and every cyclotron project presented the question, Who would most benefit, basic biomedical researchers or clinicians? The answer depended partly on size: smaller production cyclotrons—42 inches was fairly standard—produced enough isotopes for basic research and for small-scale clinical use. Larger medical cyclotrons—60 inches and over—produced enough for routine radiation therapy. The larger the machine, the more likely it

[36]Weaver to Fosdick, 23 Nov. 1937, RF 205D 12.178. John L. Heilbron and Robert W. Seidel, *Lawrence and His Laboratory: A History of the Lawrence Berkeley Laboratory*, vol. 1 (Berkeley: University of California Press, 1989), especially chs. 4–5.

was to be attached to a hospital radiological Institute and clinic, and the more Weaver worried about clinicians co-opting funds and projects intended for basic research. By the late 1930s Weaver was as wary of radiological entrepreneurs as he had been of nuclear physicists a few years before.

For cyclotron builders, mass producing radioisotopes and tending biomedical researchers brought both costs and benefits. Research on nuclear reactions made quite limited demands on accelerators. Sufficient amounts of data could be produced by operating a few hours per day or a few days per week, and with small currents of one or two microamperes. In contrast, production of isotopes required continuous operation at full power, with currents of 100–200 microamperes. As demand for radioisotopes increased, it became necessary to operate around the clock with two or three shifts of physicists in attendance.[37] Isotope production meant less time for doing physics, but it also provided pay for more graduate students and valuable experience for a large number of apprentice cyclotron builders. The extra resources were substantial: typically in Weaver's projects, half the money went directly to physics departments for operating the cyclotron.[38] Also, producing isotopes put a premium on designing machines for maximum beam intensity, which was an advantage in experiments on nuclear reactions. Experience gained in long production runs helped transform a temperamental machine into a foolproof instrument that every physics department or hospital could use routinely. Making radioisotopes did strain the human resources of physics departments, but between about 1936 and 1940 it was vital to scaling-up accelerator technology and creating a new research community.[39]

The most important use of isotope production, however, was the leverage it provided for getting money from foundations to build new and larger machines. Obviously, physics departments could not be expected to absorb the large costs (salaries, electric power, overhead) of serving biomedical groups; that could only be done by agencies that operated across disciplinary boundaries, like the RF and other foundations. The argument was compelling for Weaver, who needed to persuade his board that a project was uniquely its responsibility. At this stage in the scaling-up of accelerators it was in physicists' self-interest to cooperate with biomedical researchers. Accelerators and isotopes were essential to research fronts in both disciplines, and that made accelerator pro-

[37]Weaver diary, 3 Apr. 1939, RF 205D 12.180. Tate, "Request for funds," 12 June 1941, RF 226D 2.23.

[38]Weaver to Lawrence, 16 Mar. 1939, RF 205D 12.180.

[39]Lawrence to Weaver, 10 Nov. 1937, RF 205D 12.180. Heilbron and Seidel, *Lawrence and His Laboratory*, ch. 8.

jects ideal vehicles for getting physicists and biologists to work together. Conflict between the disciplinary aims of physicists and biomedical scientists was at a minimum. So too was the risk to Weaver that physicists would profess an interest in biomedical problems while doing pure physics.

There was also a congruence in 1936–1939 between the cost of accelerators and Weaver's resources. Circa 1938, a small medical cyclotron (forty-two inches) could be built for between $35,000 and $50,000 (less if physicists' labor was not counted). Annual operating costs, by some estimates, was as low as $8,500.[40] This was just about the size of Weaver's larger projects, and since Weaver insisted that local institutions bear half the cost (usually for a building and operating expenses), a system of regional accelerator centers was quite within Weaver's capacity. The larger sixty-inch medical cyclotrons were about three times more expensive, about the size of Weaver's rare capital grants to institutions. Beyond that size, cyclotrons were outside Weaver's program and cost more money then the entire annual budget of the program in vital processes. (The RF's contribution of $1 million plus to Lawrence's 184-inch cyclotron was a once-only withdrawal from endowment.)

Breaking the Ice: Copenhagen and Paris

It is no accident that the RF's first two cyclotron projects were European and that they were more Tisdale's doing than Weaver's. As an old IEB hand, Tisdale was more comfortable with the idea of aiding centers of physical science and less wary of physicists' motives. The Bohr and Joliot projects were meant as much to give continental physicists access to Lawrence's cyclotron technology as to produce radioisotopes for biomedical research.[41]

Both Bohr and Joliot professed keen interest in biology though in different ways. Bohr's interest, initially, was theoretical and philosophical: what fascinated him was the prospect that concepts of quantum physics, like complementarity, might have analogies in biology.[42] The prospect of material support, however, gradually gave his enthusiasm a more practical shape. Weaver told Bohr about his program in vital processes during his European tour with Lauder Jones in 1932. Bohr did

40A. L. Hughes, "Report. . . " 11 Oct. 1938, RF 1.2 228D 1.6.

41Heilbron and Seidel, *Lawrence and His Laboratory*, ch. 7. Heilbron, "The first European cyclotrons," *Rivista di Storia della Scienza* 3 (1986): 1–44.

42Gerald Holton, *Thematic Origins of Scientific Thought* (Cambridge, Mass.: Harvard University Press, 1973), ch. 4. Kay, "The secret of life: Niels Bohr's influence on the biology program of Max Delbrück," *Rivista di Storia della Scienza* 2 (1985): 487–510.

not begin to think seriously of doing experimental work in biology, however, until there was someone in his group who could do the experiments, namely, James Franck. Then the combination of new talent, new ideas, and new funding proved irresistible. By 1934 Bohr was full of enthusiasm for a cooperative project, though he had, as Tisdale noted, "a very meagre knowledge of biology." By 1935, Tisdale reported that Bohr would talk of little else and was planning to devote himself full time to biology in the next few years. It was not instrumentation and experiments that inspired Bohr so much as organization and communication. He dreamed of making his institute a nerve center of a new biology to the same way that he had made it a center of the new physics, through international conferences, exchanges and fellowships.[43]

Joliot, too, professed to be giving up nuclear physics for biology but with less enthusiasm than chagrin. Although it was his discovery of "artificial transmutation" that had triggered the hunt for new isotopes, Joliot soon found himself pushed to the margins as the enterprise became more dependent on expensive machinery and large research teams. He told Tisdale that he no longer had any hope of competing "with the Rutherfords, the Lawrences, etc., who seem to have a great deal of capital behind them." Discouraged and bitter, Joliot saw biological applications as his best bet, suited to his talent for innovation, he thought, and likely to become important in the future. Disappointed that his recent Nobel Prize had not brought fortune along with fame, he told Tisdale that he was putting all his effort into getting backing for an institute for research in physics and biology.[44] In France, he complained, young physicists had no chance to get state support, but he thought that the prospects of getting an institute were greater in biology and medicine.[45] Joliot made no effort to disguise the fact that he regarded biology as a consolation prize.

Not surprisingly, Weaver was "very doubtful" about Joliot's proposal, suspecting that "the biological slant was thrown in to create atmosphere." He thought it unlikely that Joliot really intended to work on biology; more likely he wanted an accelerator to get back in the race in nuclear physics. Even if the RF gave money only for the "biological and medical fraction of the bet," Weaver still thought the odds were long: "I

[43]Aaserud, *Redirecting Science,* ch. 5. I am grateful to Dr. Aaserud for the opportunity to read this chapter before publication. Tisdale diary, 10 Apr., 20 Oct. 1934, 24, May 1935; Tisdale to O'Brien, 10 Apr. 1934; Bohr to Weaver, 13 Apr. 1933; Tisdale to Weaver, 30 Apr. 1934; all in RF 713D 4.46.

[44]Tisdale to Weaver, 15 Feb., 19 Sept. 1935; Weaver diary, 4 June 1935; F. Joliot, "Plan for the creation of a specialized laboratory," 22 Jan. 1935; all in RF 500D 10.111. Heilbron, "First European cyclotrons."

[45]Tisdale to Weaver, 22 Jan. 1935; Tisdale diary, 27 Dec. 1935; Miller diary, 17 Sept. 1936; all in RF 500D 10.111–112. Hanson diary, 13–23 Apr. 1938, RF 205D 12.179.

do not warm up much and suspect that we would find ourselves supporting nuclear physics." Tisdale was taken aback: he had expected Weaver might balk at the cost of an accelerator project but seemed almost shocked that Weaver would doubt Joliot's sincerity.[46] In fact, Joliot's application did contain suspiciously few specifics about biological collaborators and biological problems. A personal visit to Paris changed Weaver's mind about Joliot, but he remained doubtful about his project.[47]

Weaver felt no such doubts about Bohr's intentions, and he told Tisdale they could take "a pretty generous and broadminded attitude."[48] Indeed, he seemed less nervous about the possibility of bootlegging than Tisdale or George von Hevesy, who volunteered his assurance that Bohr did not want a cyclotron "to permit him to . . . compete with the Rutherfords, Lawrences, and others who are working in the field of pure physics."[49] Bohr's reputation and evident enthusiasm made such assurances unnecessary to Weaver. Also, Bohr had more definite ideas than Joliot about his chemical and biological collaborators.

Bohr pinned his hopes on von Hevesy and physiologist August Krogh, who he hoped would devote his last few years before retirement to Bohr's project. (James Franck, who Bohr had hoped would take charge of the biological side of the project, emigrated to the United States in 1934.) Weaver was doubtful about Krogh: "Does [he] swing enough weight with the biologists," he wondered, "so that they will have to grant (whatever their reservations about a physicist and a physical-chemist) that . . . biology is properly and adequately represented?" Weaver and Tisdale also knew that Krogh wanted to begin a long-term study of the physiology and ecology of marine organisms, a project that had already been declined by the RF. Tisdale reported that Krogh had abandoned that plan, however, and was now converted to using deuterium and radiosodium to study the transport of water and metal ions across animal membranes.[50] Weaver was willing to take a risk with Krogh in order to secure von Hevesy and Bohr.

Joliot had no biologist of Krogh's caliber. His colleagues in the medical departments of the Radium Institute were eager to participate, but

[46]Weaver to Tisdale, 21 Jan. 1935; Tisdale to Weaver, 5, 15 Feb. 1935; all in RF 500D 10.111.

[47]Tisdale diary, 4, 27 June 1935; Weaver to Tisdale, 1 Oct. 1935; all in RF 500D 10.111.

[48]Weaver to Tisdale, 5 June 1934, RF 713D 4.46.

[49]Tisdale diary, 10 Apr. 1934; Tisdale to Weaver, 27 Feb. 1935; both in RF 713D 4.46–47.

[50]Tisdale diary, 8 Apr., 20, 29 Oct. 1934; Tisdale to Weaver, 16 Nov. 1934, 27 Feb. 1935; Weaver to Tisdale, 21 Jan. 1935; all in RF 713D 4.46–47.

their interests were more on the clinical side, especially cancer research. The closest equivalent to Krogh was Antoine Lacassagne, who planned to use radioisotopes to study the localized effects of radiation on tissues and organs. The biologists of the Rothschild Institute were also keen, though their plans were not very specific. (Boris Ephrussi hoped that radiophosphorus, assimilated into chromosomes, would make more specific mutations than did external X-rays and thus reveal the chemical structure of the genetic material.)[51] Weaver was never keen on the Paris project, but Tisdale was, and his persistence paid off.

The grants to Bohr and Joliot provided money to build cyclotrons, pay the salaries of Krogh and Lacassagne, and cover additional expenses for radioisotope research. Matching funds were readily forthcoming from private patrons in Denmark, and in France from the new Socialist government of Leon Blum, who had made science a top social priority. Conveniently for Joliot, his wife, Iréne Joliot-Curie, was Blum's undersecretary for research.[52]

In fact, the two European cyclotrons were hardly used at all for biomedical researches. There were delays. Making these temperamental machines work as they did at Berkeley was always a frustrating experience. Apparently innocent minor changes in design resulted in inexplicable failures and months of de-bugging. Two of Lawrence's young cyclotroneers had finally to be called in, which cost Weaver a little money and good deal of time and effort arranging things with Lawrence.[53] Even then, the machines at Paris and Copenhagen machines did not begin to operate until early 1939, just in time for the discovery of uranium fission (Joliot was ecstatic, Weaver was not) and the outbreak of war.[54]

The biomedical researches in both projects were done, in fact, with isotopes produced from old-fashioned radium-beryllium sources. These weak sources of neutrons could produce isotopes only in tiny amounts. Most departments of physics had such sources, but the larger ones were owned by hospitals and were in continual use for cancer therapy. They were jealously guarded by clinicians, who made them

[51]Joliot, "Plan for the creation of a specialized laboratory," 22 Jan. 1935; Tisdale to Weaver, 15 Feb. 1935; O'Brien to Gregg, 15 Feb. 1935; all in RF 500D 10.111.

[52]Tisdale to Weaver, 27 Feb. 1935; Weaver to Tisdale, 23 Feb. 1937; Tisdale diary, 27–28 Jan. 1936; all in RF 713D 4.47. Tisdale to Weaver, 8 Jan. 1937; Weaver to Tisdale, 16 Jan., 17 Feb. 1937; Weaver diary, 5, 9 May 1937; Tisdale diary, 27 Dec. 1935; Miller memo, 17 Sept. 1936; all in RF 500D 10.112.

[53]Tisdale diary, 10 Mar., 12 Apr. 1937; Weaver diary, 5 May 1937; Weaver to Tisdale, 17 May 1937; Tisdale docket, 25 Aug. 1937; all in RF 500D 10.112. Bohr to Weaver, 4 Apr. 1937, RF 713D 4.47. Heilbron and Seidel, *Lawrence and His Laboratory*, pp. 329–348.

[54]Bohr to Tisdale, 6 Jan. 1939, RF 713D 4.48. Tisdale diary, 31 Jan. 1939, RF 500D 10.113.

available to basic researchers for only two weeks in the year. It was precisely to relieve that bottleneck that Weaver decided to support the building of cyclotrons. While they were being built, however, the advantage lay with people who enjoyed access to radium sources and who knew enough to pick problems that could be done with the few isotopes that were available. It was not the "unlimited applications" of the isotope prophets that proved them right, but the one or two problems that happened to coincide with what was doable with precyclotron isotope technology.

By far the most important researches were done with phosphorus-32, which was easy to make from sulfur-32 and had a half-life just long enough for biological work. It was about the only useful radioisotope of the elements of biological interest, since radioisotopes of hydrogen, oxygen, nitrogen, and carbon were still unknown. Phosphorus was long known to be an important component of lipids, of course, but the major use of radiophosphorus was in the newer field of carbohydrate metabolism. The role of phosphate intermediates in the breakdown of sugars was one of the hottest research fronts in biochemistry in the early 1930s, and radiophosphorus was ideally suited to unraveling the intermediary steps in the metabolic chain. This coincidence made it possible to do important work with the limited resources that were available at the time. Not everyone, however, combined the requisite knowledge of biochemistry and chemical physics. Particular individuals and contexts were crucial.

Radioisotopes: George von Hevesy

On the European side, the crucial person was von Hevesy, who, with a fellowship from the RF's program for refugee scholars, arrived at Copenhagen in 1934. (Officially he was a visiting professor, but he had no intention of returning to Germany.) Officially in Brönsted's chemical institute, von Hevesy was really a member of Niels Bohr's circle. (Bohr told Tisdale that it was von Hevesy who gave his philosophical interest in biology a definite shape and urgency.)[55] He was one of the very few people in the world who was experienced in biological uses of isotopes. His earlier work at Freiburg on the absorption and transport of lead was the first to use isotopes as "tracers" of physiological processes. His knowledge of chemistry enabled him to move easily into intermediary metabolism, which was even better suited to the tracer method than bulk transport. And he was experienced in collaborative research with

[55]Tisdale diary, 10 Apr., 20, 29 Oct. 1934; Bohr and Brönsted to Jones, 11 Oct. 1933; Jones diary, 30 Oct. 1933; Mason diary, 1 May 1933; all in RF 713D 4.46.

biomedical scientists. That experience had not always been a happy one—he complained about the difficulties of working with physiologists who knew no chemistry—but he had learned from his mistakes.[56] His experience in managing cooperative research enabled him quickly to put his personal stamp on the Copenhagen project, not just in providing analytical services but in setting the research agenda.

Von Hevesy had come to Copenhagen with plans for a broad program of researches in physical chemistry using various kinds of instruments, especially X-ray diffraction. He hoped that the RF would continue supporting that work, as it had been doing for some years at Freiburg. Weaver made it clear that he was terminating old IEB projects in pure physics and chemistry, but he invited Hevesy to redesign his application to put the emphasis on biological uses of physical chemistry. Von Hevesy was easily persuaded to tie his application to Bohr's cyclotron project, and that almost guaranteed that he would concentrate on radioisotopes. Thus, with gentle pressure from Weaver, von Hevesy returned to his earlier interest in isotopic tracers but now in a broader context and in the new field of intermediary metabolism.[57]

Von Hevesy also made sure that he would not have to depend on clinicians for his supply of isotopes. In 1935, taking advantage of the celebration of Bohr's fiftieth birthday, he arranged for Bohr's Danish patrons to present him with a large radium source to be used exclusively for research in pure physics and biology.[58] In the next few years von Hevesy turned out some twenty papers of fundamental importance on phosphorous metabolism: on the "rejuvenation" of bone, measured by turnover of phosphate; on the metabolism and distribution of lecithin, an important lipid in cell membranes; on turnover of carbohydrate in muscle; metabolism and energy production in embryogenesis, and so on. Von Hevesy found himself in a new field hitherto inaccessible and, for a time, with hardly any competition. More than anything else, his brilliant papers demonstrated to Weaver, and indeed to the world, what radioisotopes could do in biochemistry and physiology.

Von Hevesy also reached out to other groups. Research on circulation in plants was undertaken with biochemist Kai Linderstrøm-Lang at the Carlsberg Laboratory; with physiologist Einar Lundsgaard on metabolism in perfused organs; and with clinicians on sulfur metabolism in mice, phosphorus metabolism in teeth, and other topics. Von

56George von Hevesy, "Historical sketch of the biological application of tracer elements," *Cold Spring Harbor Symposium* 13 (1948): 129–150.

57Weaver to Tisdale, 25 May 1934; O'Brien diary, 10 Apr. 1934; Tisdale diary, 28 Apr. 1934; all in RF 713D 4.46. Aaserud, *Redirecting science,* ch. 5.

58Tisdale diary, 29 Oct. 1934, 22 May 1935, RF 713D 4.46–47.

Hevesy was adept in these collaborative relations. In projects involving radiophosphorus he himself selected the problems, then sought out appropriate biomedical experts to help with animals, dissection, and sampling. He encouraged his collaborators to suggest problems that were suited to the tracer method. In other cases biomedical scientists came to him with problems which required his expertise with isotopes. In such cases the work was done outside his group with von Hevesy providing analytical services. (This was the case with Lundsgaard and Linderstrøm-Lang.) In all projects, however, von Hevesy made it a point to thoroughly understand the biological problems, and that was crucial to his remarkable success.[59]

August Krogh, in contrast, did not make use of the tracer method to expand his research program. He was preoccupied with one specialized problem, the transport of water across animal membranes. While biding his time until the cyclotron could provide him with a broader array of novel isotopes, he worked almost exclusively with heavy water, which was commercially available.[60] The achievements of the Paris group were even more limited, partly because of their limited access to the clinicians' radium source, but more important because there was no one at the Radium Institute like von Hevesy to take charge. (Joliot was absorbed by building his cyclotron and never became personally engaged with biological problems.)[61]

Medical Cyclotrons: The United States

Weaver seemed more reluctant to sponsor accelerator projects in the United States, perhaps because cyclotron technology was accessible, through Lawrence's growing network of disciples, to any department with the money to buy one. Copenhagen and Paris could be justified as unique opportunities; on the U. S. side, it was harder for Weaver to pick a few projects without laying the RF open to claims from others with seemingly equal claims. And how to prevent clinicians from co-opting cyclotron projects for purely clinical research and therapeutics? The advent of the large medical cyclotrons was a golden opportunity for ra-

[59]Tisdale diary, 22, 23 May 1935, 27–28 Jan., 29 Oct., 3 Nov. 1936, 17 Jan. 1938; Hevesy, "Brief summary of the physico-biological researches," 13 Sept. 1938, RF 713D 4.47–48.

[60]Von Hevesy, "Brief summary," 13 Sept. 1938; Tisdale diary, 29 Oct. 1934, 24 May 1935; Miller diary, 25 Jan. 1935; all in RF 713D 4.47.

[61]Joliot, "Rapport," 20 Jan. 1939; Tisdale diary, 31 Jan. 1939; both in RF 500D 10.112–113. Bohr, at least, organized conferences on physics and biology and spread the word of von Hevesy's achievements. Tisdale diary, 22 May 1935; Bohr to Tisdale, 6 Jan., 30 Sept. 1939; all in RF 713D 4.47–48.

diologists. Combining foundation grants and research in radiation biology with income-producing therapeutic service was an appealing strategy for enhancing the importance of their specialty. Entrepreneurial radiologists with visions of grand institutes were not unfamiliar to Weaver in the late-1930s, and he dealt with them as warily as he did with entrepreneurial physicists. Together they were the Scylla and Charybdis of Weaver's isotope projects as accelerators moved from the domain of nuclear physics, through biochemistry and biophysics, to becoming centerpieces of radiological institutes. It is no accident that Weaver undertook his four American projects at the point where the risk of cooption by nuclear physicists had just receded and before the risk of cooption by radiologists had become too serious. It was a narrow window of opportunity.

The medical cyclotron projects were as complex as any that Weaver undertook. Experienced physicists were required full time to operate the cyclotron and make the isotopes, and radiochemists had to be there to fashion irradiated target material into usable form for biochemists and physiologists. Biophysicists trained in physics, radiochemistry, and physiology were essential, everyone agreed, but the point was moot since there were few of them and no departments of biophysics to produce them. (True hybrids like von Hevesy were rare.) Thus, the main thing was to make sure that there was good communication between the physicists and radiochemists and the various groups who used the isotopes they produced.

Weaver's role in these projects was to ensure that they were properly organized and to provide money for parts of the package that were not provided locally. In some cases the RF paid for the big machine, in others it provided research expenses, or both. In the Minnesota project, Weaver provided $36,000 to build a high-pressure Van de Graaff accelerator and to get biological research started.[62] At Rochester a small cyclotron was already in operation in the physics department; there the RF provided $35,000 over three years for biomedical (and some physical) research and for a support group of physicists and chemists to operate the machine, prepare labeled compounds, and build radiation counters.[63] The grant of $60,000 to Washington University was largely for construction of a large medical cyclotron, since a local endowment was already available for biological and medical applications.[64] Lawrence's emergency request for $30,000 to complete the new medical

[62]J. T. Tate, "Request for funds," 13 Feb. 1937, pp. 21–22, RF 226D 2.21. Heilbron and Seidel, *Lawrence and His Laboratory,* ch. 6, covers all U. S. cyclotrons.

[63]Weaver diary, 30 July 1936; S. L. Warren to Hanson, 13 Jan. 1938; Warren to Gregg, 23 June 1939; all in RF 200D 161.1977.

[64]P. A. Shaffer to Weaver, 8 Nov. 1938, RF 1.2 228D 1.6.

cyclotron was equally for development of his larger (and safer) machine and for biomedical research.[65] (A second grant was made in 1939 to enlarge the biomedical work.)[66]

These projects reveal what a successful appeal for funds looked like. One essential ingredient was a group of reputable physicists who were genuinely interested in working with biomedical scientists. What most impressed Weaver, however, was biomedical researchers who already had experience in using radioisotopes. Thus, the initial advantage lay with groups who had ready access to other sources of isotopes and who had already learned to organize cooperative projects. The Minnesota project was the closest of the four to Weaver's ideal, mainly because of participation by the Mayo Foundation and Clinic. Comparable to the Rockefeller Institute in its outstanding research staff, the Mayo Foundation pursued the same strategy of building clinical research on a strong base of biochemistry and biophysics. The Mayo group had the additional advantage that it was attached to the University of Minnesota as a kind of extramural graduate faculty, thus facilitating collaboration with physicists and chemists.

The key person at Mayo was Frank C. Mann, a physician with marked organizational skills and a keen interest in basic research. His Institute of Experimental Medicine included two biochemists, Jesse Bollman and Eunice Flock, and two biophysicists, Edward Baldes and Julia Herrick. It was a young group (all under forty) and equally strong in physical and biomedical science. Bollman had an M.D. degree but had also studied biochemistry; Flock had a Ph.D. in physiology and specialized in intermediary metabolism. Baldes, with Ph.D.s in both physics (Harvard) and physiology (London), worked on circulation and nerve-muscle physiology. Herrick was a Ph.D. biophysicist and specialized in radiation biology.[67]

Most important, Mann's group had access to the Mayo Clinic's large radium source and had already begun to use radiophosphorus and radiosodium in researches on metabolism and transport. They had a good working arrangement with chemists from the university, who used the Mayo source to make isotopes for their own work on exchange of ions at the surface of inorganic crystals. In return they provided radiochemical services for Mann's group. This arrangement was nonexpandable, however. The clinic's radium source was in constant use for cancer therapy except for two weeks each year. Meanwhile, the demand

[65]Lawrence to Weaver, 14 Oct., 10 Nov. 1937; Weaver diary, 20 Oct., 23 Nov. 1937; all in RF 205 12.178.

[66]Weaver to Lawrence, 6, 16 Mar. 1939; Lawrence to Weaver, 11 Mar. 1939; all in RF 205D 12.180.

[67]Tate, "Request for funds," 13 Feb. 1937, RF 226D 2.21.

for access to isotopes was increasing rapidly in other science departments, especially among the biochemists and plant physiologists of the School of Agriculture. The solution was obvious: an accelerator could produce in hours what was available from the Mayo source in a year. Physicist John Tate was eager to have his department join the high-energy physics club and was more than willing to supply other groups with isotopes and provide routine service. Thus most of the elements of a cooperative project were already in place, and Weaver had only to put the whole package together.[68]

The importance of biomedical leadership is also apparent in the Rochester project. The initiative there came from Stafford Warren, head of the Department of Radiology. Warren was already receiving support from Weaver for research on "diathermy" (artificial fever induced by microwave radiation). Warren's project was wholly clinical (Weaver had inherited it from Gregg) and Weaver was as eager to terminate it as Warren was to keep it alive.[69] It was in these circumstances that he came to Weaver with a plan for a large project in radioisotopes. The circumstances did seem favorable. Physicist Lee DuBridge was interested, and George Whipple was using radioiron supplied by Lawrence in his research on anemia, for which he had recently (1934) won the Nobel Prize. In Warren's group a young biophysicist, William F. Bale, had done preliminary experiments with phosphorus. Despite these hopeful signs, however, Weaver did not warm to the project.[70]

It is likely that Weaver had doubts about Warren's real intentions, though he never said so. An aggressive grant-getter, Warren displayed little interest in the basic problems of radiation biology, as Weaver had hoped he might. He was a clinician, and the modest results of the diathermy work were quite out of proportion with his ambition to acquire an endowment of $10 million for an institute of biophysics and radiology. Talk of "institutes" combined with an average record of achievement always made Weaver run the other way. Also, Warren's group of radiologists lacked the experience and fundamental interests of Frank Mann's group. Bale was green, just out of graduate school, and though Warren had a list of chemists allegedly eager to take part, a visit by Weaver revealed that this group existed only on paper. He

[68]Weaver diary, 30 Oct. 1936, 9 Nov. 1936, 18 Jan. 1937; Tate, "Request for funds," 13 Feb. 1937, pp. 11–17; Tate, "Request for funds to aid in the continuation of a cooperative program," in G. S. Ford to RF, 12 June 1941; Bollman to J. E. Williams, 11 Apr. 1941; all in RF 226D 2.21–23.

[69]Lambert to S. Warren, 7 May 1935; Weaver diary, 9 Sept. 1935; Hanson diary, 4 June 1937; Gregg, "Appraisal," June 1938; all in RF 200A 114.1398.

[70]Hanson diary, 4 June 1937; Weaver diary, 3 Nov. 1937; Warren to Gregg, 17 Dec 1937; all in RF 200D 161.1977.

also found that Warren and Bale planned to use radiophosphorus to study the effects of radiation on normal and cancerous tissues, not to do basic biochemistry and biophysics.[71]

After a good deal of negotiation and deflating of grandiose hopes, Weaver did finally approve a modest grant, though not for the large medical cyclotron that DuBridge wanted, or for Warren's radiological institute. What tipped the scales is not clear, but differences between the initial and revised proposals suggest that Weaver got DuBridge to be more actively involved. Warren continued to lobby Weaver and Hanson for advice on raising funds for his institute, undeterred by their studied lack of interest.[72]

The conflict between clinical and biophysical uses of isotopes was the chief issue in the Washington University and Berkeley projects, owing to the greater size of the cyclotrons proposed. Neither DuBridge's nor Tate's machines were large enough for routine clinical therapy, and that limited the potential for radiological bootlegging. The scaled-up models of Lawrence and Alfred L. Hughes, however, were intended for clinical use as well as basic biomedical research. Such machines were much more expensive to build and operate. DuBridge estimated the initial cost of a sixty-inch machine at $100,000–$200,000 and annual expenses of $20,000–$40,000.[73] Almost of necessity these large machines had to be attached to hospitals and clinics, not physics departments.

At Berkeley, too, scaling-up was pushing cyclotron projects beyond the scope of Weaver's program. Ernest Lawrence's new Radiation Laboratory was the prototype of the new sixty-inch medical cyclotron facility, in which clinical research and physics were done side by side under one roof.[74] Lawrence assured Weaver that his true interest was fundamental biological research. Perhaps it was, but his proposal and Weaver's on-site observations suggest that Lawrence was driven to emphasize clinical applications by the practical necessities of building a larger machine. It was not just the extra cost in money. At every stage of scaling-up, Lawrence had discovered, he could justify the additional expense only by pointing to something qualitatively new, something that the larger machine could do that smaller ones could not. The practical necessities of fund raising required that some visible threshhold be crossed.

[71]Weaver diary, 4 July, 9 Sept. 1935, RF 200A 114.1398. Weaver diary, 1 Apr. 1938, 9 Feb. 1940; Warren to Hanson, 13 Jan. 1938; all in RF 200D 161.1977–1978.

[72]Hanson diary, 18 Jan. 1938, 8 Nov. 1940; Weaver diary, 20–21 Feb. 1940; all in RF 200D 161.1977–1978.

[73]Dubridge, "Nuclear physics research program," 20–21 Feb. 1940, RF 200D 161.1978.

[74]Heilbron and Seidel, *Lawrence and His Laboratory*, ch. 8.

In the mid-1930s, it was the threshhold between pure physics and biomedical research; in 1938, between basic research and clinical therapy.[75] Medical foundations might be impressed by Lawrence's argument that his new cyclotron could produce more bang for the buck than radium (for cancer therapy), but it made Weaver nervous.[76] Weaver's uneasiness about the clinical emphasis of the new facility was balanced, however, by Lawrence's demonstrated interest in basic biomedical research (he was then supplying isotopes to some dozen research groups) and by his unique prestige as the inventor of the cyclotron. So (as with Bohr) Weaver did not inquire too closely into Lawrence's real interest in biology but simply enjoined him to use the RF's grant exclusively for basic research.[77]

In the Washington University project Weaver helped divert a local endowment from clinical to basic research. Income from the $750,000 endowment of the Mallinckrodt Radiological Institute had till then been used to subsidize radiation treatments. But patient fees had since grown to cover expenses, and the new medical dean, biochemist Philip Shaffer, was eager for the Medical School to compete in "the astonishing developments" in radioisotopes. For Weaver it was a singular opportunity to create an endowed center for isotope research, and he drove a hard bargain. Shaffer, in deanly fashion, hoped to keep part of the fund in reserve for other projects, but Weaver insisted that the entire income be earmarked for basic biomedical research.[78] Besides the strong departments of radiology and physics, there was also the prospect of participation by Weaver's favorite biophysicist, Francis Schmitt. The physicist too put their best foot forward. In preparation for their application to the RF, Alfred L. Hughes personally made an exhaustive inspection of American accelerators, gathering information on cost, technical problems, and uses. His long report made an incontrovertible case for investing in a large medical cyclotron rather than smaller and cheaper models. Hughes's authoritative report dispelled any lingering doubts Weaver had about the large medical cyclotrons and smoothed the passage of the proposal through the RF's board.[79]

[75]Weaver diary, 25 Jan., 29 Oct. 1937; Lawrence to Weaver, 14 Oct., 10 Nov. 1937; all in RF 205D 12.178.

[76]Lawrence to Weaver, 29 Nov. 1937; Weaver diary, 29 Oct. 1937; both in RF 205D 12.179.

[77]Dockets 38013 and 39042; Weaver to Lawrence, 6 Mar. 1939; Lawrence to Weaver, 10 Nov. 1937, 11, 16 Mar. 1939; all in RF 205D 12.178 and 12.180.

[78]Shaffer to Weaver, 8 Nov. 1938, 24 Mar., 19 Apr., 16 May 1939; Weaver to Shaffer, 9 May 1939; all in RF 1.2 228D 1.6–7.

[79]Hughes, "Report," 11 Oct. 1938; Shaffer to Weaver, 8 Nov. 1938; Docket 39249; Weaver to Shaffer, 4 Apr. 1939; Schmitt to Shaffer, 15 Oct. 1938; Lawrence to Hughes, 11 Feb. 1939; all in RF 1.2 228D 1.7.

It was clear, however, that the RF could invest in only a few such large machines. As the design of medical cyclotrons stabilized, more institutions would become capable of organizing clinical projects comparable to Lawrence's or Hughes's. Weaver needed a rationale for limiting cyclotron projects, and in the late 1930s he began to think in terms of a system of regional centers. He depicted the cyclotron at St. Louis as an instrument for the Middle West, comparable to Lawrence's Rad Lab and, on the East coast, the forty-two inch medical cyclotrons at MIT and Harvard. The grant to Manne Siegbahn in 1943 was likewise justified as a Swedish national cyclotron.[80] This conception of regional centers helped justify grants for these expensive machines and at the same time closed the door to future claims. Thus an appeal by DuBridge in 1940 for a large medical cyclotron got nowhere.[81] The window of opportunity for medical cyclotrons opened and shut within a few short years.

The changing meaning of cyclotrons as they scaled-up is especially evident in the RF's grant for Ernest Lawrence's 184-inch cyclotron in 1941.[82] At $1,210,000 it was Weaver's largest project and one of the grandest—he liked to compare it to the IEB's grant for the 200-inch telescope at Mt. Palomar. It was not, however, a medical cyclotron and not a part of the program in "vital processes." It was presented to the trustees as a special project in pure physics. This was why, for Weaver, the size of the new machine was a practical advantage. It made the project unique, guaranteeing that a large gift to Lawrence would not trigger an avalanche of proposals from competing groups. Weaver warned Lawrence not to scale-down his plans in the hope that the RF and other foundations would be more willing to back a cheaper project. In fact, he told Lawrence to double his request from $750,000 to $1.5 million, on the grounds that a unique international facility "built for all science" would be easier to sell than a smaller one.[83]

Weaver realized that the 184-inch machine could not be sold as a medical cyclotron, as Lawrence hoped to do. Weaver instinctively felt that the RF's trustees might respond to a forthright appeal for a breakthrough into high-energy physics, but was unlikely to spend $1 million on a risky machine just to make more radioisotopes. Accustomed to selling a mix of physics and biomedical application, Lawrence was doubtful at first, but Weaver's eyes had been opened by a talk with Her-

[80]Siegbahn to Hanson, 26 Oct. 1944, 6 Feb. 1945; Hanson to Siegbahn, 5 Dec. 1944; all in RF 800D 6.54.

[81]DuBridge, "Nuclear physics research program," 20–21 Feb. 1940, RF 200D 161.1978.

[82]Heilbron and Seidel, *Lawrence and His Laboratory*, ch. 10.

[83]Weaver diary, 27 Oct. 1939; Hanson diary, 6 Nov. 1939; Weaver to Fosdick, 25 Jan. 1940; F. B. Jewett to Weaver, 16 Feb. 1940; all in RF 205D 12.181–183.

bert Gasser, who advised him to argue the case as one in nuclear physics, not physiology.[84] Doubtless Gasser's advice reflected his belief that RF officers should not have "programs," but whatever his motives he shared Weaver's view that the RF should bear the whole burden and get the whole credit for making Lawrence's grand scheme a reality. As Weaver wrote to Fosdick, it was "the biggest opportunity that has come along in eight years or that is likely to come along in eighteen more!"[85]

Stage-managed with exacting care by Weaver, the RF's grant for the big cyclotron was a showpiece of science patronage. But it also demarcates the end of the time when cyclotron projects could be accommodated within Weaver's program in "vital processes." Physicists returned to physics, and radioisotopes came into routine use, mass produced in the reactors of the Atomic Energy Commission. In the world of big physics and big biomedical research, there was a smaller role for Weaver and his kind of science management.

Instruments and Interdisciplinary Relations

Radioisotope technology gave cyclotron projects a singular and effective social organization: highly centralized on the physics side, and on the biomedical side diverse and decentralized. In the Minnesota project, isotopes made by Tate's physicists were distributed to Frank Mann's group at the Mayo Clinic, Maurice Visscher in physiology, George Burr in biochemistry, Elvin Stakeman in plant pathology, and several teams of inorganic and physical chemists. Each worked on specialized problems, relying on the physicists and radiochemists for technical service.[86] So too at Berkeley, most of the fundamental research with isotopes was done by satellite groups linked to the Rad Lab by service and supply. These included David Greenberg and Carl Schmidt in biochemistry (protein metabolism), Dennis Hoagland in agriculture (plant nutrition), Israel Chaikoff (lipid metabolism in tumors), and others in the departments of surgery and medicine working on cerebrospinal fluid, Addison's disease, and thyroid. Further afield, Lawrence supplied Henry Borsook's group of protein biochemists at Caltech, Whipple at Rochester, and others.[87] The network supplied by

[84]Weaver to Lawrence, 13 Feb. 1940; Lawrence to Weaver, 21 Feb. 1940; Weaver diary, 8 Mar. 1940; Weaver to K. T. Compton, 25 Mar. 1940; Gasser to Weaver, 19 Mar. 1940; all in RF 205D 13.183–184.

[85]Weaver to Fosdick, 25 Jan. 1940, RF 205D 12.182. Staff conference, 12 Mar. 1940, RF 205D 12.184.

[86]Tate, "Request for funds," 13 Feb. 1937, RF 226D 2.21.

[87]Heilbron and Seidel, *Lawrence and His Laboratory*, ch. 8. Lawrence to Weaver, 11 Mar. 1939, RF 205D 12.180.

DuBridge was likewise scattered among various departments of the medical school.[88]

But if the social makeup of isotope projects was decentralized and eclectic, the problems chosen were remarkably consistent from one project to another. This was not Weaver's choice but a reflection of how radioisotopes intersected with currently active research fronts. Whereas spectroscopy was a general tool for identifying chemical structures, isotopes suddenly opened up specific problems like intermediary metabolism, in which much hard work by biochemists had hitherto yielded meager results.[89] Not surprisingly, the most important and accessible problems were usually the ones selected, and the limited range of radioisotopes (before the invention of carbon-14) made for still greater concentration of effort. Studies of phosphorus and sulfur metabolism were the most popular—the isotopes were easy to make, and the problems were significant ones for biochemists. The physiology of membranes and transport were the staple of isotope research on the biophysical side, for similar reasons. Isotopes of sodium, potassium, and calcium were available, and it did not take special imagination to use them on mainstream problems of cell physiology. For a few years, isotope technology gave a programmatic coherence to Weaver's projects in that field, even though the choice of collaborators and problems was almost wholly opportunistic.

In their lack of direct control of problems, Weaver's isotope projects resembled the GEB's departmental grants-in-aid funds of the 1920s. In their coherence of subject matter, however, they came close to achieving what Weaver had hoped to do in institutional grants to centers of general physiology. Instruments, Weaver discovered, were an ideal vehicle for his style of science management, far better than grants for disciplinary communities or for particular scientific problems. Organizing projects around instruments enabled Weaver to delegate the choice of problems to scientists without completely giving up control to the hazards of disciplinary logrolling, as happened to Thorkelson and Mason with grants to departments and NRC committees. Weaver did not have to risk everything on predicting which problems would pay off or what groups of experts would excel. Nor did physicists and biologists have to struggle to discover problems in which both sides took an equal interest. Selecting instruments, rather than problems or disciplines, gave researchers complete freedom within definite boundaries, and gave Weaver a degree of control that was not too heavy-handed and not too loose. More would have invited criticism from his clients and the trustees; less would have put him at their mercy.

88 Warren to Gregg, 23 June 1939, RF 200D 1651.1977.

89 Kohler, "Schoenheimer."

It is easy to see why Weaver showed so little interest when enthusiastic planners like Philip Shaffer urged him to consolidate the "more or less incidental or sporadic studies" of the cyclotron satellites into a central institute, where, Shaffer felt, the new methods could be explored in a more planned and rational way.[90] Weaver avoided any such superorganization. The combination of central instrument and decentralized user groups—a kind of scientific line and staff organization—may be the closest Weaver ever came to the ideal vehicle for an active but hands-off management.

Weaver's interest in instrumentation has struck some observers in a different way. Pnina Abir-Am, for example, has described it as "technology transfer"—the term is borrowed from critiques of multinational capitalism, with the implications fully intended of technological imperialism, dependency, and exploitation. In this view physical scientists (and Weaver) were colonizers and exploiters, biologists the colonized and exploited, and physical instruments were the vehicles for domination. This was, it is alleged, the only possible relation that Weaver could conceive and the only possible one in a relation mediated by laboratory technologies.[91]

A broad look at all Weaver's projects reveals a more complex picture. The relations between physical and biomedical scientists could and did take varied forms. Colonization was the rarest, not surprisingly since it required a drastic change in professional commitments and life-style. A service relation was more common, because it was familiar (to physical chemists at least) and not too demanding. But the experience was anything but hegemonic; more often the relation was loose to the point of being ineffectual, as it was in the spectroscopy projects. Usually, sponsorship entailed a kind of trade relation, in which both parties gained something they would not otherwise have had. There is little evidence of one side dominating the other, in part because Weaver took pains to prevent either the physical or biomedical partners in a project from using it for parochial disciplinary ends. He selected instrumentations and designed projects specifically to prevent such abuse.

No doubt instruments were vehicles for transfer of practices and values between the natural sciences (as commerce is of cultural values). But it was meant to be, and usually was, a two-way street. The discourse of power and hegemony seems too blunt an instrument to dissect the subtle and varied social experiences of collaboration and patronage. Working with spectroscopists or accelerator physicists or X-ray crystallographers does not seem to have oppressed biomedical scientists or

[90]Shaffer to Weaver, 19 Apr. 1939, RF 1.2 228D 1.7.

[91]Abir-Am, "The discourse of physical power and biological knowledge in the 1930s."

perverted their modes of practice. It is striking how readily unfamiliar kinds of instrumentation were assimilated into ongoing lines of biomedical practice. It is striking, too, how varied and flexible Weaver was in his dealings with different groups of scientists—biomolecular boundary crossers, discipline or specialty communities, invisible colleges, one-idea men and women.

The record of Weaver's activities reveals two things: a constant vision of a scientific culture in which chemists, physicists, and mathematicians, as well as biologists, could work on problems of "vital processes"; and a pragmatic and flexible approach to translating that vision into practice in the field.

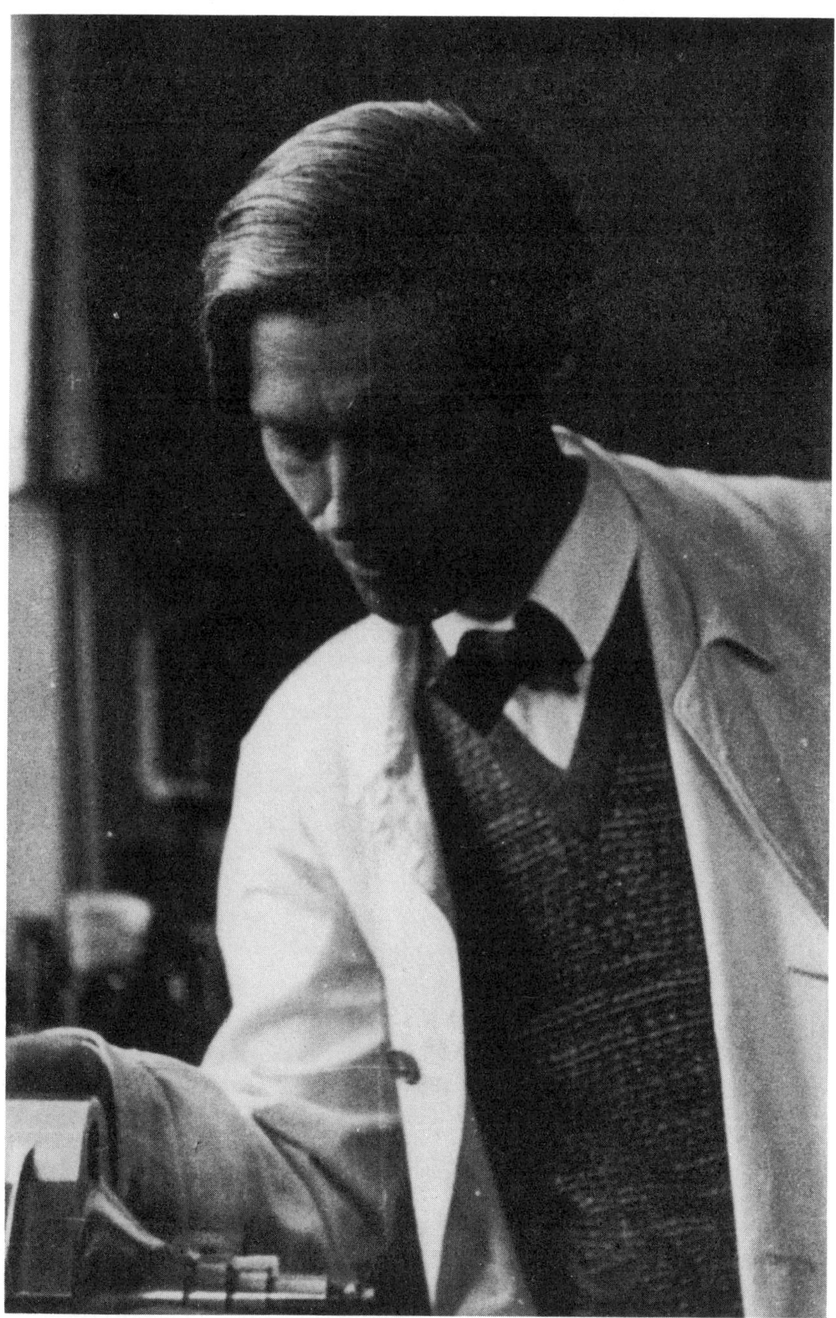

25. The Svedberg, early 1930s. Courtesy of the Rockefeller Archive Center.

26. Ultracentrifuge, University of Uppsala, early 1930s. Courtesy of the Rockefeller Archive Center.

27. Linus Pauling, Caltech, late 1930s. Courtesy of the Rockefeller Archive Center.

28. Electron diffraction apparatus, Caltech, late 1930s. Courtesy of the Rockefeller Archive Center.

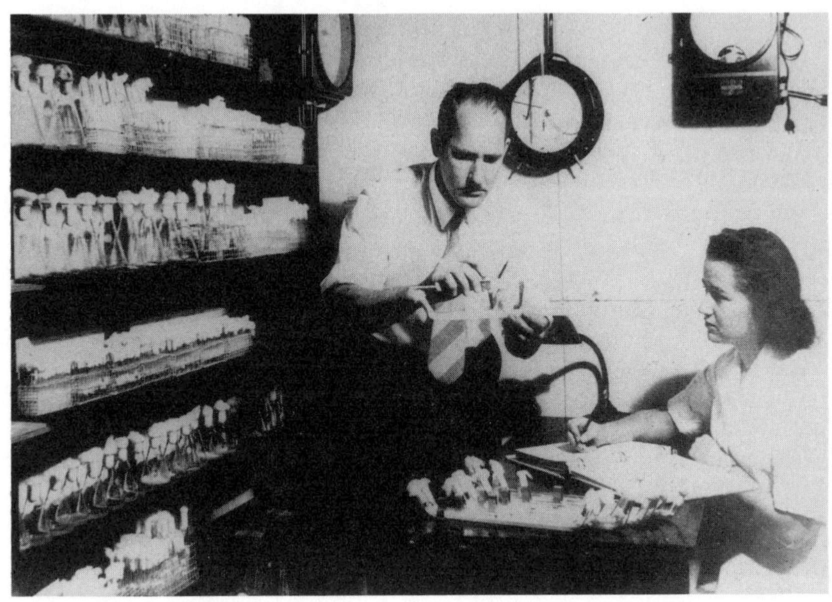

29. George Beadle and assistant measuring growth of *Neurospora,* Stanford University, 1943. Courtesy of the Rockefeller Archive Center.

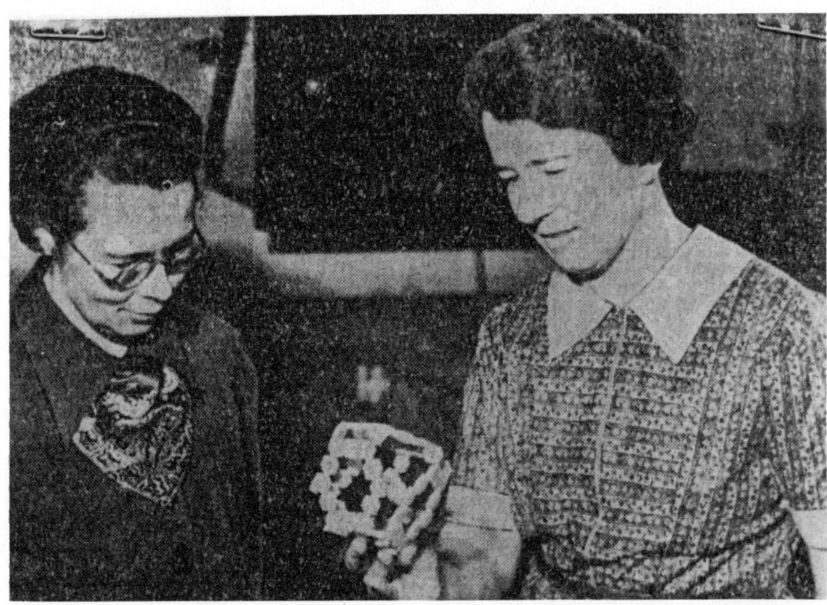

30. Dorothy Wrinch holding cyclol model, 1938. Courtesy of the Rockefeller Archive Center.

31. Spectroscopic biology laboratory of Thorfin R. Hogness, University of Chicago, 1935. Courtesy of the Rockefeller Archive Center.

32. Oscilloscope, Washington University, late 1930s. Courtesy of the Rockefeller Archive Center.

33. George von Hevesy, Copenhagen, late 1930s. Courtesy of the Rockefeller Archive Center.

34. George von Hevesy, Niels Bohr, and high voltage accelerator, Copenhagen, late 1930s. Courtesy of the Rockefeller Archive Center.

35. Sixty-inch medical cyclotron, University of California, 1939. Courtesy of the Rockefeller Archive Center.

Part IV

Conclusion

Chapter Fourteen

Partners in Science in Perspective

The history of foundation patronage from 1900 to 1940 charts the evolution of a partnership that became steadily more complex and pervasive. Sponsored research became more extensively practiced. The privilege of a few in 1900, by 1940 grant-getting was an essential skill for making academic careers in most disciplines. Sponsored research was also a more intensive experience for those who engaged in it. Small grants-in-aid were not very demanding of recipients in the early 1900s; the project grants of the 1930s involved an intellectual give-and-take with foundation managers. Doing science became a more complex activity, in which there were many more essential actors. We saw the strains of an unfamiliar and unwelcome sharing of authority in Woodward's contretemps with his academic clients. The institutional grants of the 1920s were designed to make Rose and Trowbridge partners in institution building, but also to keep them clear of decisions about research agendas and practices. Weaver's project grants of the 1930s took him into the inner sanctum of scientific careers, laboratory practices, and community organization. Weaver's cycle of visits, proposals, conferences, reports, and renewals was the social machinery of an active, working partnership. Experience gradually eroded fears on both sides that traditional prerogatives would not be respected, and produced forms of sponsorship that were ever more intimate and interactive.

The science programs of the Carnegie, Rockefeller, and other foundations were an evolving social system, which changed with experience and political-economic circumstances. The nineteenth-century system of grants-in-aid to needy academics, which fit a culture of scarcity, broke down under the strain of the new wealth coming to higher education and research from the large foundations. The new system of patronage rested on assumptions of the need for manpower and infrastructure, in part because of the demographic disruptions of World War I. The system of large institutional grants flourished in the prosperous 1920s, a part of the postwar enthusiasm for institution building and community development on the grand scale. The return (in a new way) to individual project grants in the 1930s was partly an accommodation to the scarcity of the Depression years. At the same time, each major change in the system of patronage was conditioned by the experiences and the limitations of the system that preceded it. Carnegie and Rockefeller grants to the National Research Council were designed to remedy flaws perceived in the Carnegie Institution's grants. The science programs of the International and General Education Boards flourished in space created but not occupied by the general purpose foundations. Disenchantment with Rose's system building gave force to Fosdick and Mason's vision of a foundation for the advancement of knowledge in specific fields of science.

Management, with a Personal Touch

Looking back from the end of the twentieth century, the creation of this system of financing science is part of the inexorable trend toward large-scale organization–the "organizational synthesis," which may well be this century's hallmark.[1] There is a definite family resemblance between the Carnegie and Rockefeller boards and other large twentieth-century institutions, like business corporations, government agencies, universities, hospitals, and so on. The large foundations were cultural transmission belts, carrying business methods and managerial values from the world of large corporations into academic science. Robert Woodward was perhaps the most self-conscious and deliberate in doing that, but business ideals were no less implicit in the activities of Trowbridge, Thorkelson, and Weaver. William Howell observed how a byproduct of extramural sponsorship was "a sort of competitive struggle for tangible results which gives to scientific research something of the character of a business proposition." Applauding sponsorship as an incentive to produce good work, Howell lamented the loss of academic idealism and independence that also seemed to come with it.[2]

The system of foundation patronage is part of the "organizational synthesis," certainly, but a peculiar part, less bureaucratic than others,

[1]Louis Galambos, "Technology, political economy, and professionalization: Central themes of the organizational synthesis," *Bus. Hist. Rev.* 57 (1983): 471–493. Alfred D. Chandler, Jr., *The Visible Hand* (Cambridge, Mass.: Harvard University Press, 1977). William H. McNeill, *Pursuit of Power* (Chicago: University of Chicago Press, 1982). Steven Wheatley, *The Politics of Patronage*, pp. ix–xv.

[2]Howell to S. Flexner, 10 Nov. 1934, pp. 2–3, RF 915 4.41.

and retaining more of the personal, voluntaristic spirit of nineteenth-century institutions. Compared to corporate or government organizations, even the Rockefeller Foundation was small and informal, and most foundations operated like family charities—philanthropic mom-and-pop shops. One of the most striking things about foundation patronage was its continuing emphasis on individuals, even in institutional grants: recall Woodward's "research associates" scheme, or Rose's vision of a system of mentors. Weaver's projects were as much investments in individuals as in specific research projects. The Rockefeller Foundation was remarkably unbureaucratic: it had no application forms, and the only iron-clad rule was that a university official had to approve every grant application (preferably a president, but a dean would do).

The personal quality of Weaver's practice is apparent in the how-to guide he wrote in 1946 for new recruits to foundation work. It was not a handbook of procedures but personal tips on field practice, learned in the school of hard knocks.[3] The essence of making projects, for Weaver, was knowing people and their work at firsthand, in the field. Depending on correspondence, he quipped, was like "buying out of a catalogue."[4] The first rule was to be a good listener: "When seeking information," Weaver urged, "officers should, in the language of radiation theory, be good absorbers and very poor emitters."[5] Good listening meant not being influenced by either smooth or maladroit salesmanship. Good people should not be turned away if they happened to have unappealing personalities, and charmers should be treated with particular care. "When Rob Millikan or Jim Angell used to show up in their prime," Weaver recalled, "we would lock the safe, throw the key out the window, and put cotton in our ears . . . [and] the race of Millikans and Angells has by no means died out."[6] Formal interviews in the office, Weaver found, were less useful than getting younger researchers alone, "with Herr Geheimrat not present . . . in their own labs, sitting on stools" or on their back porches, talking informally about themselves and their work.[7]

The great danger was that personal relationships might become not simpler and more natural but strained and "shadowed by some hint of personal indebtedness."[8] It was easy to give in to "the sense of power that goes with being a Scientific Santa Claus with a pack full of blank checks." There were tricks for keeping relations business-like, the

[3]Weaver, "N. S. notes on officers' techniques," 11 Jan. 1946, RF 915 2.13.

[4]Ibid., p. 13.

[5]Ibid., p. 2.

[6]Ibid. pp. 5–6.

[7]Ibid., p. 17. Weaver to Fosdick, 4 Aug. 1937, RF 205D 5.77.

[8]Ibid., pp. 15–16.

most important of which was to act always as an agent of the foundation, never as a personal advocate. In the Paris office, for example, it had been the custom to refer to a fictitious "Fellowship Committee," to remove from approvals "the sticky honey of personal favor, and [from] . . . declinations the unpleasant taint of personal responsibility."[9] The essence of Weaver's style of patronage, however, was his personal knowledge of people and fields and his willingness to make personal judgments based on that knowledge. He could and did make mistaken judgments. That did not trouble him particularly: mistakes were a sign that officers were seeking out imaginative and interesting projects. The worst mistake an officer could make, Weaver felt, was to be paralyzed by fear of making mistakes, and on the whole he thought that he could have made a few more than he did.[10]

Weaver was alert to the little subterfuges that academics used to sell their ideas. He quickly acquired an immunity to aggressive salesmen, who would say of everyone in his group, "That man is simply a human dynamo, he lives for nothing but research, etc." At the same time, Weaver was remarkably open to scientific enthusiasts, always listening and keeping an open mind until he could decide if they were cranks, saints, or green talents who needed coaching.[11] He took an unfailing interest in human personality and behavior, noting idiosyncracies of character and the varied ways that individuals presented themselves. He was bemused by a distinguished chemist who suavely offered him the choice of continuing an annual grant to his department or capitalizing it (Weaver called his attention to "other possibilities"). He was no less bemused by those who seemed to forget why they had come to call, like the phychologist R. S. Woodworth, who pleaded for the Committee on Child Development by describing its work as not very exciting but worthy of being continued. Weaver listened patiently to James McKeen Cattell repeatedly express his dislike of foundations. He listened unmoved to the egotistical Herbert McLean Evans, when Evans "call[ed] to make acquaintance of WW and deliver a somewhat oratorical speech concerning endocrinology in general and H. M. Evans in particular."[12] Weaver could be a little vain about his hobnobbing with the truly great—swapping professional jokes with Albert Einstein was duly recorded—but he was quick to see through those whose representations were not backed up by real accomplishment.[13]

Weaver most enjoyed listening to people who were spontaneously

[9]Weaver, "N. S. notes," p. 4.

[10]Weaver, "Notes," pp. 50–51.

[11]Weaver diary, 3 Feb. 1933; 10, 23 Oct. 1934; 21 July, 9 Sept., 27 Dec. 1935; all in RF.

[12]Ibid., 29 Nov. 1932; 30, 31 Mar. 1933; 16 Apr. 1935.

[13]Ibid., 27 Apr. 1934.

enthusiastic about their work. Herman Schneider, and enthusiast for workers' education, came to plead for a grant to the University of Cincinnati and ended up spending several hours in Weaver's office and "entertain[ing] WW with an original and stimulating discourse on engineering, education, economics, oriental philosophy etc."[14] Like Thorkelson, Weaver was at heart a teacher, and like all good teachers he was an avid student, both of science and of human nature. Nothing gave him more pleasure than smoothing the way for young and talented scientists. Obliged to be polite to the pompous and pretentious, Weaver took his revenge by recording their antics in his diary. He seemed immune to cynicism and relished being part of the kaleidoscopic human scene.

Woodward, Greene, Embree, Rose, Trowbridge, Thorkelson, and Weaver were a variety of "research entrepreneur," to extend the term coined by Charles Rosenberg to describe the first-generation directors of agricultural experiment stations.[15] These scientist-administrators stood between state legislators and farm groups who saw science as practical service, and agricultural chemists and biologists who saw service as a basis for fundamental research. The research entrepreneurs invented ways for scientists to get the resources that came with service, but without sacrificing disciplinary goals. They negotiated a trade-off of authority, in which scientists controlled their workplaces and the process of production, and sponsors defined the larger contexts in which the public value of their science would be judged. The same process was carried out wherever science was practiced: in government agencies by bureau chiefs like John Wesley Powell, Harvey Wiley, or Charles Walcott; and in industry by directors of research like Willis Whitney, Frank Jewett, or Charles Stine.[16]

In colleges and universities, too, a balance between research and undergraduate teaching was negotiated by department heads and reform presidents like Charles W. Eliot and Arthur Hadley. In universities, however, a species of middle manager of research did not appear, no doubt because academic research did not generate income. Deans of graduate schools occupied an institutional position analogous, say, to

[14]Ibid., 28 Mar. 1935.

[15]C. E. Rosenberg, "Science, technology, and economic growth: The case of the agricultural experiment station scientist, 1875–1914," *Agric. Hist.* 45 (1971): 1–20. Reprinted in *No Other Gods* (Baltimore: Johns Hopkins University Press, 1976), ch. 9.

[16]A. Hunter Dupree, *Science in the Federal Government* (Cambridge, Mass.: Harvard University Press, 1957), pp. 211–214, 292–293. George Wise, "A new role for professional scientists in industry: Industrial research at General Electric, 1900–1916," *T&C* 21 (1980): 408–429. Wise, *Willis R. Whitney, General Electric, and the Origins of U.S. Industrial Research* (New York: Columbia University Press, 1985). David A. Hounshell and John K. Smith, Jr., *Science and Corporate Strategy* (Cambridge: Cambridge University Press, 1988).

station directors, but they did not acquire the same authority because they did not control resources. Those who did have money to spend, interestingly, also began to exhibit managerial behavior. Charles S. Slichter is exemplary, when as graduate dean he got the power to allocate faculty grants-in-aid. It is no accident that three of Slichter's students and protégés—Thorkelson, Mason, and Weaver—turn up as foundation managers, or that Trowbridge, upon retiring from the IEB, became a graduate dean. In the absence of intramural managers of academic research, that role came to be occupied by foundation officers. They became managers of research in universities, rather like bureau chiefs, station heads, and directors of industrial R&D divisions, only extramurally.

Foundation officers mediated between academic scientists and the business and civic leaders who tended Carnegie's and Rockefeller's philanthropic trusts. They sought alliances with men of affairs, like Elihu Root or Raymond Fosdick, who had come to see science as a national good. There were comparable alliances in industry with technically alert executives; in agricultural stations with specialty farmers, agricultural journalists, and agribusinessmen; and in government with politicians who realized that science had become an issue with which political careers could be made.[17] On the other side, foundation managers allied themselves with the small but growing number of academics—like Hale, Millikan, Lillie, Richards, Morgan, or Compton—who realized that organization and management were good ways to keep ahead of the pack in the increasingly crowded and competitive world of basic research. So too in government, industry, and agricultural stations, research managers worked with a new generation of scientists who discerned in practical application not a threat to disciplinary purity but new opportunities for fundamental research on problems that academics had overlooked. A new and fruitful partnership was born.

Many of the key issues of foundation patronage follow from the institutional separation of the funding and the doing of research. Certainly it was the root of Robert Woodward's tribulations with the grants program (as overseer of the CIW departments he was much more like his industrial or governmental brethren). Institutional separation also complicated the relation between foundation managers and their academic clients, because the missions of foundations and universities, while similar, were not identical. For some forty years, foundations were like rich cousins to university scientists: definitely family but beyond the reach of unconditional family claims.

[17]Leonard S. Reich, *The Making of American Industrial Research* (Cambridge: Cambridge University Press, 1985). C. E. Rosenberg, "The Adams Act: Politics and the cause of scientific research," *Agric. Hist.* 38 (1964): 3–12 (reprinted in *No Other Gods,* ch. 10).

Being extramural also gave foundation officers a kind of freedom to concentrate resources that academic insiders could seldom have. As Weaver observed, it was difficult for most university administrators to be selective and programmatic in allocating internal resources (the Charles Slichters were exceptional). To preserve harmony the good, average, and even the not-so-good all had to be aided, and every discipline, whether it was productive and seminal or narrow and played out. The Rockefeller Foundation, being extramural, could avoid village politics: "It has the opportunity to choose only the very best men, the very best special facilities, and only those researches which most significantly key together," Weaver wrote. "This is admittedly an ideal, but it is also a duty; and the success of the officers should be measured in terms of the degree to which this ideal is approached."[18] As Walter Cannon remarked, in re Henry Dakin's plea for laissez-faire in 1934, foundation officers were bound to influence the course of science, like it or not, and they had best do so in an openly programmatic and considered way. Weaver did not approve of the way Frederick Keppel doled out the Carnegie Corporation's funds to projects that caught his fancy. It was like peering at a world of varied opportunities through a hole in a board, he thought: what Keppel saw was interesting but perhaps less interesting that other things he chose not to see.[19] Having a definite program, Weaver believed, was one of the social obligations of private endowments.

Foundation officers were so close to being academic insiders that they had always to remind themselves that they were not. Weaver observed that officers were frequently asked by recipients of grants to advise on the management of projects, often in the most friendly and flattering terms. He almost always declined to do so: not because it was improper for officers to advise but because the proper moment for advice was before the grant was made: "A grant once made should be like an adolescent son or daughter: the parent is deeply concerned and may even be worried; but he should have the good sense to realize that his personal contribution is essentially over."[20] Yet that is only half the story. True, Weaver was most actively managerial in negotiating projects, but projects became part of a social system of patronage in which Weaver and his colleagues took a continuing and active part, especially as grants came up for renewal. Recall, too, his judicious but continuing engagement in the careers of the protomolecular biologists.

Foundation managers also had to work within existing academic in-

[18]Weaver memo, 5 Nov. 1936, p. 8, RF 915 1.8. See also Gregg to K. T. Compton, 28 May 1941, RF 224D 2.17.

[19]Weaver, oral history, pp. 401–403.

[20]Weaver, "N. S. notes on officers' techniques," p. 11.

stitutions. Nothing they did or could do altered the traditional form of disciplines and departments. Rose and Weaver both failed to create separate institutional space for general physiology or biophysics. Weaver's experiences reveal how little leverage he had to develop specialties like bio-organic chemistry or mammalian genetics. Foundation managers shrank from any radical redesign of institutions. They were impervious to appeals to endow research professorships or research "institutes" separate from college-based departments. They steered clear of schemes for consolidating departments into divisional superorganizations. Weaver was instinctively suspicious of presidents who pointed to the "marvelous atmosphere of cooperation and collaboration," as would anyone, he thought, "who carries scar tissue won in ancient academic battles."[21] Institutions were not changed by patronage; behavior and practices within institutions were.

The enduring effects of foundation programs should be sought in the greater variety of participants in research and in the more complex organization of scientific work. Sponsorship made possible, indeed essential, a quickened pace of production and made academics more aware of using resources efficiently. It put a premium on agility in getting quickly into and out of productive or fashionable lines of research, encouraging strategies of cream skimming and school research that before had been common only in a few fast-paced research fronts. Applying for fellowships and grants encouraged a more self-conscious management of careers. The strategic use of new laboratory technologies and cross-disciplinary connections became more commonplace. In short, modes of practice that were customary in research institutes, government bureaus, and agricultural experiment stations invaded university departments via projects sponsored by foundations. This accumulated experience was one reason, perhaps, why scientists took so readily to their new federal sponsors after 1945.

I do not want to overstate the case. There were other systems of patronage that also encouraged scientists to adopt a more managed style of practice. Geological surveys had influenced the work of academic geologists and geographers since the mid-nineteenth century. Industrial consulting had long been a significant influence on chemists' working habits and choice of problems. Since the early 1900s federal grants for agricultural research had been distributed on a project basis; applications for specific projects were screened and approved by managers in the U.S. Department of Agriculture. In the 1930s industries, especially food and drug industries, were beginning to give academics money to do research relevant to their in-house R&D. Scientists in many sectors

[21]Ibid., p. 26.

of the academic world were getting used to working with sponsors, especially scientists in disciplines connected to the production of goods or services. The large foundations were not leaders in the twentieth-century managerial revolution. But they did lead, for two decades or so, in extending managerial habits from business and government to the natural sciences in universities.

Contrast the record of Weaver's productive grant-giving with the failure of contemporary efforts to give federal agencies a role in sponsoring research in universities. Karl Compton got nowhere in his schemes for public endowment for basic research through the Science Advisory Board (1933–1935). Lyman Briggs's plans to establish a program of extramural grants in the National Bureau of Standards went down to defeat year after year in Congress. Conservatives in the National Academy of Sciences subverted Compton's efforts. Briggs was thwarted by congressmen who believed in service for pay, and by land-grant universities who saw a threat to their privileged access to public resources. Neither side was able to conceive a basis for a system of patronage that gave both a measure of authority, and both were overly sensitive to threats to established prerogatives. Academic scientists were reluctant to give up laissez-faire, fearing bureaucratic strings attached to federal grants. Politicians and public administrators could not afford politically to compromise ideals of strict procedure and accountability. Every effort to create a system of federal sponsorship bogged down in ritual displays of authority. It seemed better to do nothing than to compromise.[22]

The politics of patronage lagged behind practice in the 1930s. While scientists and potential patrons bickered and rehashed "old rows," as Fosdick put it, a new system of patronage was being put into practice, most notably by the Rockefeller and other foundations, but also by other agencies with special incentives and opportunities. In the late 1930s, competitive programs of grants for research on farm surpluses were established by the U.S. Department of Agriculture, and for research on cancer by the new National Cancer Institute.[23] Industries also sponsored academic research indirectly through graduate fellowships (which had the advantage of training researchers for careers

[22]Carroll W. Pursell, Jr., "The anatomy of a failure: The Science Advisory Board, 1933–1935," *Proc. Amer. Philos. Soc.* 109 (1965): 342–351. Lewis E. Auerbach, "Scientists in the New Deal: A pre-war episode in the relations between science and government in the United States," *Minerva* 3 (1965): 457–482. Pursell, "A preface to government support of research and development: Research legislation and the National Bureau of Standards, 1935–1941," *T&C* 9 (1968): 145–164. Robert Kargon and Elizabeth Hodes, "Karl Compton, Isiah Bowman, and the politics of science in the great depression." *Isis* 76 (1985): 301–318.

[23]Dupree, *Science in the Federal Government*, pp. 355–367.

in industrial R&D).[24] These were only harbingers of things to come, however. It took a hot and a cold war to wear down ideological and political barriers to government patronage of basic research in universities.[25]

In the 1930s, the institutional similarities of foundations and universities—in their histories, aims, personnel, and cultural values—made it much easier for them than for other potential patrons to create a system of sponsorship in the natural sciences. Foundation patronage was more flexible and forgiving than federal or industrial patronage. Foundation managers of science were more free to improvise and adapt to local opportunities. This practical experience of sponsorship was a far more powerful solvent of academic and philanthropic doubts than any ideologizing.

Toward Big Science?

It is hard not to think of foundation programs as a prologue to the postwar system of federal patronage, but it may be that we should resist the natural inclination to seek continuities and look also for differences.[26] The system of federal patronage, which evolved mainly out of military research programs of World War II, differed in important ways from prewar foundation programs. The most important difference was the pervasive presence of peer review. This system gave a smaller role to program managers, who in some cases have become little more than executive secretaries of peer review committees. The peer review system increased bureaucratic oversight in small things, while in matters of consequence it bound sponsors to academic and disciplinary agendas. Would contemporary scientists, accustomed to the buffer of peer committees, tolerate for long the personal interventions of people like Weaver, Thorkelson or Trowbridge? I doubt it. Will histo-

[24]Richard H. Shryock, *American Medical Research Past and Present* (New York: Commonwealth Fund, 1947), pp. 141–144, 158. R. T. Major, "Contributions of the chemical industry to science," *Sci. Monthly* 51 (1940): 158–164. Francis Boyer, "The pharmaceutical manufacturer and academic research," *New Eng. J. Med.* 228 (1943): 529–532. Callie Hull and M. Mico, "Research supported by industry through scholarships, fellowships, and grants," *J. Chem. Educ.* 21 (1944): 180–190.

[25]Paul Forman, "Behind quantum electronics: National security as basis for physical research in the United States, 1940–1960," *Hist. Stud. Phys. Sci.* 18 (1987/1988): 149–229.

[26]Daniel J. Kevles, "The National Science Foundation and the debate over postwar research policy, 1942–1945," *Isis* 68 (1977): 5–26. Nathan Reingold, "Vannevar Bush's new deal for research: Or the triumph of the old order," *Hist. Stud. Phys. Sci.* 17 (1987): 299–344.

ries of federal patronage revolve around the vision and personalities of program managers? It seems unlikely.[27]

Except for the NRC fellowship and grants-in-aid committees, peer review is strikingly absent from the historical record of foundation patronage. The possibility was mooted from time to time, usually by scientists, but no one in foundations seemed to like it. Woodward angrily rejected the idea when Chamberlin suggested that it might reduce conflict with his constituents, recall, and Rose and Trowbridge's expert advisors proved a mixed blessing. In his review of Weaver's program in 1934, Simon Flexner proposed that an advisory board be "intercalated between the officers . . . who mature plans, and the trustees, who act on them."[28] Doubtless he hoped that heavy-caliber experts would keep Weaver in his place. But Fosdick simply ignored the whole issue. Weaver parried Fosdick's suggestion that specialists be brought in to advise, arguing that outsiders would not know how the foundation worked and could not help but push their own professional interests.[29] The actions of the 1934 panel, so obviously self-interested, were not a good auger of what such a group might do if installed permanently within the RF. Formal sharing of control between managers and scientific constituents did not seem to be a viable option.

As it recedes in time, the relationship between foundation managers and academic scientists may look more and more like a singular episode, a chapter that was opened suddenly by the first World War and closed forever by the second.

In 1951 Weaver gave the Rockefeller Foundation's new president, Chester Barnard, a plan for dramatically scaling-back support of the natural sciences in the United States. The war, Weaver pointed out, had so dramatically changed the funding of basic science that there was no longer any real need for the foundation to take part. Of the roughly $24 million available for sponsored research in basic biology in the United States, he estimated, no less than $20.7 million came from five federal agencies. The NIH alone gave $13.7 million, voluntary health agencies

[27]But see Forman, "Behind quantum electrodynamics," p. 209, and C. Stuart Gilmore, "Federal funding and knowledge growth in ionospheric physics, 1945–1981," *Soc. Stud. Sci.* 16 (1986): 105–133, especially pp. 123–127.

[28]Dakin to Flexner, 16 Nov. 1934, p. 3; Cannon to Flexner, 21 Nov. 1934, p. 6; Howell to Flexner, 10 Nov. 1934, p. 3; Flexner to panel, 19 Nov. 1934, p. 5; Fosdick to Flexner, 19 Nov. 1934, pp. 1–2; all in RF 915 4.41. Howell suggested a compromise, in which Weaver would handle larger projects and a panel of experts would allocate fellowships and grants-in-aid.

[29]Tisdale to Weaver, 25 Mar. 1934, RF 915 1.1.

$1.7 million, the Rockefeller Foundation $1.2 million, and $0.35 million from other foundations. Of the thirteen largest funding agencies, only the RF and the Commonwealth Fund existed when the trustees approved the program in vital processes in 1933. The message was almost too obvious: it was time to move on.[30] Weaver had already disengaged himself from the natural sciences program and was spending most of his time on the new challenge of agricultural programs in Latin America.[31] Agriculture was for Weaver after 1945 what vital processes had been in the 1930s. It took over a decade for the natural sciences program to be wound down, but the interesting story was over by 1945, maybe even by 1940. As federal subsidies burgeoned in the 1950s and 1960s, the large foundations gradually returned to the programs in health, general education, and welfare that had been their stock in trade before 1916. The dynamic partnership between scientists and foundation managers, so characteristic of the interwar decades, had become routinized. The age of "big science" had begun.

In assessing the historical meaning of foundations in the history of science, we need to look both forward to "big science" and backward to the time when foundations emerged out of the civic and voluntary reform movements of the late nineteenth century. The interwar style of patronage had as much affinity with the small-scale, personalized social relations of the nineteenth century as with the larger, more impersonal bureaucracies of the postwar world. The personal touch of Thorkelson, Trowbridge, and Weaver was crucial to their success in creating a new system of sponsorship. These managers of science operated in a transitional period: that is, a period in which old and new were so equally mixed that fundamentally new things could be done in ways that seemed reassuringly traditional. In the large foundations, voluntarism, and a tendency to disregard differences between private and public authority, lasted well after they had gone out of style in the larger worlds of politics and profit making. In the Rockefeller Foundation especially, the nineteenth and twentieth centuries overlapped for a few decades. And in that time and place, a handful of migrants from academia helped create new modes of organization, funding, and practice in one of the most individualistic of cultural enterprises.

[30] Weaver to Barnard, 27 Sept. 1951; "Liquidating grants," 16 Jan. 1953; both in 915 2.14 and 2.16. Weaver, oral history, pp. 581–588.

[31] Weaver, oral history, pp. 653–665. Weaver, *Scene of Change,* pp. 94–104. Deborah Fitzgerald, "Exporting American agriculture: The Rockefeller Foundation in Mexico, 1943–1953," *Soc. Stud. Sci.* 16 (1986): 457–484.

Abbreviations

CC	Carnegie Corporation, New York, N.Y.
CDW	Charles D. Walcott Papers, Smithsonian Institution, Washington D.C.
CIT	California Institute of Technology, Pasadena
CIW	Carnegie Institution of Washington, Washington D.C.
CUL	Columbia University Library, Archives Division, New York, N.Y.
ECP	Edward C. Pickering Papers, Pusey Library, Harvard University, Cambridge, Mass.
EGC	Edwin G. Conklin Papers, Firestone Library, Princeton University, Princeton, N.J.
ERE	Edwin R. Embree Papers, RAC.
FTG	Frederick T. Gates Papers, RAC.
GEB	General Education Board, RAC.
GEH	George Ellery Hale Papers, CIT Archives, microfilm.
IEB	International Education Board, RAC.
IHB	International Health Board, RAC.
JMC	James McKeen Cattell Papers, Library of Congress, Washington, D.C.
JSB	John Shaw Billings Papers, New York Public Library, New York, N.Y.
KTC	Karl T. Compton Papers, Massachusetts Institution of Technology Archives, Cambridge, Mass.
LC	Library of Congress, Washington, D.C.
NAS	National Academy of Sciences Archives, Washington, D.C.
RAC	Rockefeller Archive Center, North Tarrytown, N.Y.
RAM	Robert A. Milliken Papers, CIT Archives, microfilm.
RBF	Raymond B. Fosdick Papers, Harvey Mudd Library, Princeton University, Princeton, N.J.
RF	Rockefeller Foundation, RAC.
SF	Simon Flexner Papers, American Philosophical Society, Philadelphia, Pa.
TWR	Theodore W. Richards Papers, Pusey Library, Harvard University, Cambridge, Mass.
UC	Presidents' Papers, University of Chicago Archives, Chicago, Ill.
UW	University of Wisconsin Archives, Madison, Wis.

Notes: If box and folder numbers or labels are not specified, documents are filed by name of correspondent.

Citations to collections in the Rockefeller Archive Center include series, box, and folder (e.g., 205 45.2), but omit record group numbers except where omission may be misleading. RF project files are in R.G. 1 (1.1 for those initiated before 1945 and 1.2 for those initiated after 1945 or unprocessed); Rockefeller family papers in R.G. 2; program and policy (900 series) in R.G. 3 (including officers' conferences); IHB papers in R.G. 5; officers diaries in R.G. 12 (12.1 for RF officers, 12.9 for IEB); board minutes in R.G. 16.

Index